MAPPING AMERICA'S NATIONAL PARKS

PRESERVING OUR NATURAL AND CULTURAL TREASURES

Foreword by Ken Burns and Dayton Duncan

MAPPING AMERICA'S NATIONAL PARKS

PRESERVING OUR NATURAL AND CULTURAL TREASURES

Esri Press
REDLANDS | CALIFORNIA

U.S. NATIONAL PARK SERVICE

Cover credits: Aldo Leopold Wilderness Research Institute, Chad Lopez, Kass Green & Associates, NPS, USGS.

Esri Press, 380 New York Street, Redlands, California 92373-8100

ISBN: 9781589485464
Library of Congress Control Number: 2020952262

For purchasing and distribution options (both domestic and international), please visit esripress.esri.com.

CONTENTS

A
Map of
Lewis and Clark's Track
Across the Western Portion of
NORTH AMERICA,
from the
Mississippi to the Pacific Ocean;
By Order of the Executive of
The United States
in 1804, 5 & 6.
Copied by Samuel Lewis from the Original Drawing of W. Clark.
MISSOURI RIVER
RIVER LA PLATTE
Longitude West from Washington
Longitude West from London
Sho-sho-nes
Snake Indians
Callahpo-e-wah
Ricaras
London Published April 1814 by Longman Hurst Rees Orme & Brown Paternoster Row.

FOREWORD

When Lewis and Clark returned from their historic exploration of the American West in 1806, they brought back news of lands teeming with wildlife (many never before described for science), vast prairies, a diverse array of American Indian people, and mountains much, much bigger than Thomas Jefferson imagined. The myth of a fabled Northwest Passage—a waterway across the continent that explorers since the time of Columbus had been seeking—died with their report.

Proof of it was made clear with a map William Clark put together and published in 1814, based on his own calculations of their route using dead reckoning. But it also contained information gathered from American Indians (some drew maps in the dirt to explain the terrain) and from consultations with fur trappers who visited him in St. Louis in intervening years. It is one of the seminal maps of the United States' own westward journey to the Pacific—and also, we think, a lovely work of art.

On it, Clark included information he got from a former member of the expedition, John Colter, who had remained for years in the West and became a legendary mountain man. In one extraordinary journey, Colter traversed the headwaters of the Yellowstone River and traveled into the Teton range of what is now Wyoming. Clark's map tracing Colter's route shows a large lake and places denoted as "Hot Spring Brimstone" and "Boiling Spring" and a river called "Stinking Water." For years, Easterners derided the mountain men's tales of sulfurous geysers and boiling mud pots as "Colter's Hell"—as seemingly fanciful as the myth of a Northwest Passage.

Later explorations—bringing back photographs, paintings, and yes, more detailed maps—proved that such a place did, in fact, exist. In 1872, it became the world's first national park: Yellowstone.

As this beautiful and important book demonstrates, maps are still intertwined with everything associated with "America's best idea"—from helping millions of visitors navigate their own journeys of discovery to helping the National Park Service better manage the wildlife and landscapes it protects and preserves for future generations. As Honorary Park Rangers—and lifelong lovers of maps—we're thankful to Esri for continuing this great tradition.

Ken Burns and Dayton Duncan,
The National Parks: America's Best Idea, documentary and book (2009)

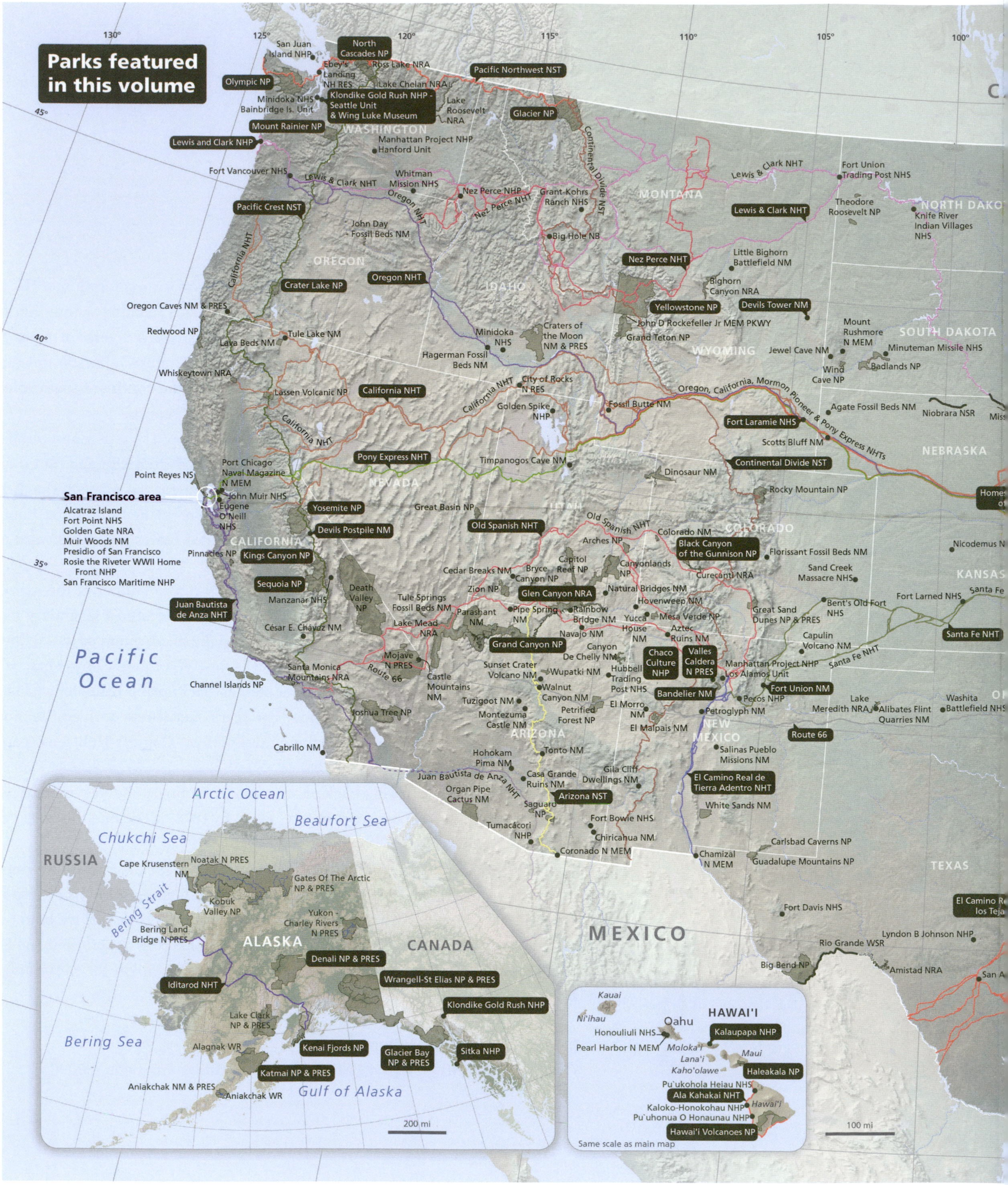
Parks featured in this volume
130°
125°
120°
115°
110°
105°
100°
45°
40°
35°
San Juan Island NHP
North Cascades NP
Ebey's Landing NH RES
Ross Lake NRA
Olympic NP
Lake Chelan NRA
Pacific Northwest NST
Minidoka NHS Bainbridge Is. Unit
Klondike Gold Rush NHP - Seattle Unit & Wing Luke Museum
Lake Roosevelt NRA
Glacier NP
Mount Rainier NP
WASHINGTON
Lewis and Clark NHP
Manhattan Project NHP Hanford Unit
Continental Divide NST
Fort Vancouver NHS
Lewis & Clark NHT
Whitman Mission NHS
Nez Perce NHP
Grant-Kohrs Ranch NHS
MONTANA
Lewis & Clark NHT
Fort Union Trading Post NHS
Oregon NHT
Nez Perce NHT
Pacific Crest NST
John Day Fossil Beds NM
Big Hole NB
Lewis & Clark NHT
Theodore Roosevelt NP
NORTH DAKO
Knife River Indian Villages NHS
California NHT
OREGON
Nez Perce NHT
Little Bighorn Battlefield NM
Crater Lake NP
Oregon NHT
IDAHO
Bighorn Canyon NRA
Yellowstone NP
Devils Tower NM
Oregon Caves NM & PRES
John D Rockefeller Jr MEM PKWY
Redwood NP
Tule Lake NM
Minidoka NHS
Craters of the Moon NM & PRES
Grand Teton NP
Mount Rushmore N MEM
SOUTH DAKOTA
Lava Beds NM
WYOMING
Jewel Cave NM
Minuteman Missile NHS
Hagerman Fossil Beds NM
Wind Cave NP
Badlands NP
Whiskeytown NRA
City of Rocks N RES
Lassen Volcanic NP
California NHT
California NHT
Oregon, California, Mormon Pioneer & Pony Express NHTs
Golden Spike NHP
Fossil Butte NM
Agate Fossil Beds NM
Niobrara NSR
California NHT
Fort Laramie NHS
Scotts Bluff NM
NEBRASKA
Pony Express NHT
Timpanogos Cave NM
Port Chicago Naval Magazine N MEM
Point Reyes NS
Continental Divide NST
NEVADA
Dinosaur NM
San Francisco area
Alcatraz Island
Fort Point NHS
Golden Gate NRA
Muir Woods NM
Presidio of San Francisco
Rosie the Riveter WWII Home Front NHP
San Francisco Maritime NHP
John Muir NHS
Eugene O'Neill NHS
Rocky Mountain NP
Yosemite NP
Great Basin NP
UTAH
Old Spanish NHT
Old Spanish NHT
Devils Postpile NM
Colorado NM
COLORADO
CALIFORNIA
Arches NP
Black Canyon of the Gunnison NP
Nicodemus NHS
Pinnacles NP
Kings Canyon NP
Capitol Reef NP
Florissant Fossil Beds NM
Canyonlands NP
Sand Creek Massacre NHS
Cedar Breaks NM
Bryce Canyon NP
Curecanti NRA
KANSAS
Sequoia NP
Death Valley NP
Zion NP
Natural Bridges NM
Glen Canyon NRA
Fort Larned NHS
Juan Bautista de Anza NHT
Manzanar NHS
Tule Springs Fossil Beds NM
Hovenweep NM
Parashant NM
Pipe Spring NM
Rainbow Bridge NM
Yucca House NM
Mesa Verde NP
Great Sand Dunes NP & PRES
Bent's Old Fort NHS
Santa Fe
César E. Chávez NM
Lake Mead NRA
Aztec Ruins NM
Navajo NM
Santa Fe NHT
Grand Canyon NP
Capulin Volcano NM
Canyon De Chelly NM
Chaco Culture NHP
Valles Caldera N PRES
Pacific Ocean
Mojave N PRES
Sunset Crater Volcano NM
Manhattan Project NHP Los Alamos Unit
Santa Fe NHT
Santa Monica Mountains NRA
Route 66
Hubbell Trading Post NHS
Wupatki NM
Castle Mountains NM
Walnut Canyon NM
Fort Union NM
Channel Islands NP
Bandelier NM
Pecos NHP
Lake Meredith NRA
Washita Battlefield NHS
Tuzigoot NM
Alibates Flint Quarries NM
El Morro NM
Petrified Forest NP
Petroglyph NM
Joshua Tree NP
Montezuma Castle NM
ARIZONA
NEW MEXICO
El Malpais NM
Route 66
Cabrillo NM
Tonto NM
Salinas Pueblo Missions NM
Hohokam Pima NM
Gila Cliff Dwellings NM
Casa Grande Ruins NM
El Camino Real de Tierra Adentro NHT
Juan Bautista de Anza NHT
Organ Pipe Cactus NM
Arizona NST
Saguaro NP
White Sands NM
Fort Bowie NHS
Tumacácori NHP
Chiricahua NM
Carlsbad Caverns NP
Coronado N MEM
Chamizal N MEM
Guadalupe Mountains NP
TEXAS
El Camino Real de los Tejas
Fort Davis NHS
MEXICO
Lyndon B Johnson NHP
Rio Grande WSR
Big Bend NP
Amistad NRA
San A
Arctic Ocean
Beaufort Sea
Chukchi Sea
RUSSIA
Cape Krusenstern NM
Noatak N PRES
Gates Of The Arctic NP & PRES
Bering Strait
Kobuk Valley NP
Yukon - Charley Rivers N PRES
Bering Land Bridge N PRES
ALASKA
CANADA
Denali NP & PRES
Wrangell-St Elias NP & PRES
Iditarod NHT
Klondike Gold Rush NHP
Lake Clark NP & PRES
Bering Sea
Alagnak WR
Kenai Fjords NP
Glacier Bay NP & PRES
Sitka NHP
Katmai NP & PRES
Aniakchak NM & PRES
Aniakchak WR
Gulf of Alaska
200 mi
Kauai
Ni'ihau
Oahu
HAWAI'I
Honouliuli NHS
Kalaupapa NHP
Pearl Harbor N MEM
Moloka'i
Lana'i
Maui
Kaho'olawe
Haleakala NP
Pu'ukohola Heiau NHS
Ala Kahakai NHT
Kaloko-Honokohau NHP
Hawai'i
Pu'uhonua O Honaunau NHP
Hawai'i Volcanoes NP
Same scale as main map
100 mi

Library of Congress, Geography and Map Division

INTRODUCTION

TOM PATTERSON, NPS HARPERS FERRY CENTER FOR MEDIA SERVICES (RETIRED)

Maps and the US National Park Service (NPS) mission are inextricably linked. The founding legislation of the NPS, the Organic Act of 1916, famously describes the dual, and sometimes conflicting, mission of the agency:

> *....to conserve the scenery and the natural and historic objects and the wildlife therein and to provide for the enjoyment of the same in such manner and by such means as will leave them unimpaired for the enjoyment of future generations.*

To achieve both objectives—enjoyment and preservation—the NPS relies heavily on maps. Consider enjoyment. A map is the first information that visitors receive from the friendly ranger when entering a park. Its primary function is orientation and navigation; the map guides them to the visitor center, points of interest, and various facilities that make for a comfortable and safe visit. But that is only part of what the map does. It can also tell interpretive stories about things not readily apparent—for example, troop movements at a site that today looks more like a manicured park than a battlefield, previous geologic events, and future climate change. As an interpretive tool, the visitor map helps connect the tangible with the intangible, allowing people to better understand and appreciate the park they are in.

The visitor map is the public-facing side of NPS cartography. Even more important are the maps and geospatial analysis that the public largely does not see, which serve the critical preservation needs of the NPS mission. It is a delicate balance. Without adequate preservation, future generations will not enjoy the parks as fully as we do today.

The chapters in this book showcase a disparate selection of maps used for managing parks effectively, including public safety, natural and cultural resource protection, working with communities and partners, and planning for fire. The common thread uniting these maps is the research and science underpinning the data they depict, and the use of GIS to analyze this data to reveal spatial insights not otherwise obvious. Be it viewshed analysis for a proposed cell tower or cataloging sensitive archeological sites, GIS gives park managers the means to make informed decisions.

The maps in this book also showcase the diversity of park sites. As shown in the previous map, the more than 400 sites administered by the NPS are spread across 11 time zones—from the eastern Caribbean to the western Pacific—and span 80 degrees of latitude from arctic Alaska to the South Pacific. Within this huge sweep of geography are the inspiring landscapes, unique cultures, and historical events that have shaped the American experience. And what better way to visualize all of this than with an assemblage of points, lines, polygons, and labels on graphical devices called maps.

1 CARTOGRAPHY IN THE NATIONAL PARK SERVICE

BRENDAN M. BRAY, NPS HARPERS FERRY CENTER FOR MEDIA SERVICES

THE EVOLUTION OF PARK MAPS

National Park Service (NPS) cartographic maps are a common feature found in park brochures, exhibits, and other interpretive media. Just as every park is different, so too are park maps. These maps are custom designed to emphasize park symbols of national significance and critical safety and stewardship information, such as roads, trail heads, campgrounds, or restrooms. Regardless of the size and location of the park, maps are often the first piece of information that visitors receive as they begin their experience.

NPS maps are specially designed to offer quick utility and convey complex geographic information in an easy-to-read, visually appealing way. But these maps don't just look beautiful; they are highly technical and aim to seamlessly integrate layers of geographic information with the aesthetic look and interpretive themes inherent in these special places.

NPS cartographers at the Harpers Ferry Center for Interpretive Media use a variety of GIS tools to generate accurate maps using satellite and aerial imagery, national land cover data, and georeferenced park features. Cartographers also employ various artistic techniques, such as hand-drawn waves crashing along coastlines or cubed buildings in towns and villages to capture recognizable features of a park. On these next pages, you will see examples of pieces created to orient and immerse visitors in a park experience without simultaneously overwhelming them with information.

In the site maps for Salem Maritime and Booker T. Washington (see chapter 3, "Visitors and resource protection") historic sites, cartographers used a combination of GIS and graphic design software to generate axonometric, pseudo-3D projections to show buildings on a standard street grid base layer. In this photo-realistic view of the park, buildings seemingly leap off the page and illustrate the architectural complexities of ornate historic structures. In the Grand Canyon trail map (see chapter 2, "Recreation"), cartographers used terrain texture shading and natural land colors to create a natural look and feel to help hikers easily find their way across complex terrain and understand the physical demands of descending below the canyon rim. The *Stehekin* map (see chapter 2, "Recreation") is a full 3D natural land-color representation of an incredible landscape where cartographers combined elevation data with natural features and texture shading to generate this eye-popping map that is highly artistic and extremely informative. The *Congaree National Park* map is a planimetric map built with lidar-derived elevation data. This high-resolution data teases out every bend and meander of hydrology and unveils a complex network of drainages within the park boundary.

More than 300 million people visit US national parks every year. National parks will continue to see rapid growth in visitation over the coming decades. To meet this growing demand, NPS cartographers will continue to explore new ways to blend traditional cartography, graphic design, and digital mapping to generate highly useful and beautiful products for our visitors.

RUSSELL CAVE NATIONAL MONUMENT 3D PARK MAP

JOSEPH MILBRATH, NPS HARPERS FERRY CENTER FOR MEDIA SERVICES

This 3D map of the park was created for a new park brochure in 2016. The map shows the park's location along the valley and ridge topography of northeast Alabama. National Agricultural Imagery Program (NAIP) imagery was draped on a digital elevation model (DEM) in 3D software, Natural Scene Designer, and 3D trees were "planted" where forested areas were present in the imagery. Geospatial data was used for the park's trails, roads, and buildings.

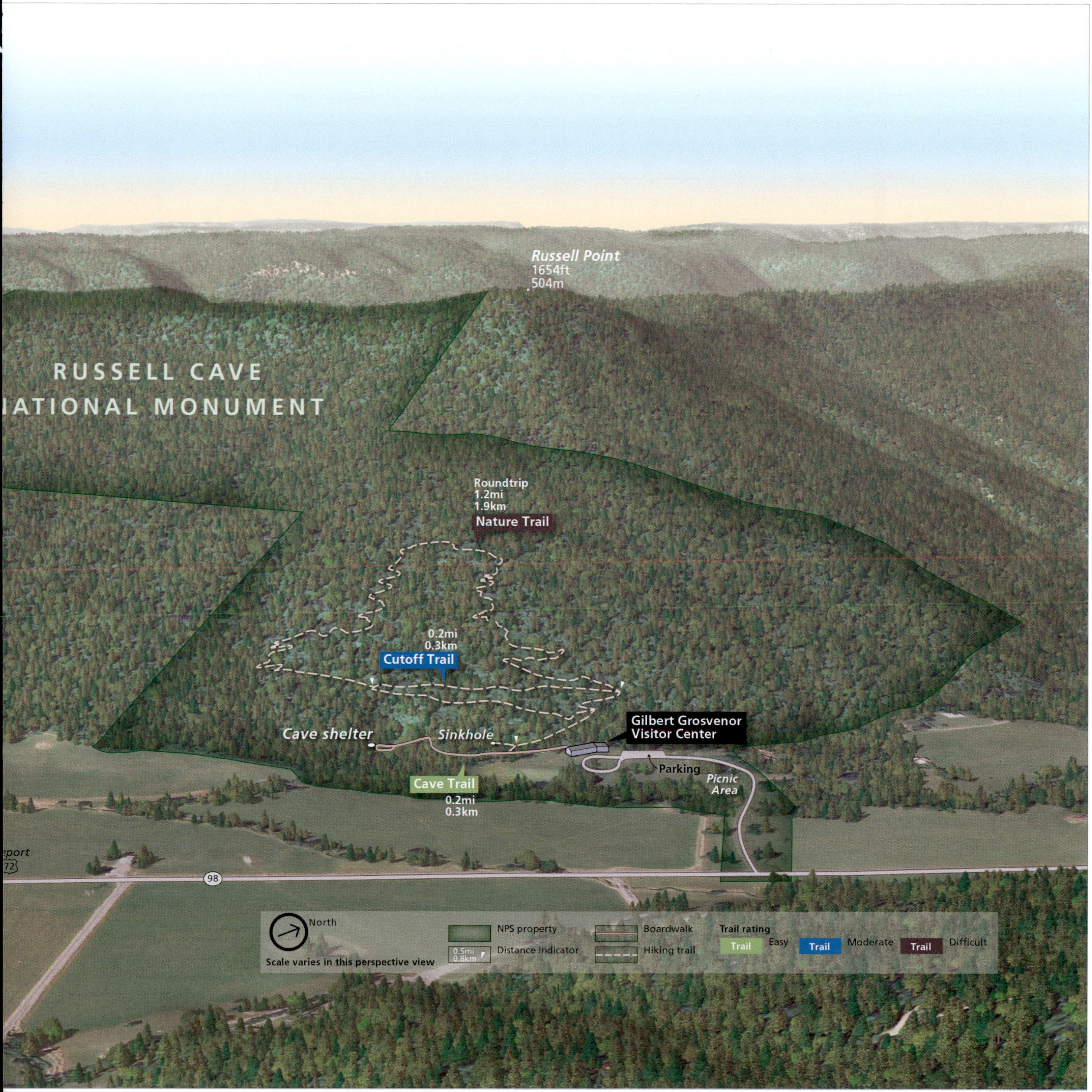

This 3D map shows the park's location along the valley and ridge topography of northeast Alabama. Data sources: USGS, NPS, and NAIP.

VALLES CALDERA NATIONAL PRESERVE PARK MAP

JOSEPH MILBRATH, NPS HARPERS FERRY CENTER FOR MEDIA SERVICES

The map of Valles Caldera National Preserve was created for the preserve's first brochure in 2017. The map's relief was built by blending a United States Geological Survey (USGS) DEM with lidar data collected to study volcanic flows in the Banco Bonito portion of the preserve. National land cover data was used to determine forested and meadow areas. The two land cover classifications illustrate the park's inverted *Sky Island*, where high elevations are forested, and low elevations contain open meadows. The map also includes geospatial data of the park's rivers, roads, and trail networks.

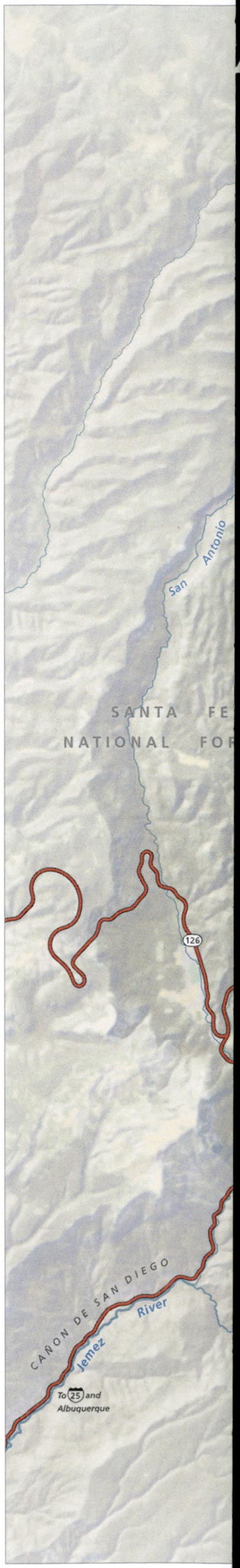

The map of Valles Caldera National Preserve was created for the preserve's first brochure in 2017. Data sources: USGS and NPS.

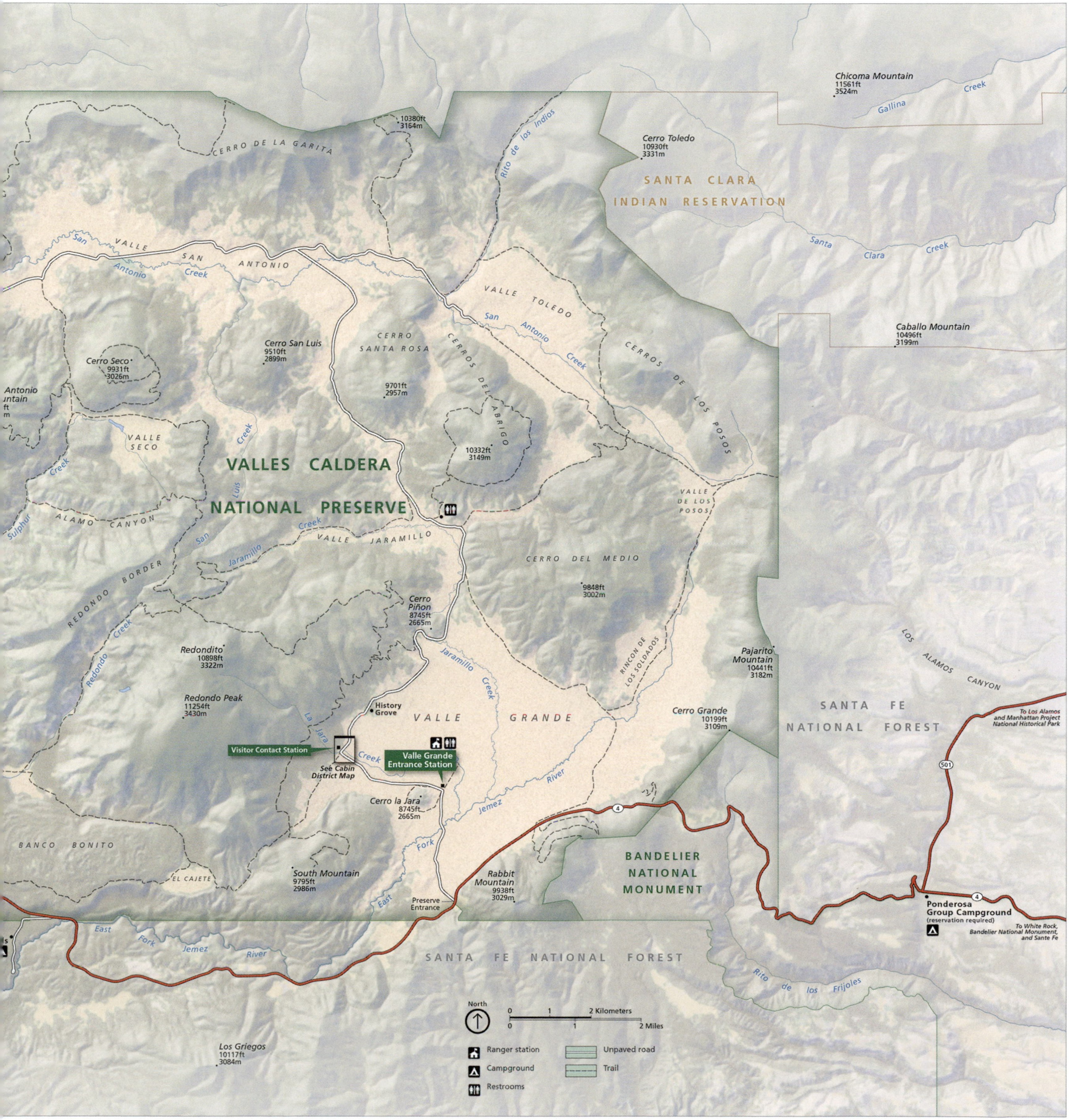
Chicoma Mountain
11561ft
3524m
Gallina
Creek
10380ft
3164m
CERRO DE LA GARITA
Rito de los Indios
Cerro Toledo
10930ft
3331m
SANTA CLARA
INDIAN RESERVATION
Santa
Clara
Creek
San
VALLE
SAN
ANTONIO
Antonio
Creek
VALLE TOLEDO
San
Antonio
Creek
Caballo Mountain
10496ft
3199m
CERRO
SANTA ROSA
CERROS DEL ABRIGO
CERROS DE LOS POSOS
Cerro San Luis
9510ft
2899m
Cerro Seco
9931ft
3026m
Antonio
Mountain
9701ft
2957m
VALLE
SECO
Creek
10332ft
3149m
VALLES CALDERA
NATIONAL PRESERVE
San Luis
Creek
VALLE
DE LOS
POSOS
Sulphur
ALAMO CANYON
Creek
San
VALLE JARAMILLO
Jaramillo
CERRO DEL MEDIO
9848ft
3002m
REDONDO BORDER
Cerro
Piñon
8745ft
2665m
Creek
Redondo
Redondito
10898ft
3322m
Jaramillo
Creek
RINCON DE LOS SOLDADOS
Pajarito
Mountain
10441ft
3182m
LOS ALAMOS CANYON
Redondo Peak
11254ft
3430m
History
Grove
VALLE GRANDE
Cerro Grande
10199ft
3109m
SANTA FE
NATIONAL FOREST
To Los Alamos
and Manhattan Project
National Historical Park
La Jara
Visitor Contact Station
Creek
Valle Grande
Entrance Station
See Cabin
District Map
501
River
Jemez
Cerro la Jara
8745ft
2665m
4
BANCO BONITO
Fork
BANDELIER
NATIONAL
MONUMENT
South Mountain
9795ft
2986m
EL CAJETE
Rabbit
Mountain
9938ft
3029m
East
Preserve
Entrance
Ponderosa
Group Campground
(reservation required)
4
To White Rock,
Bandelier National Monument,
and Sante Fe
East
Fork
Jemez
River
SANTA FE NATIONAL FOREST
Rito de los Frijoles
North
0 1 2 Kilometers
0 1 2 Miles
Ranger station
Campground
Restrooms
Unpaved road
Trail
Los Griegos
10117ft
3084m

KATMAI NATIONAL PARK AND PRESERVE PANORAMA

JOE MILBRATH AND JIM EYNARD, NPS HARPERS FERRY CENTER FOR MEDIA SERVICES

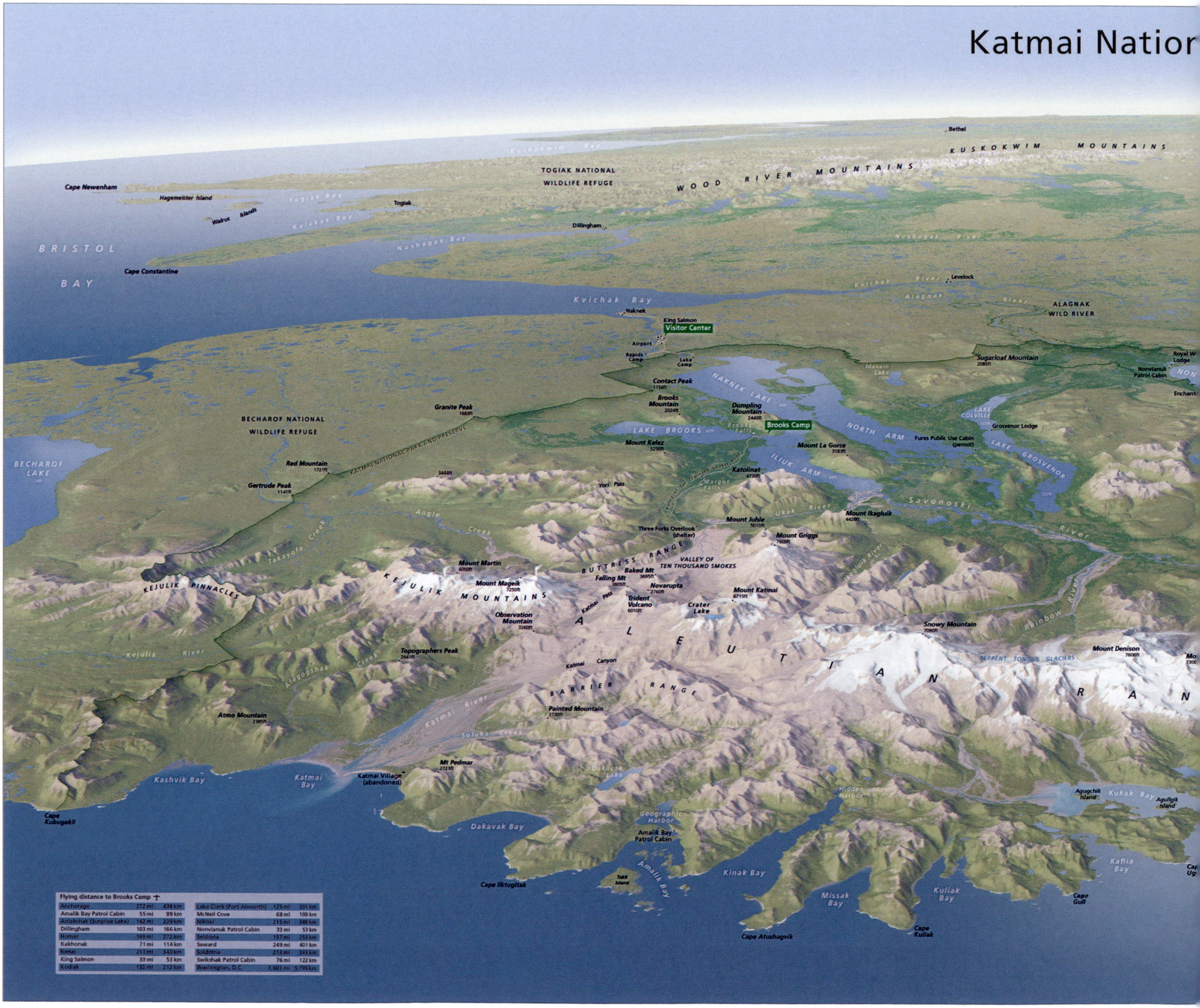

Flying distance to Brooks Camp

Anchorage	272 mi	438 km	Lake Clark (Port Alsworth)	125 mi	201 km
Amalik Bay Patrol Cabin	55 mi	89 km	McNeil Cove	68 mi	109 km
Aniakchak (Surprise Lake)	142 mi	229 km	Nikiski	215 mi	346 km
Dillingham	103 mi	166 km	Nonvianuk Patrol Cabin	33 mi	53 km
Homer	169 mi	272 km	Seldovia	157 mi	253 km
Kakhonak	71 mi	114 km	Seward	249 mi	401 km
Kenai	211 mi	340 km	Soldotna	213 mi	343 km
King Salmon	33 mi	53 km	Swikshak Patrol Cabin	76 mi	122 km
Kodiak	132 mi	212 km	Washington, D.C.	3,601 mi	5,795 km

This panorama of Katmai National Park and Preserve in Alaska was created to give visitors a perspective of the park's features outside of Brooks Falls, a popular destination where tourists can view brown bears feasting on trout and salmon. The map's relief was created using 3D software, Natural Scene Designer, and combines a DEM, Landsat imagery, and national land cover data. As a result, the terrain combines blue water features through dense forests and shrubs skyward to rocky ridgelines and glaciated peaks. Hand-drawn plumes vent from the range's active volcanic peaks, and waves crash along coastlines. The maps rivers, lakes, roads, and trails were derived from geospatial data.

This map shows a panorama of Katmai National Park and Preserve. Data sources: USGS DEM, USGS National Hydrography Dataset, NPS lands Boundary, NPS trails and roads.

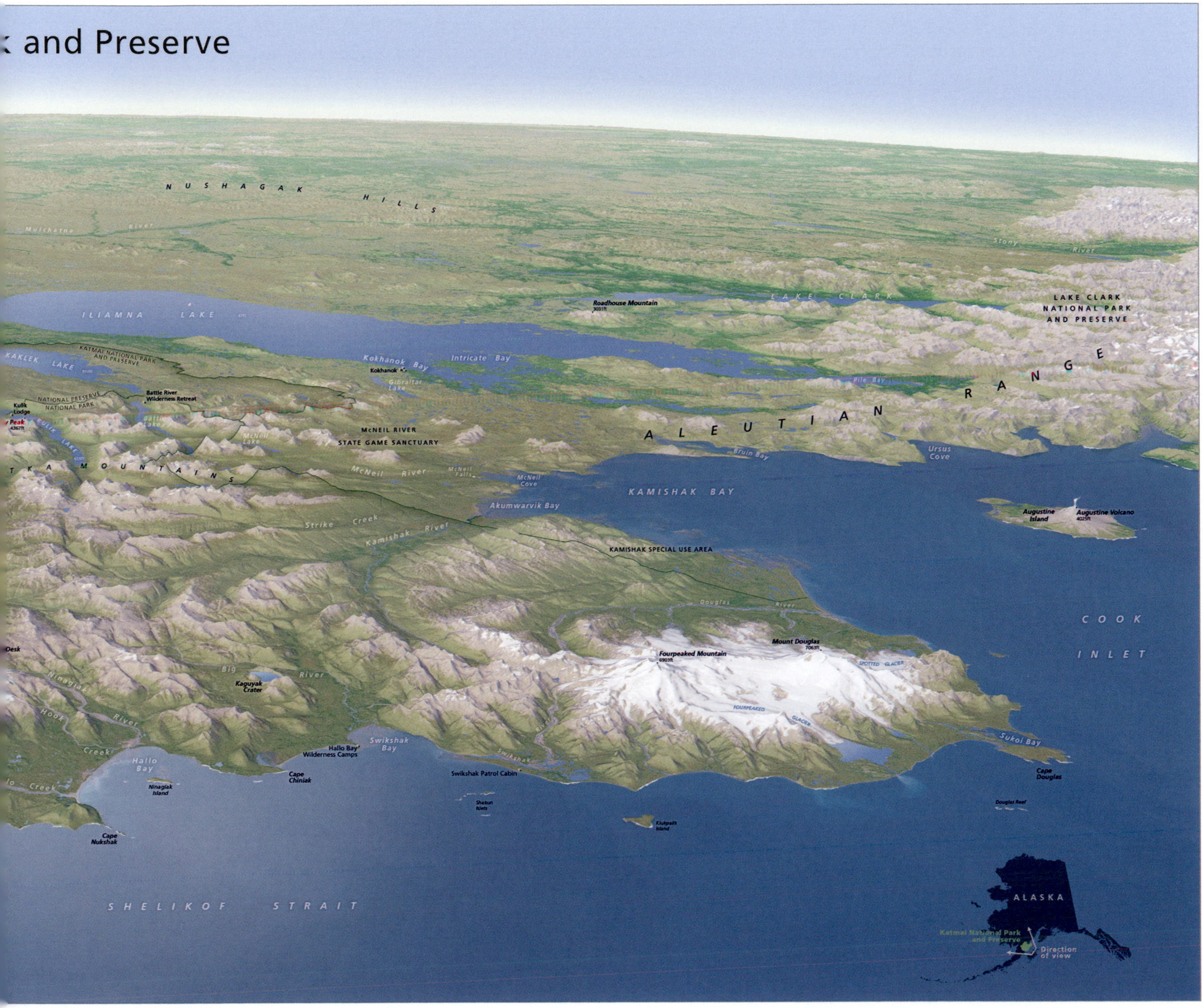

SALEM MARITIME NATIONAL HISTORIC SITE

JIM EYNARD, NPS HARPERS FERRY CENTER FOR MEDIA SERVICES

This map of Salem Maritime National Historic Site and the surrounding area in Salem, Massachusetts, was designed to be used as part of a brochure and as a wall map in the visitor center. There are three levels of buildings in the visual hierarchy, with the most important buildings shown as pseudo-3D axonometric buildings. Buildings of medium importance are shown as dark orange, and all other buildings on the map are shown in light gray. This map could be used as part of a walking tour as visitors navigate this park site and the surrounding area.

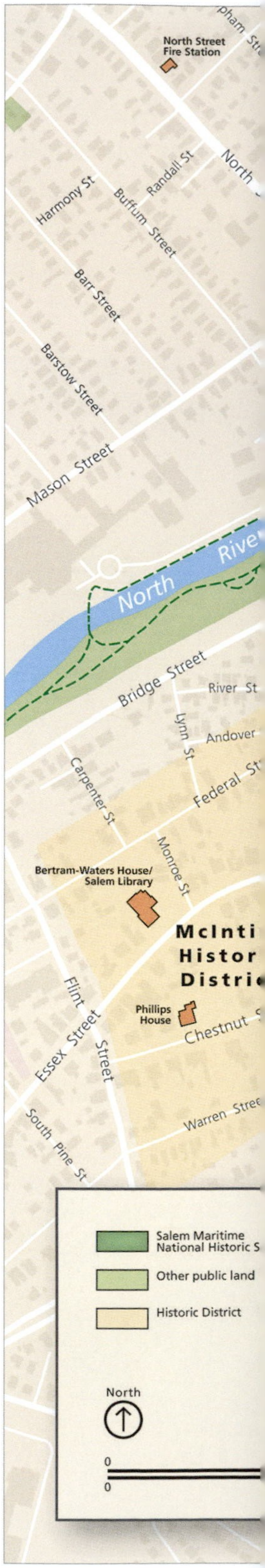

This map shows the Salem Maritime National Historic Site and the surrounding area in Salem, Massachusetts. Data sources: NPS lands boundary, NPS, and Open Street Map.

FORT UNION NATIONAL MONUMENT

JIM EYNARD, NPS HARPERS FERRY CENTER FOR MEDIA SERVICES

This 3D perspective view of Fort Union National Monument in New Mexico shows how the 19th-century fort was strategically located along the Santa Fe Trail. The wide-angle view shows the many natural resources in the area that were once utilized by the people at the fort including the forest on Turkey Mountain, adobe fields at the northern edge of the park boundary, and the water resources in the area. Overlaid on the 3D map is a map of the extensive fort network in the southwestern United States during the time period.

This 3D perspective view of Fort Union National Monument in New Mexico shows the nineteenth-century fort strategically located along the Santa Fe Trail. Data sources:USGS, NPS, and NAIP.

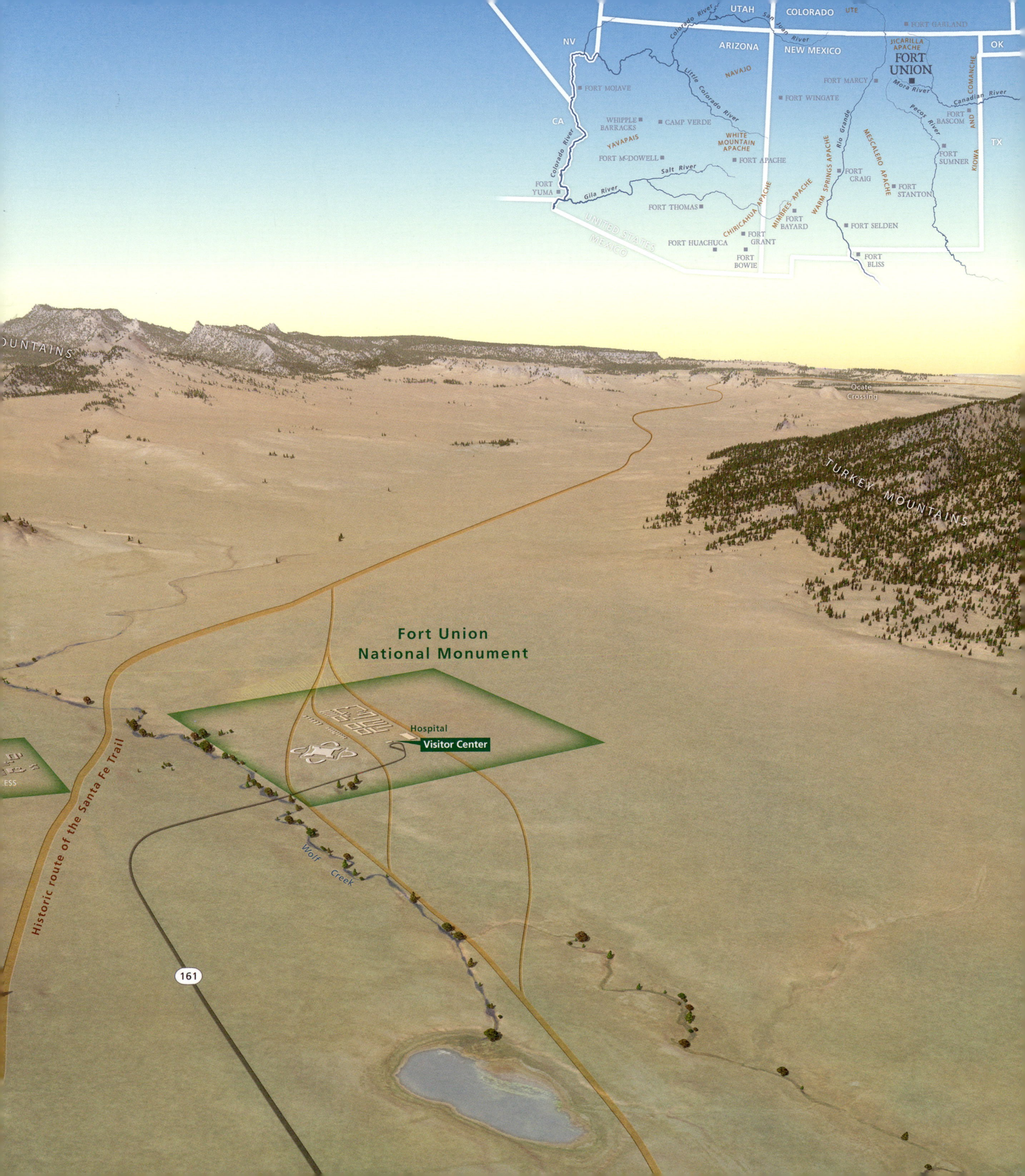
UTAH
COLORADO
UTE
FORT GARLAND
NV
ARIZONA
NEW MEXICO
JICARILLA APACHE
FORT UNION
OK
FORT MARCY
Mora River
NAVAJO
FORT MOJAVE
FORT WINGATE
Canadian River
COMANCHE AND KIOWA
CA
WHIPPLE BARRACKS
CAMP VERDE
FORT BASCOM
Pecos River
YAVAPAIS
WHITE MOUNTAIN APACHE
TX
FORT McDOWELL
FORT APACHE
FORT SUMNER
Salt River
FORT CRAIG
MESCALERO APACHE
FORT STANTON
FORT YUMA
Gila River
Colorado River
Little Colorado River
San Juan River
Rio Grande
WARM SPRINGS APACHE
MIMBRES APACHE
CHIRICAHUA APACHE
FORT THOMAS
FORT BAYARD
UNITED STATES
MEXICO
FORT HUACHUCA
FORT GRANT
FORT SELDEN
FORT BOWIE
FORT BLISS
MOUNTAINS
Ocate Crossing
TURKEY MOUNTAINS
Fort Union National Monument
Hospital
Visitor Center
Historic route of the Santa Fe Trail
Wolf Creek
161

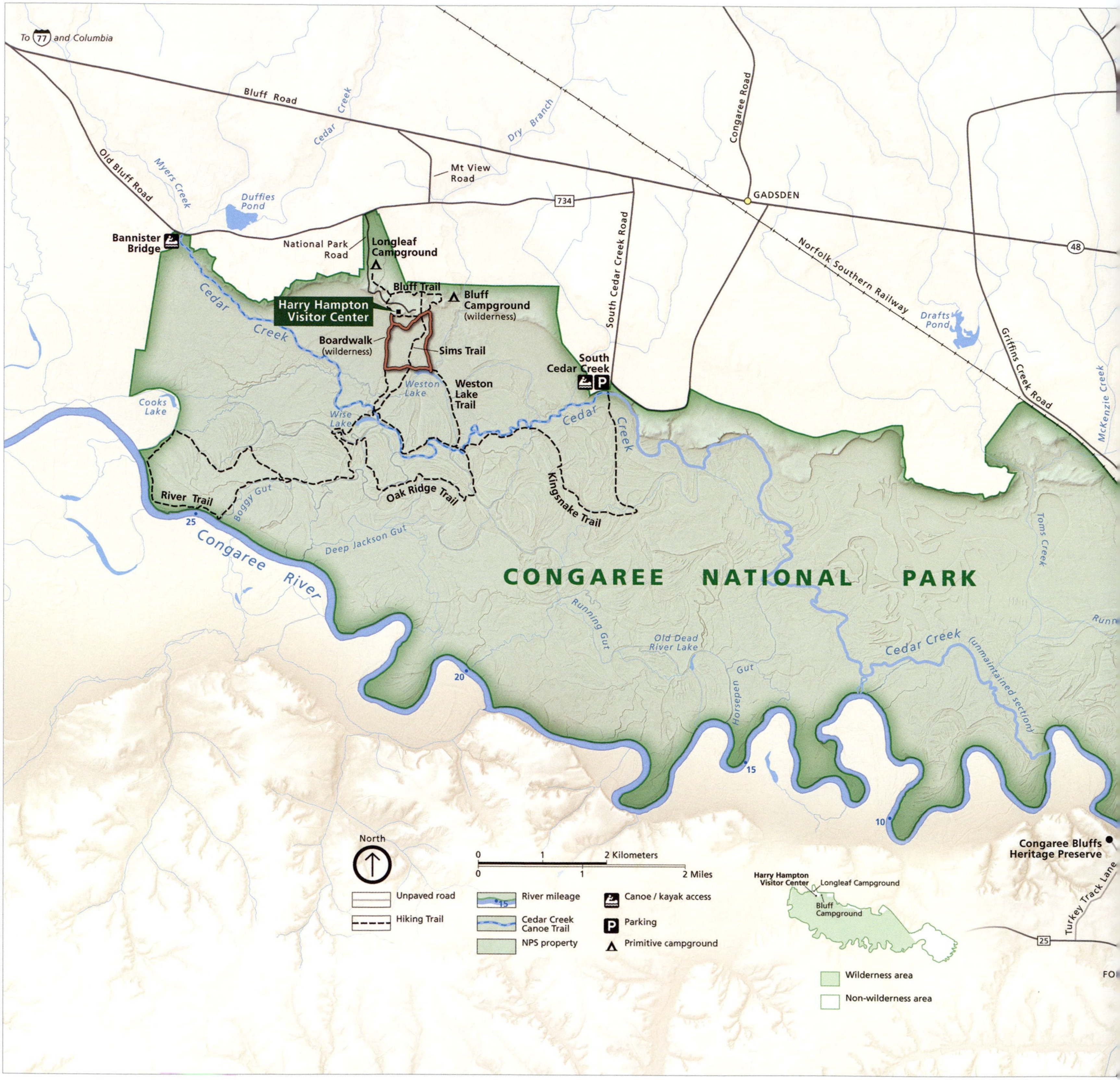

To 77 and Columbia
Bluff Road
Old Bluff Road
Myers Creek
Cedar Creek
Dry Branch
Mt View Road
Congaree Road
GADSDEN
734
48
Duffies Pond
Bannister Bridge
National Park Road
Longleaf Campground
Bluff Trail
Bluff Campground (wilderness)
Harry Hampton Visitor Center
Boardwalk (wilderness)
Sims Trail
Weston Lake
Weston Lake Trail
South Cedar Creek Road
South Cedar Creek
Norfolk Southern Railway
Drafts Pond
Griffins Creek Road
McKenzie Creek
Cooks Lake
Wise Lake
Cedar Creek
River Trail
Boggy Gut
Oak Ridge Trail
Kingsnake Trail
Deep Jackson Gut
Congaree River
25
20
15
10
CONGAREE NATIONAL PARK
Toms Creek
Running Gut
Old Dead River Lake
Horsepen Gut
Cedar Creek (unmaintained section)
Congaree Bluffs Heritage Preserve
Turkey Track Lane
25
North
0 1 2 Kilometers
0 1 2 Miles
Unpaved road
Hiking Trail
River mileage
Cedar Creek Canoe Trail
NPS property
Canoe / kayak access
Parking
Primitive campground
Harry Hampton Visitor Center
Longleaf Campground
Bluff Campground
Wilderness area
Non-wilderness area

Congaree National Park's brochure map. Data sources: USGS, NPS.

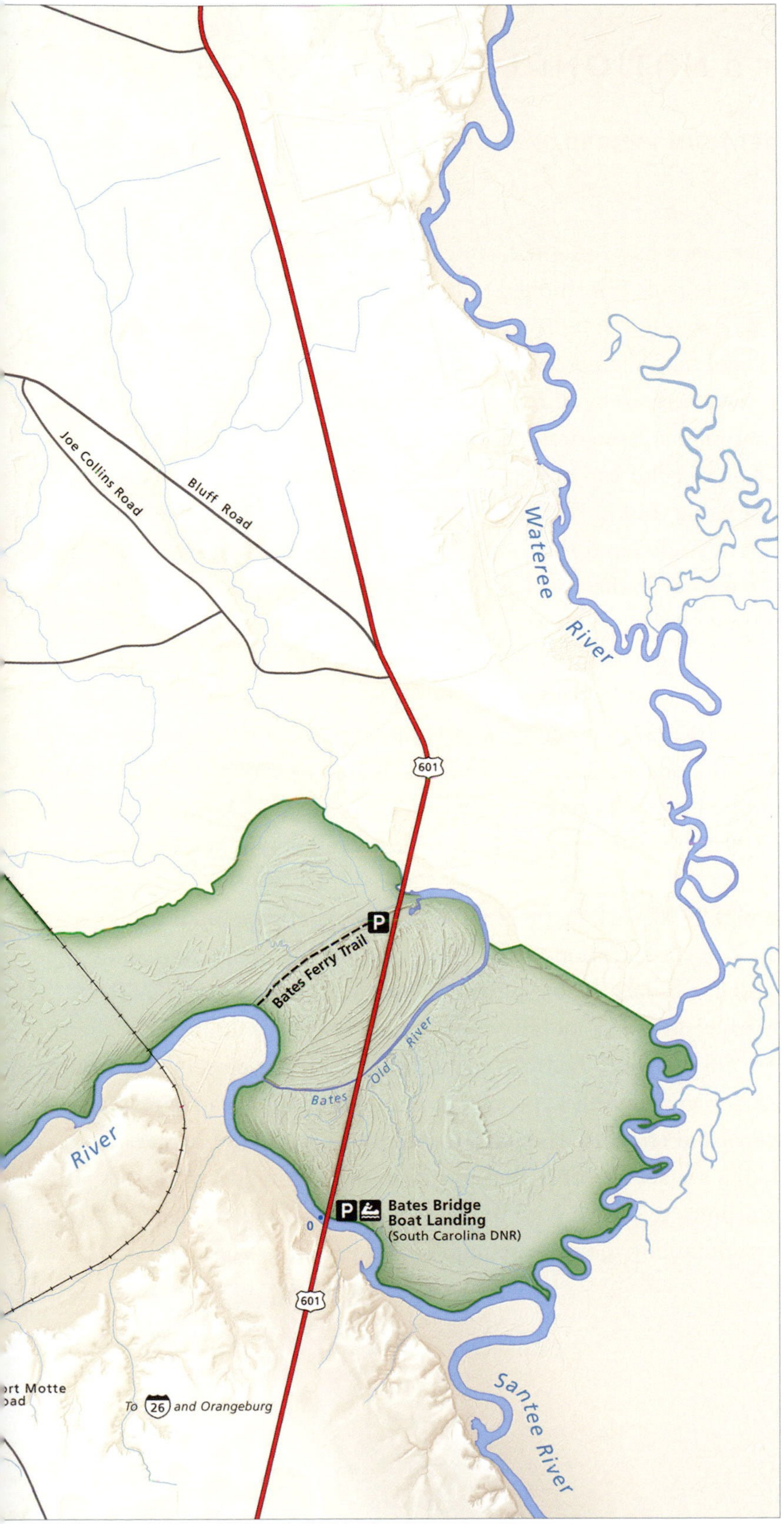

CONGAREE NATIONAL PARK

JIM EYNARD, NPS HARPERS FERRY CENTER FOR MEDIA SERVICES

Congaree National Park's brochure map uses high-resolution lidar derived elevation data to show the intricate network of waterways within this relatively flat area and highlights the wilderness area that makes up much of the park in South Carolina. Congaree National Park protects over 11,000 acres of old growth floodplain forest, the largest remaining parcel of this ecosystem that once included 52 million acres throughout the southeastern United States.

EVOLUTION OF A NATIONAL PARK SERVICE MAP

TOM PATTERSON (RETIRED), JIM EYNARD, AND JOE MILBRATH, NPS HARPERS FERRY CENTER FOR MEDIA SERVICES

National Park Service maps have oriented visitors to Yellowstone National Park's fabled beauty since the park transitioned to the NPS in 1917. This series of five maps illustrates the evolution of NPS cartography from traditional hand-drawn relief to the present-day digital maps composed of GIS data. First, the *Yellowstone* map from 1957 depicts a clear visual hierarchy of roads, points of interest, and physical features. A dominant, monochromatic, hand-drawn shaded relief softly highlights the park's topography. The first infusion of color can be found in the 1967 park map where a green park fill would become a defining trait in NPS maps. In its infancy, however, the dark-green fill overpowers the park's relief and roadways, making important features difficult to read. Symbols also begin to appear in the maps of the 1960s as spiky tents pointing visitors to campgrounds, and quadrant-circles peculiarly indicate ranger stations. In the 1979 map, NPS map symbology shifts to recognizable black pictographs, now an identifying feature of NPS maps. The map's shaded relief is restored to prominence with warm yellow highlights and cool blue shadows. In 1987, a green fill and boundary ribbon distinguishes the park from neighboring federal lands and becomes another defining characteristic of NPS maps. A black border and title banner, signature elements from designer Massimo Vignelli's retooled layout of the *Unrigid* brochure in 1977, frame a map that would go largely unchanged for nearly 30 years.

Finally, the 2018 map highlights park features with shaded relief generated using GIS data. Two important land cover classifications, meadows and forested areas, were delineated from the national land cover dataset to replicate the natural environment. With meadows clearly highlighted, visitors can identify areas to safely view wildlife from their vehicles or spotting scopes. The map also features a generalized terrain, closely resembling its hand-drawn predecessors, blended with hypsometric tints that range from snowy alpine peaks to darker dense valleys and streams. The park's road network, visitor centers, and points of interest are styled to a visual identity that has been developing over the past 50 years. NPS cartography continues to evolve with improvements in GIS technology and the availability of data, while maintaining the artistic qualities of previous generations.

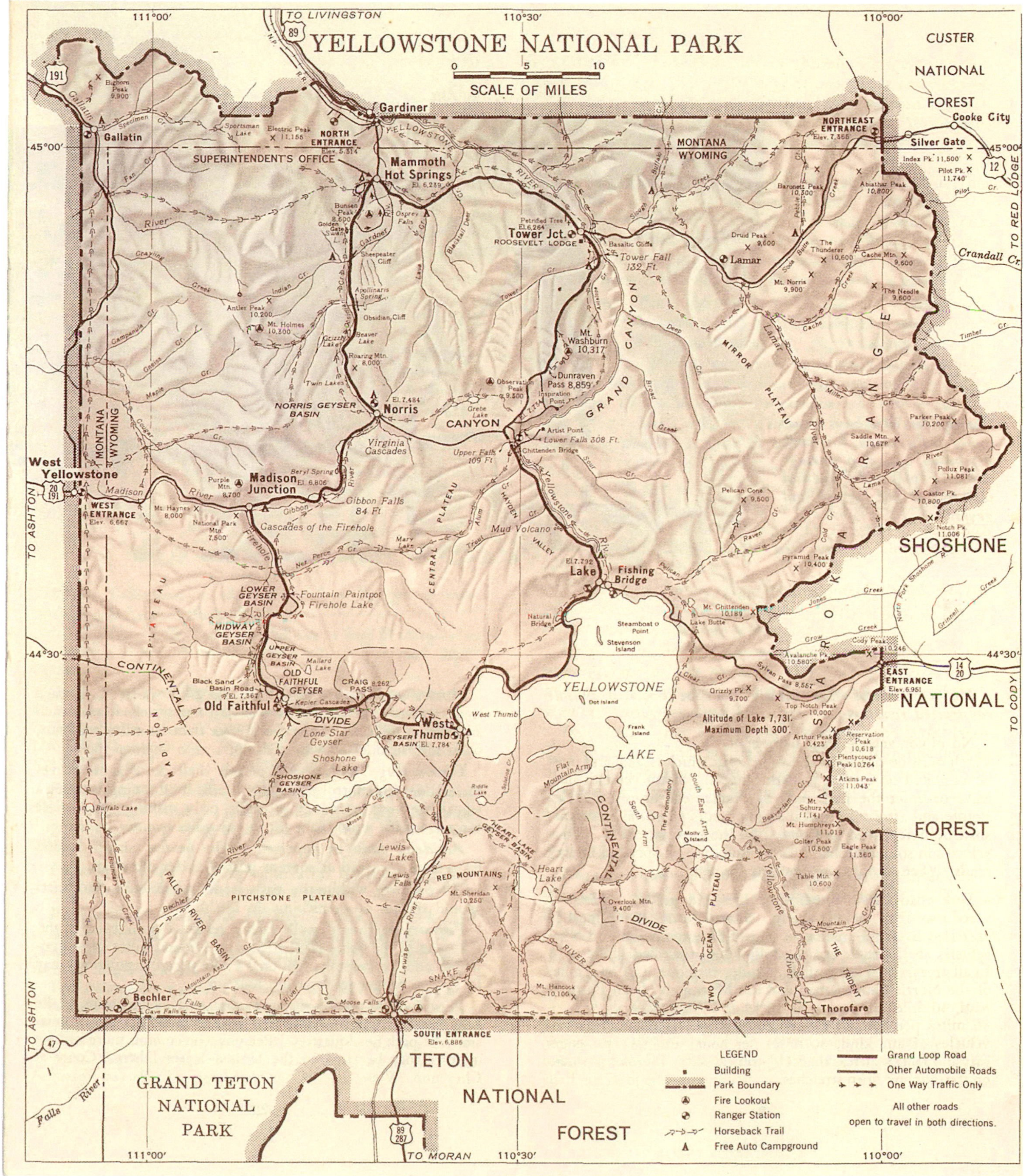

The 1957 map of Yellowstone depicts a clear visual hierarchy of roads, points of interest, and physical features.

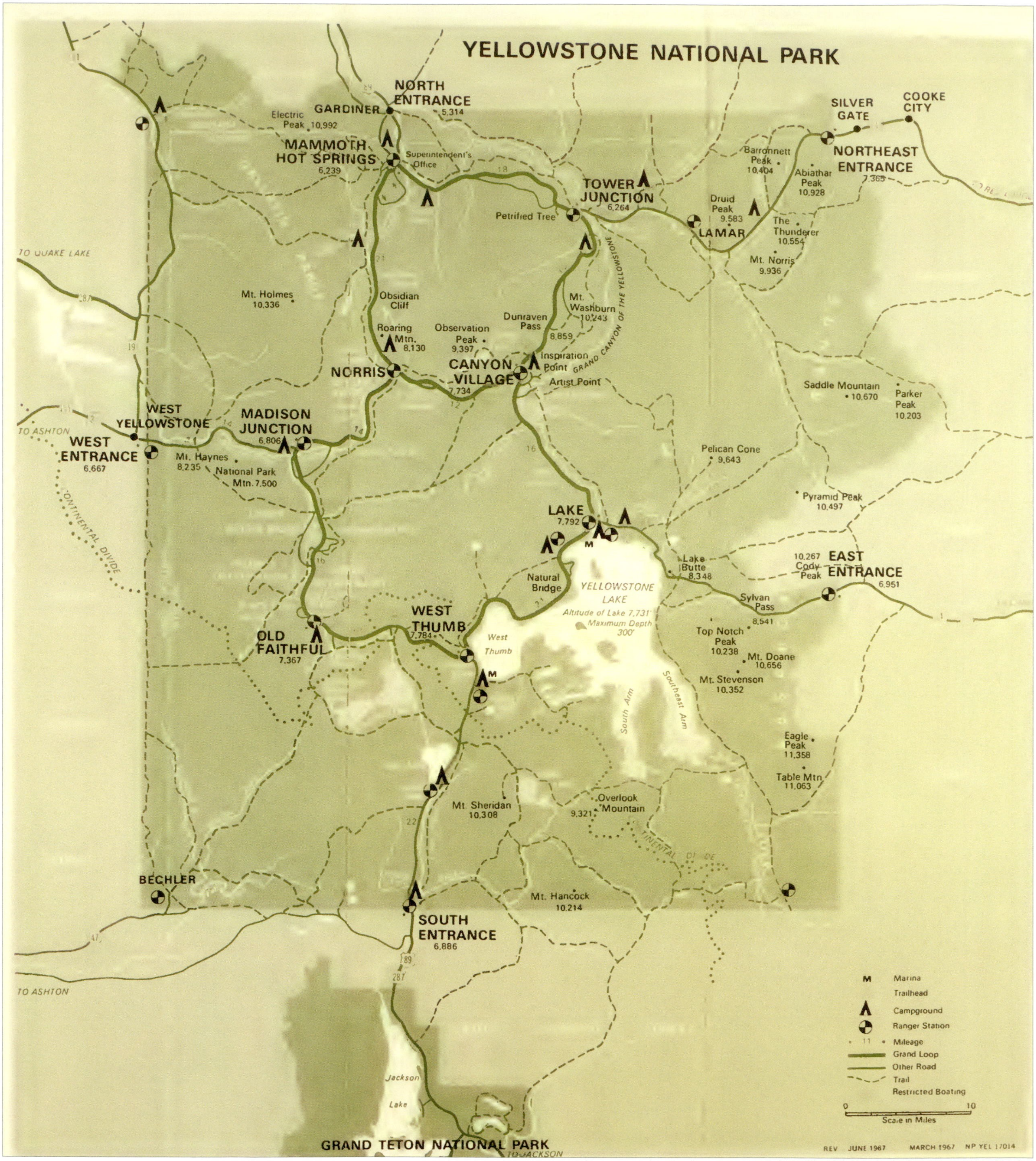

The first infusion of color can be found in the 1967 park map where a green park fill would become a defining trait in NPS maps.

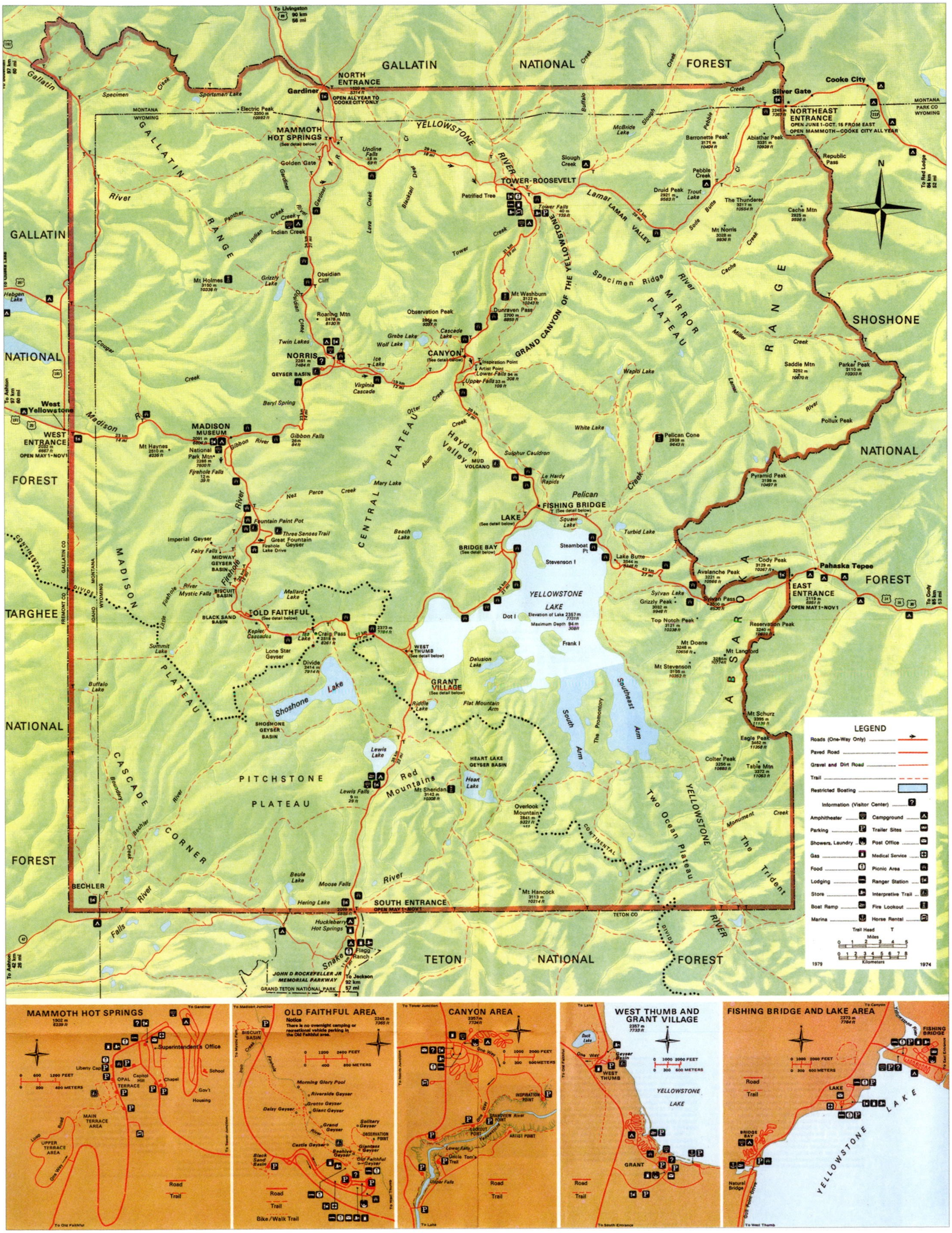

In the 1979 map, NPS map symbology shifts to recognizable black pictographs, now an identifying feature of NPS maps. Data sources on pages 15-18: NPS, Harpers Ferry Center.

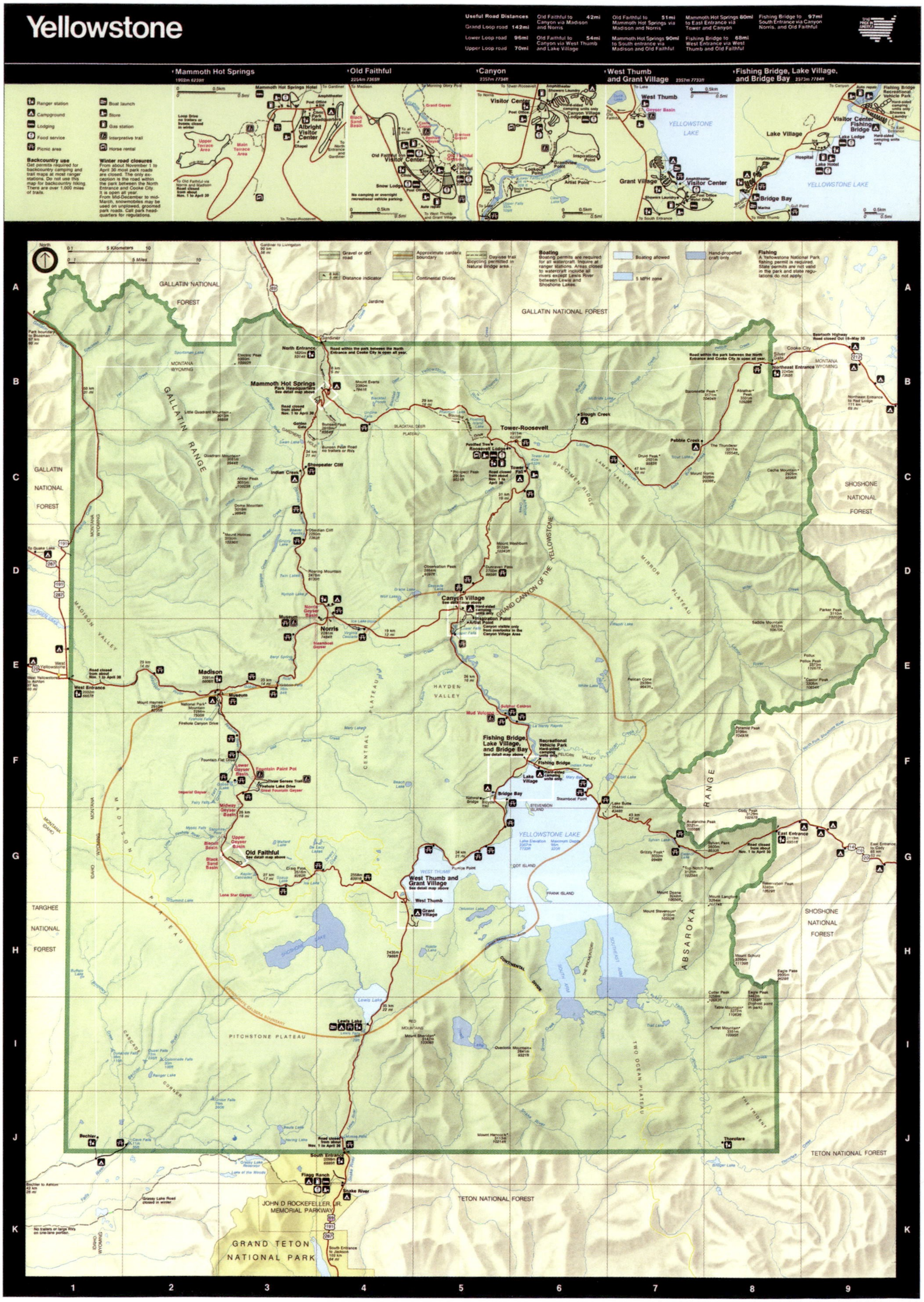

In 1987, a green fill and boundary ribbon distinguishes the park from neighboring federal lands and becomes another defining characteristic of NPS maps.

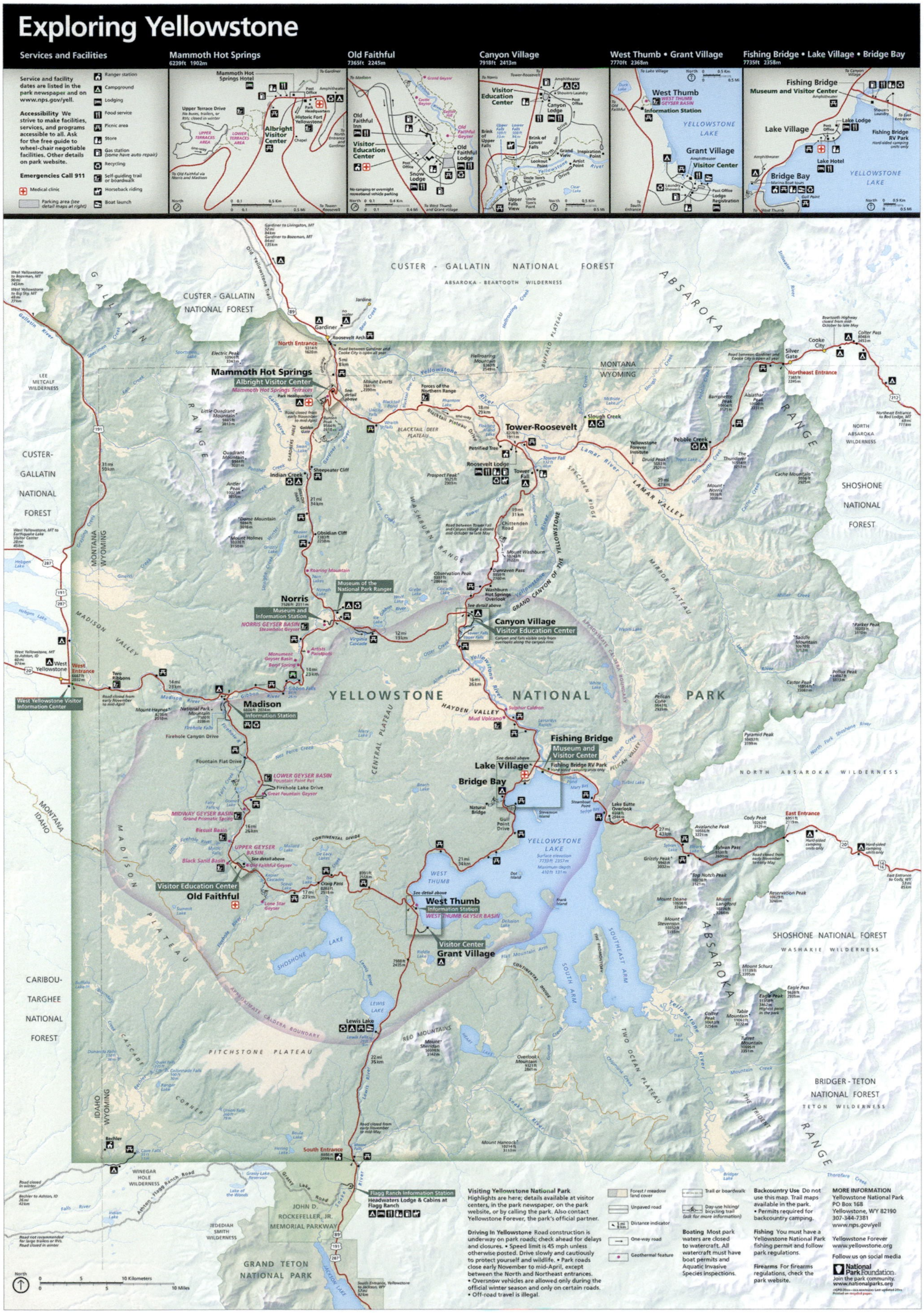

The 2018 *Exploring Yellowstone* map highlights park features with shaded relief generated using GIS data. Data sources: USGS, NPS.

2 RECREATION

SUSAN MCPARTLAND, NPS DENVER SERVICE CENTER PLANNING DIVISION

ENJOYING THE PARKS

Recreation at its core is the act of interacting with the outside world. Recreation can improve human health in physical, emotional, and spiritual ways. It also supports the values that led to and continue to support the very idea of public lands in the United States. This support is because of the connections people build while learning about or recreating in a site.

The people who visit National Park Service (NPS) sites are as varied and unique as the places they seek. So too are the recreation opportunities that await them. The long and diverse list of recreational activities available in park sites changes over time as people find new and innovative ways to interact with the outside world. Biking, kayaking, hunting, virtual touring, rock climbing, walking, bird watching, and hiking are just a few of the ways people recreate in NPS sites. No matter the form of recreation, people build connections to the cultural and natural resources that the park service protects.

As the projects in this chapter demonstrate, mapping is a powerful, dynamic, and important medium within the world of recreation. Mapping can be used for trip planning as a person seeks to understand what they can expect along a trail. It can also help connect people with opportunities that they may not have known were available to them or connect them with a recreation activity they have never tried before. Mapping can equally be used by a person who may never visit a faraway place, and yet through 3D modeling and mapping, that place becomes real and important to them. Just as GIS aids the NPS with management decisions, it also helps the public understand how and why a park site is being managed and how they can help become stewards of that place. Ultimately, mapping builds connections between people and places that represent the many stories and perspectives of this country and the world beyond.

THE JUAN BAUTISTA DE ANZA NATIONAL HISTORIC TRAIL IN SOUTHERN ARIZONA

PHIL VALDEZ JR.,[1] THE DE ANZA SOCIETY
LIZZET PINEDA, LATINO HERITAGE INTERNSHIP PROGRAM
BRIANNA WELDON, JUAN BAUTISTA DE ANZA NATIONAL HISTORIC TRAIL

The Santa Cruz Valley in Southern Arizona is the cradle of a rich chapter in American history and one of the premier places to experience the Juan Bautista de Anza National Historic Trail in the United States. This story using ArcGIS StoryMaps examines the history, environment, and recreational opportunities found in this beautiful landscape and introduces digital visitors to the resources of the area and where to visit the Anza Trail. *Disponible en español y inglés* (available in Spanish and English) can be found at https://go.nps.gov/jubasoazmap.

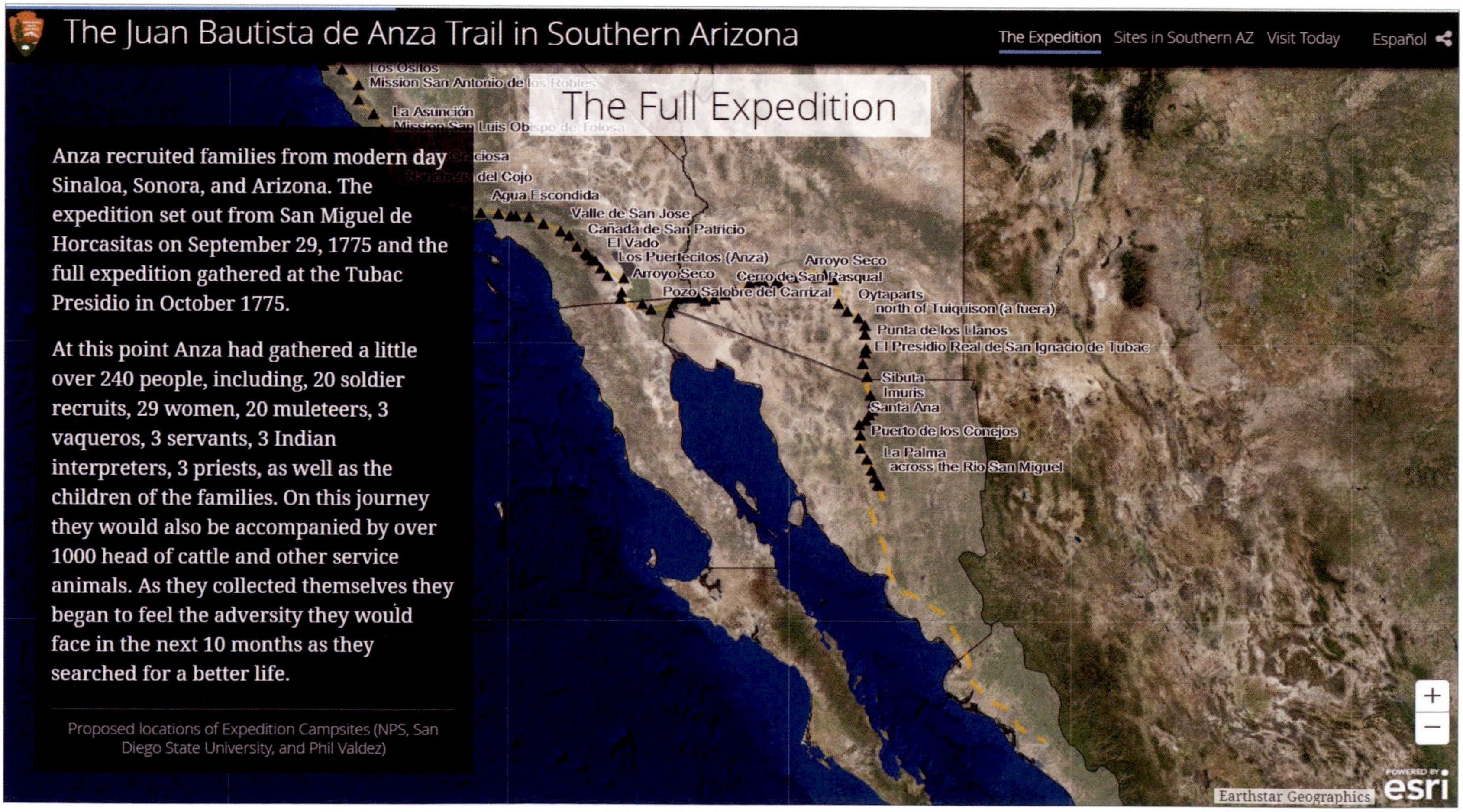
The Juan Bautista de Anza Trail in Southern Arizona
The Expedition
Sites in Southern AZ
Visit Today
Español
The Full Expedition
Anza recruited families from modern day Sinaloa, Sonora, and Arizona. The expedition set out from San Miguel de Horcasitas on September 29, 1775 and the full expedition gathered at the Tubac Presidio in October 1775.
At this point Anza had gathered a little over 240 people, including, 20 soldier recruits, 29 women, 20 muleteers, 3 vaqueros, 3 servants, 3 Indian interpreters, 3 priests, as well as the children of the families. On this journey they would also be accompanied by over 1000 head of cattle and other service animals. As they collected themselves they began to feel the adversity they would face in the next 10 months as they searched for a better life.
Proposed locations of Expedition Campsites (NPS, San Diego State University, and Phil Valdez)
Los Ositos
La Asunción
del Cojo
Agua Escondida
Valle de San Jose
Cañada de San Patricio
El Vado
Los Puertecitos (Anza)
Arroyo Seco
Arroyo Seco
Cerro de San Pasqual
Pozo Salobre del Carrizal
Oytaparts
north of Tuiquison (a fuera)
Punta de los Llanos
El Presidio Real de San Ignacio de Tubac
Sibuta
Imuris
Santa Ana
Puerto de los Conejos
La Palma
across the Rio San Miguel
Earthstar Geographics
POWERED BY esri

This story shows the full Anza Expedition. Starting on September 29, 1775, in San Miguel de Horcasitas, Sonora, Mexico, Spanish Lieutenant Colonel Juan Bautista de Anza led an ethnically diverse group of Native American, European, and African-heritage colonists 1,800 miles across the frontier of New Spain to create a settlement in present-day San Francisco, California. De Anza reached the Bay Area in late March 1776, and the rest of the expedition, including more than 100 children, arrived in late June.

Data sources: Esri, National Atlas of the United States, USGS, and NPS.

The Juan Bautista de Anza Trail in Southern Arizona
The Expedition Sites in Southern AZ Visit Today Español
Mission San Jose de Tumacácori
The travelers faced no easy task embarking on this long journey through the Sonoran desert. Today, however traveling through and even living there is much easier allowing you to explore the area and hiking the Anza trail. You can visit Tumacácori National Historical Park and participate in different activities such as earning a badge by hiking a portion of the Anza trail from Tumácacori to Tubac. Along the trail, you may experience the wildlife and see rare birds travelling along their migratory path. Stop in and learn about the expedition and the area at the museum.
On the trail to Tubac from Tumacácori (NPS)

NPS volunteers of the Anza Trail Colorguard ride horses on the Anza Trail between Tubac Presidio State Historic Park and Mission San Jose de Tumacácori National Historical Park.

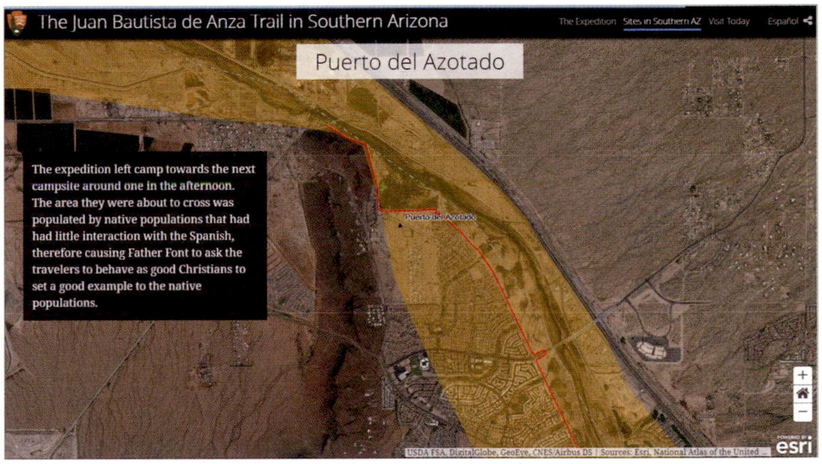
The Juan Bautista de Anza Trail in Southern Arizona
The Expedition Sites in Southern AZ Visit Today Español
Puerto del Azotado
The expedition left camp towards the next campsite around one in the afternoon. The area they were about to cross was populated by native populations that had had little interaction with the Spanish, therefore causing Father Font to ask the travelers to behave as good Christians to set a good example to the native populations.
Puerto del Azotado

The illustration shows the Puerto del Azotado campsite within the historic corridor and the present-day recreation retracement route.

The Juan Bautista de Anza Trail in Southern Arizona
The Expedition Sites in Southern AZ Visit Today Español
Puerto del Azotado
During the travel they noticed two muleteers had escaped as they felt the trip had become too much for them. Anza alerted the Native inhabitants to be on the lookout for the escaped muleteers. The expedition then halted at La Frente Negra Mountains, and at Eight at night a few Tohono O'odham approached camp with one of the fugitives. The punishment given to this muleteer was 12 slashes, he was then tied to a post where he was left overnight. This event gave the name to this campsite "Puerto del Azotado" meaning "Pass of the Slashed."
Muleteers (NPS/Bill Singleton)

These illustrations depict the muleteers that ran away at Puerto del Azotado and the expedition in the Sonoran desert. © Illustration by Bill Singleton, www.billsingleton.net.

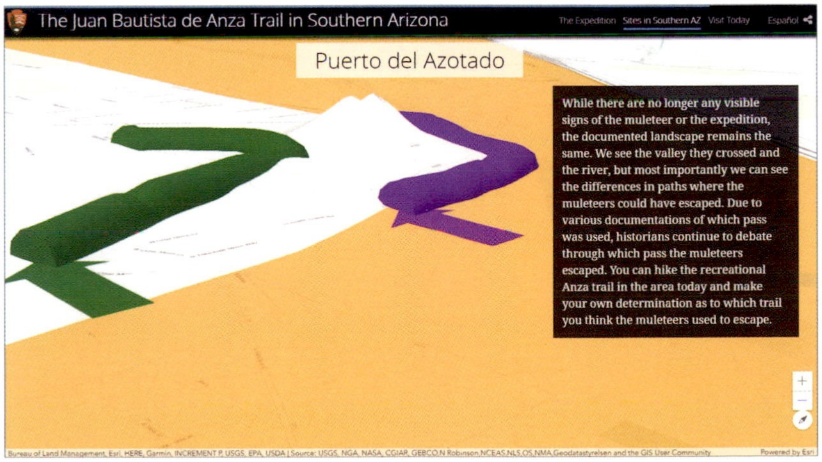
The Juan Bautista de Anza Trail in Southern Arizona
The Expedition Sites in Southern AZ Visit Today Español
Puerto del Azotado
While there are no longer any visible signs of the muleteer or the expedition, the documented landscape remains the same. We see the valley they crossed and the river, but most importantly we can see the differences in paths where the muleteers could have escaped. Due to various documentations of which pass was used, historians continue to debate through which pass the muleteers escaped. You can hike the recreational Anza trail in the area today and make your own determination as to which trail you think the muleteers used to escape.

This map shows potential escape routes of the muleteers in the area of Puerto del Azotado studied by Anza Trail volunteer historians.

NOTE

1. Dr. Phil Valdez, longtime National Park Service historian advisor and volunteer at Anza Trail, received the US Department of Interior Citizen's Award for Exceptional Service in 2015. He passed away in 2018.

A 3D MAP OF BOOKER T. WASHINGTON NATIONAL MONUMENT

JIM EYNARD, NPS HARPERS FERRY CENTER FOR MEDIA SERVICES

This large-scale 3D oblique map of this park was created for a new unigrid brochure in 2017. The map shows the reconstructed buildings and landscape of this *living history* site where Booker T. Washington was born in 1857 and lived as a slave until he was freed with the Emancipation Proclamation in 1863. Geospatial data was used for the elevation model and the locations of the buildings, paths, fences, and garden. Models were created for all the structures in 3D. The data was then brought into a 3D rendering software (Vue Infinite) to render a hyper-realistic image. Plants, trees, and atmospheric effects were created within the 3D software. While the final map is not georeferenced, it was built almost entirely with geospatial data. The ghosted buildings on the map show buildings that no longer exist at the site.

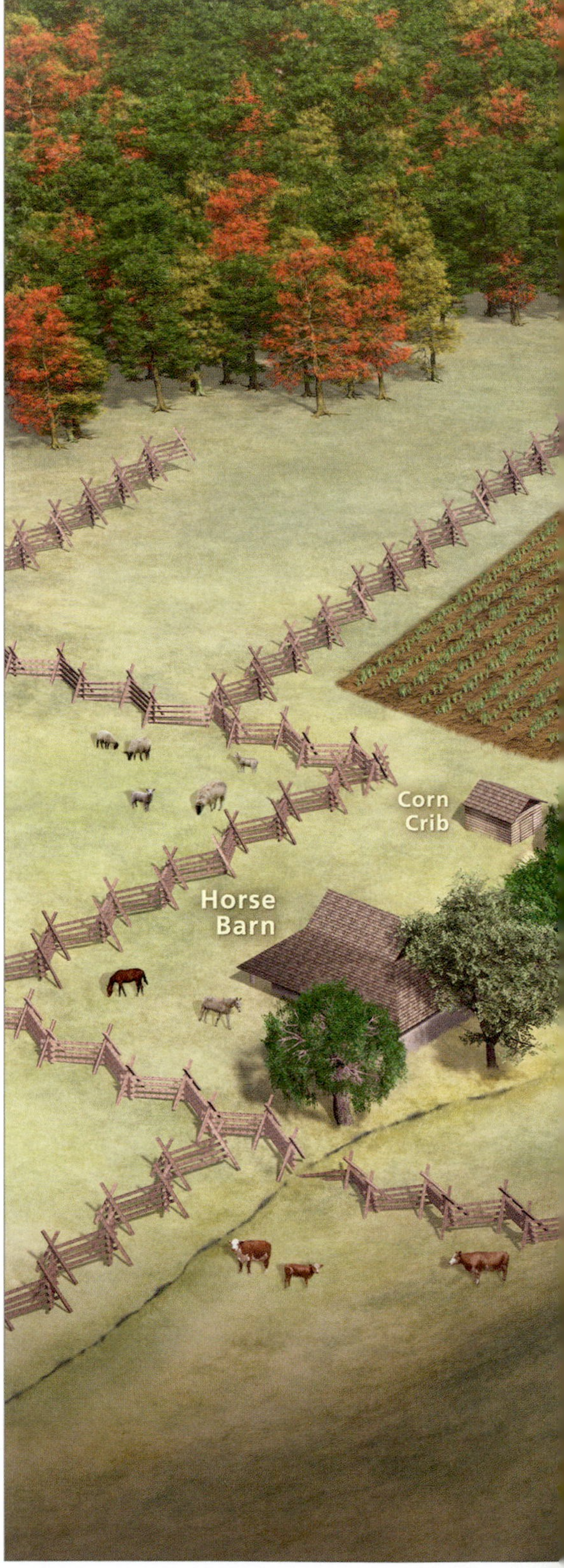

The map shows the reconstructed buildings and landscape of this living history site showing where Booker T. Washington was born in 1857. Data sources: NPS and USGS.

To
Jack-O-Lantern
Branch Heritage Trail
Tobacco
Barn
Corn
Field
Plantation
Trail
Loop
Tobacco
Field
Burroughs
House
Site
Slave
Cabin
Site
Smokehouse
Kitchen
Cabin
(where Booker
lived)
Blacksmith
Shed
Duck
Lot
Chicken
House
Plantation
Trail
Loop
Heirloom
Garden
Hog
Pens
To
Visitor
Center

DRY TORTUGAS NATIONAL PARK

TOM PATTERSON, NPS HARPERS FERRY CENTER FOR MEDIA SERVICES (RETIRED)

The 100-square-mile *Dry Tortugas National Park* map near Key West, Florida, is featured in the park's brochure. This map features aerial imagery to highlight the marine environment around Fort Jefferson and the surrounding keys. Maritime icons are used to show the locations of shipwrecks, scuba diving and snorkeling areas, and navigational buoys. The map also highlights the neighboring Tortugas Ecological Reserve and Florida Keys National Marine Sanctuary.

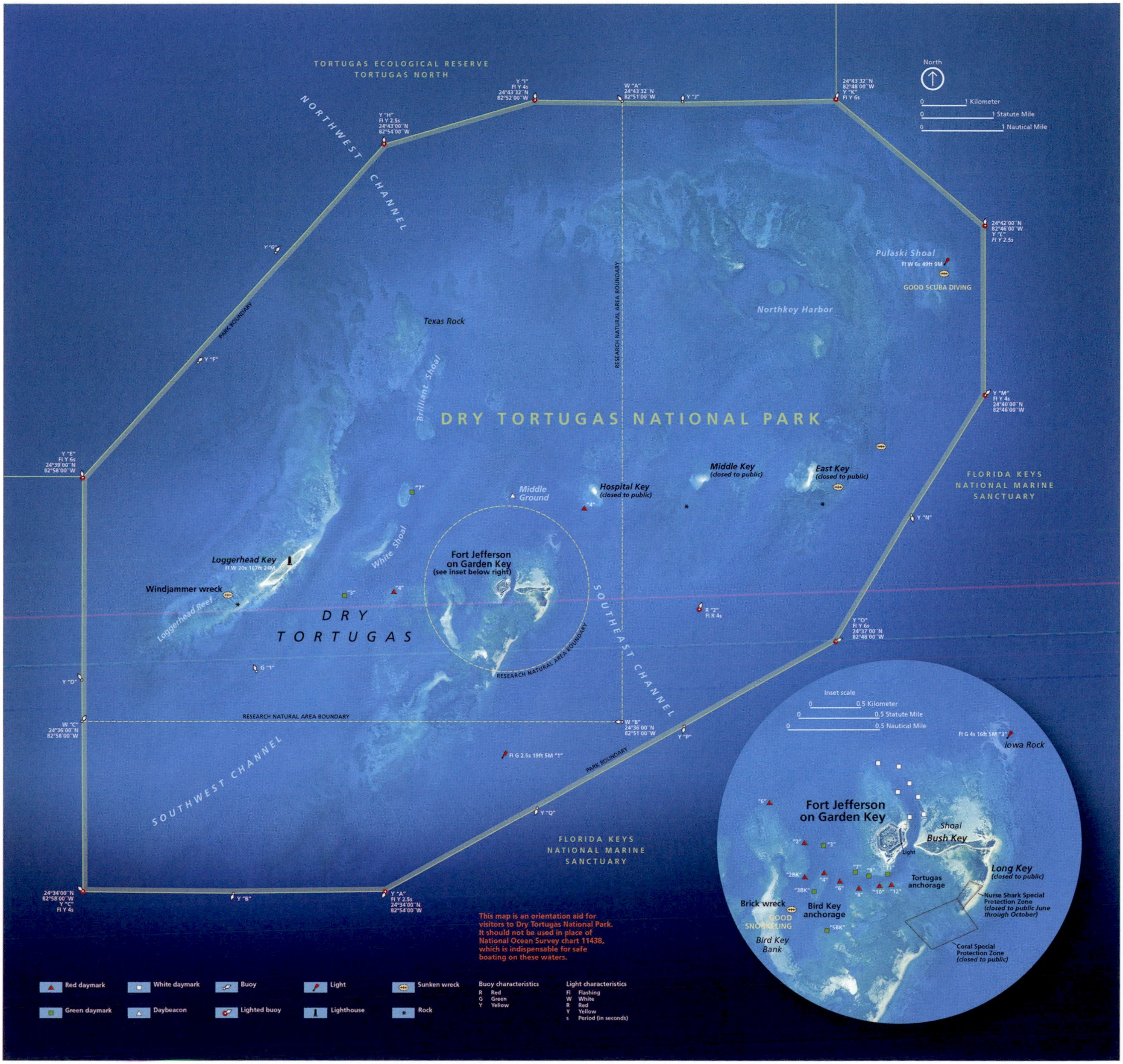

This map depicts Dry Tortugas National Park about 70 miles west of Key West, Florida, and highlights the surrounding marine environment. Data sources: NPS.

NORTH CASCADES NATIONAL PARK COMPLEX—STEHEKIN AREA

JIM EYNARD, NPS HARPERS FERRY CENTER FOR MEDIA SERVICES

The Stehekin area map was created for the new North Cascades National Park brochure published in 2018. The 3D oblique map features natural colors and a high-resolution DEM to show the mountainous terrain, snowy peaks and glaciers, lakes, and green valleys. Stehekin is a remote community in Lake Chelan National Recreation Area, part of the North Cascades National Park complex in Washington. Situated at the northwestern end of Lake Chelan, Stehekin has no road access and is accessible only by ferry, by float plane, or by hiking across mountainous terrain. The map shows the 22 miles of road in the area, an extensive network of hiking trails that connect to the Pacific Crest Trail, and other park features such as campgrounds and the Golden West Visitor Center. This Stehekin area map supplements a park map and a Highway 20 inset map also found on the park's brochure.

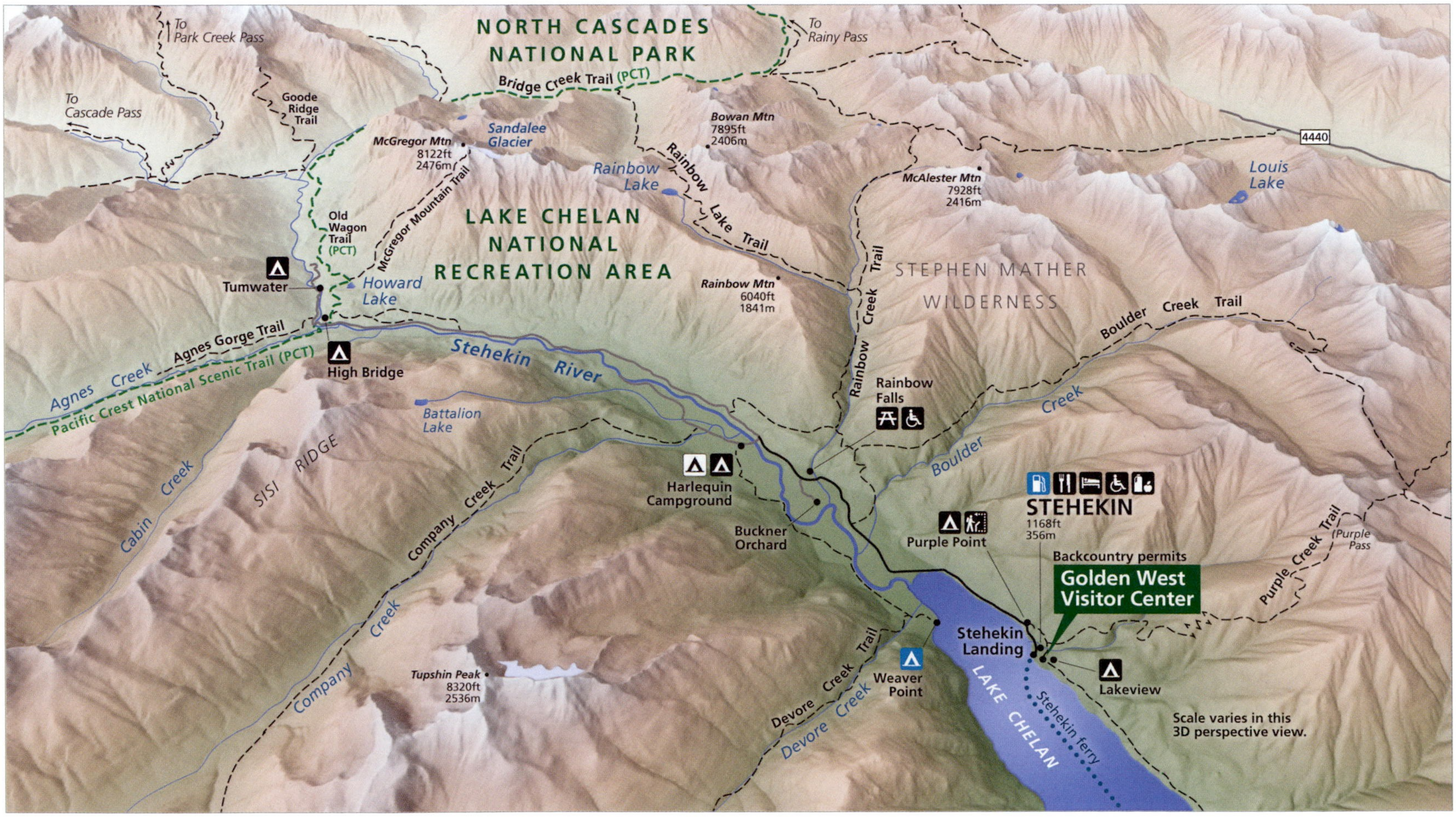

This 3D map of the Stehekin area of North Cascades National Park helps visitors navigate the remote park. Data sources: USGS and NPS.

GRAND CANYON HIKING MAP

TOM PATTERSON, NPS HARPERS FERRY CENTER FOR MEDIA SERVICES (RETIRED)

The *Grand Canyon Hiking Map* was developed in 2017 as part of a supplemental handout guide to inform visitors of the time commitment and the difficulty of hiking deep into the canyon. The hiking map highlights the popular Kaibab and Bright Angel trails, key points along each route, and blends terrain texture shading with hypsometric tints for the canyon's relief. Rock textures rendered in the Natural Scene Designer software program add subtle realistic detail. The map also features informational charts that illustrate the distance, elevation change, and time commitments hikers will endure when making the journey into the canyon.

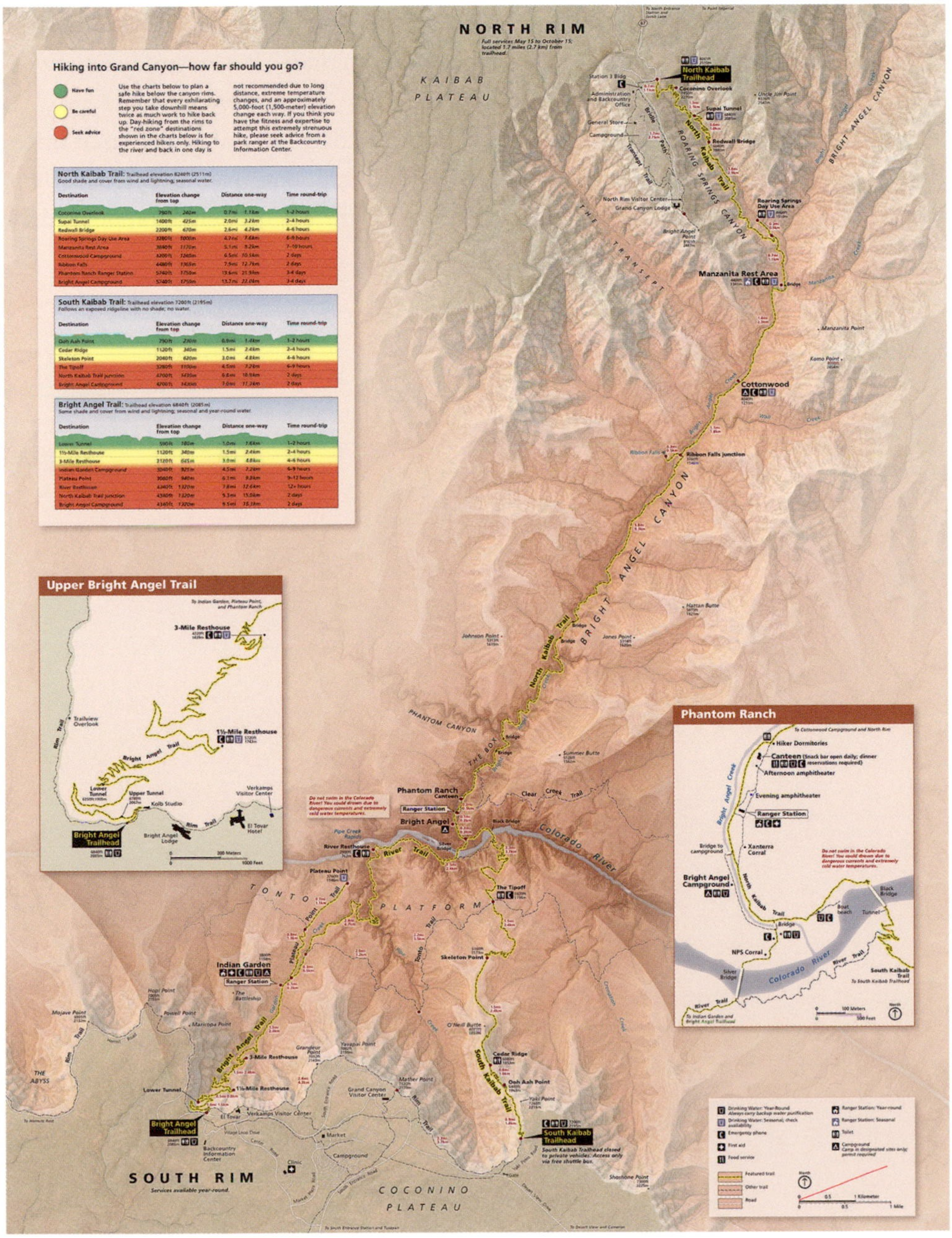

This hiking map of the Grand Canyon highlights the Kaibab Trail and Bright Angel Trail linking the North and South Rim. Data sources: USGS and NPS.

LIDAR AND THE FORTS OF PETERSBURG

DAVID W. LOWE, NPS CULTURAL RESOURCES

Petersburg, Virginia, was the location of the US Civil War's longest siege (June 1864–April 1865). For more than nine months, the Union and Confederate armies battled each other with artillery and sharpshooters from behind formidable fortifications. Each of these forts had a unique design that varied with the terrain, the role it played in the entrenched system overall, and its armament and available manpower. Federal engineers drew up precise plans of these forts. To bring these forts to life, 14 fort plans were first georeferenced and then overlain on a lidar (light detection and ranging) hillshade. Each of the forts in panel 1 of this presentation using ArcGIS StoryMaps is depicted to the same scale, so that comparisons of size and shape can be made. These overlays allow historians to identify and interpret features that survive on the ground today. *Panel 2* explains more of the process and provides an overview of the original miles of trench lines that made this siege look more like a World War I battlefield than one from the mid-19th century. The period photographs provide a glimpse of how grim the battlefield appeared at war's end.

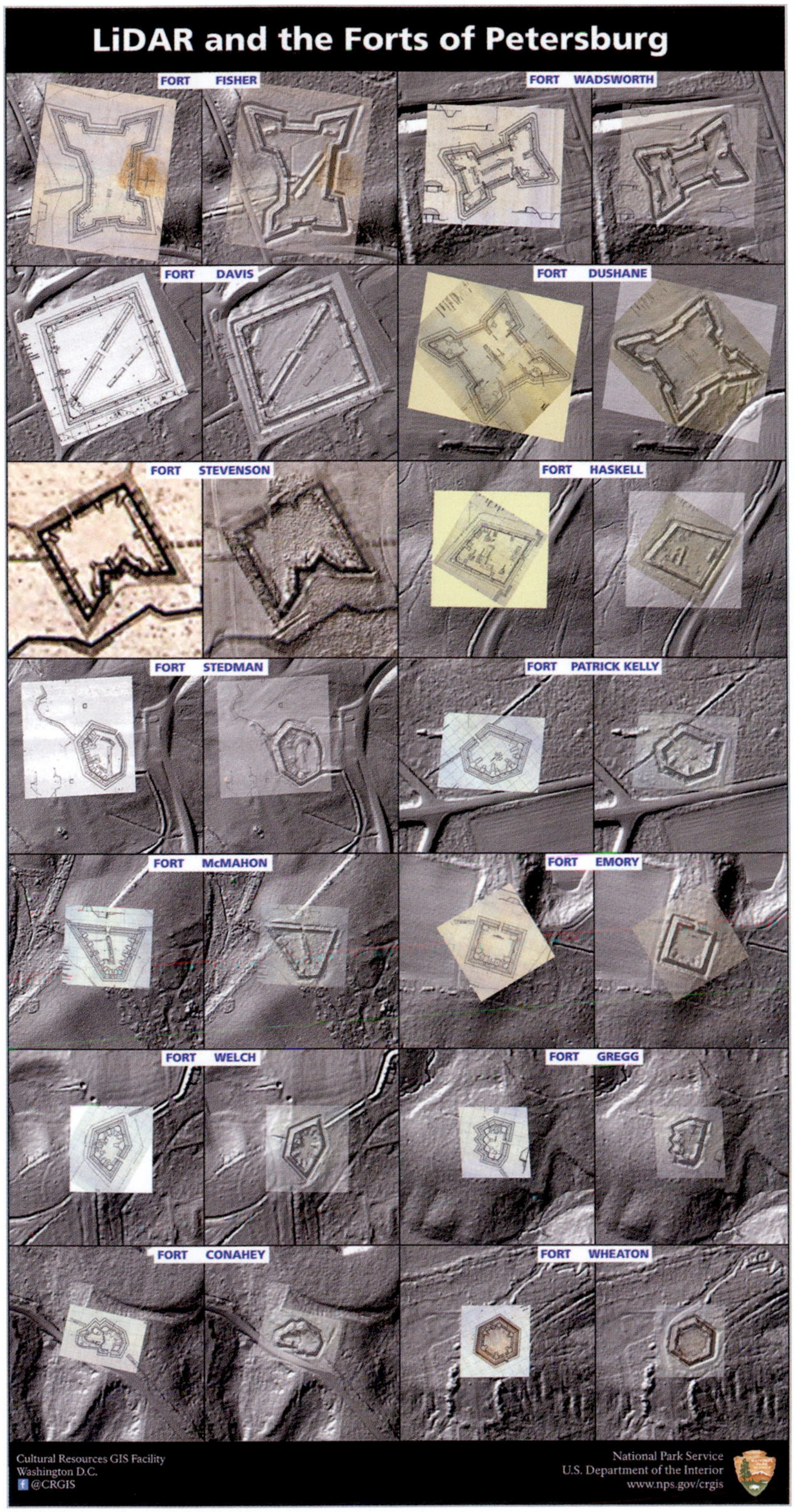

This *Lidar and the Forts of Petersburg* poster shows US engineers' plans for the Civil War forts built during the Siege of Petersburg. The poster displays fort plans coupled with transparent overlays on lidar hillshades providing a 3D view of surviving earthworks. Civil War photographs, 1861–1865, Library of Congress, Prints and Photographs Division. Data sources: NPS, Library of Congress (Prints and Photographs Division), Library of Congress (Geography and Map Division), National Archives, Esri, National Geodetic Survey, and Nathaniel Michler.

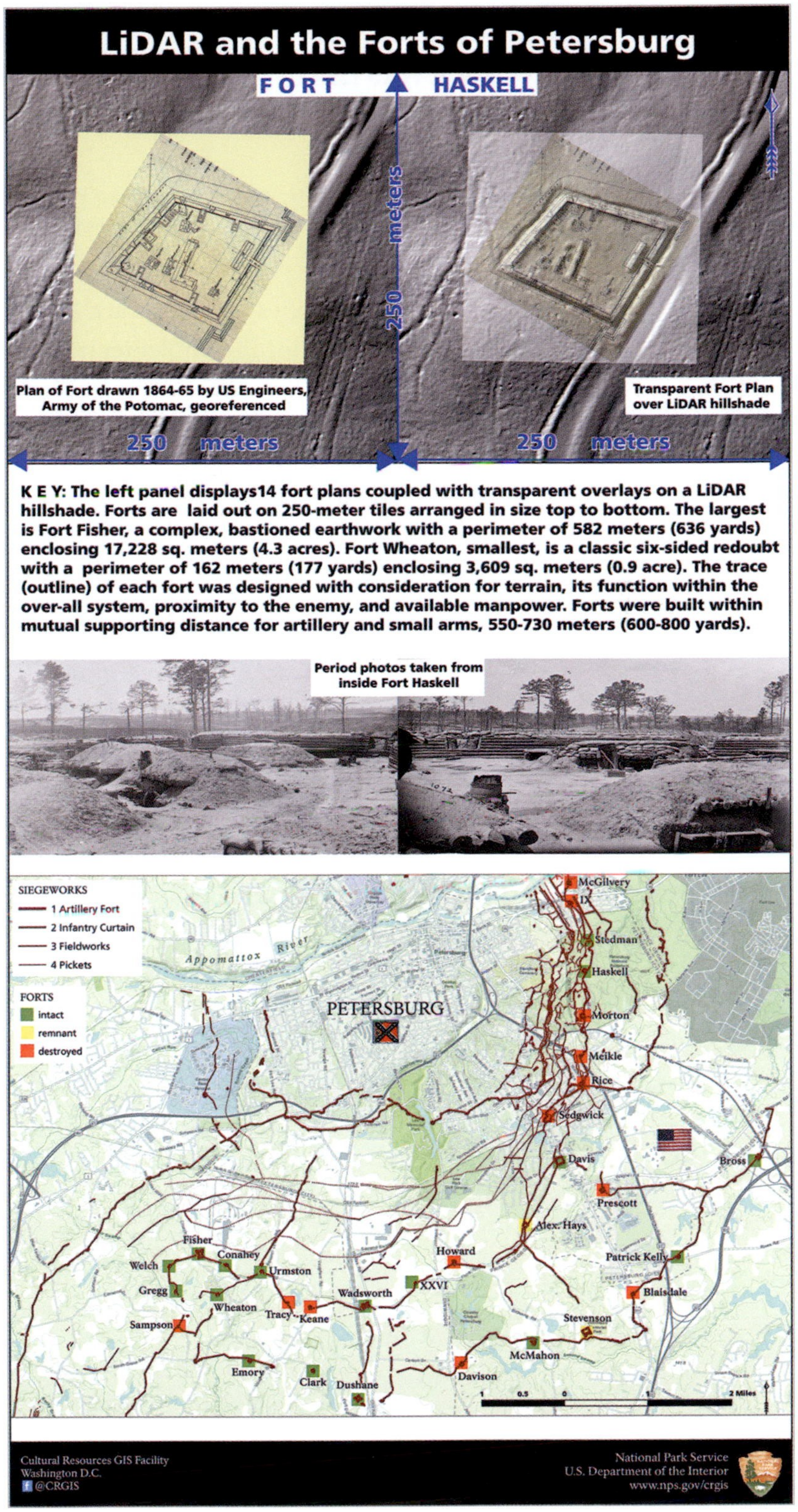

This *Lidar and the Forts of Petersburg* poster displays images, maps, and information about the project. Civil War photographs, 1861–1865, Library of Congress, Prints and Photographs Division. Data sources: NPS, Library of Congress (Prints and Photographs Division), Library of Congress (Geography and Map Division), National Archives, Esri, National Geodetic Survey, and Nathaniel Michler.

EVERGLADES NATIONAL PARK FLORIDA BAY BOATING GUIDELINES

CHRIS ANDERSON, FLORIDA FISH AND WILDLIFE CONSERVATION COMMISSION
MATT PATTERSON, EVERGLADES AND DRY TORTUGAS NATIONAL PARKS

The Everglades National Park 2016 General Management Plan calls for the implementation of pole and troll zones and slow speed corridors in Florida Bay to protect marine resources from further damage caused by motorboat propellers in seagrass beds and other shallow water habitats. The park's GIS staff, together with park management and natural resources managers, mapped pole and troll zones and slow speed and idle speed access corridors using ArcMap™. They also conducted an inventory of aids to navigation using the ArcGIS® Collector mobile application.

Boater education is a second management plan priority to protect marine resources and improve boating and safety in the park. Park staff worked closely with the Florida Fish and Wildlife Conservation Commission's cartography staff to develop boating guidelines for the Florida Bay (see the commission's map for Florida Bay) using the park's pole and troll zones, on-plane, idle speed and slow speed corridors, and aids to navigation GIS layers. The Florida Bay Boating Guide will help promote and educate the public of the new management zones and regulations in Everglades National Park.

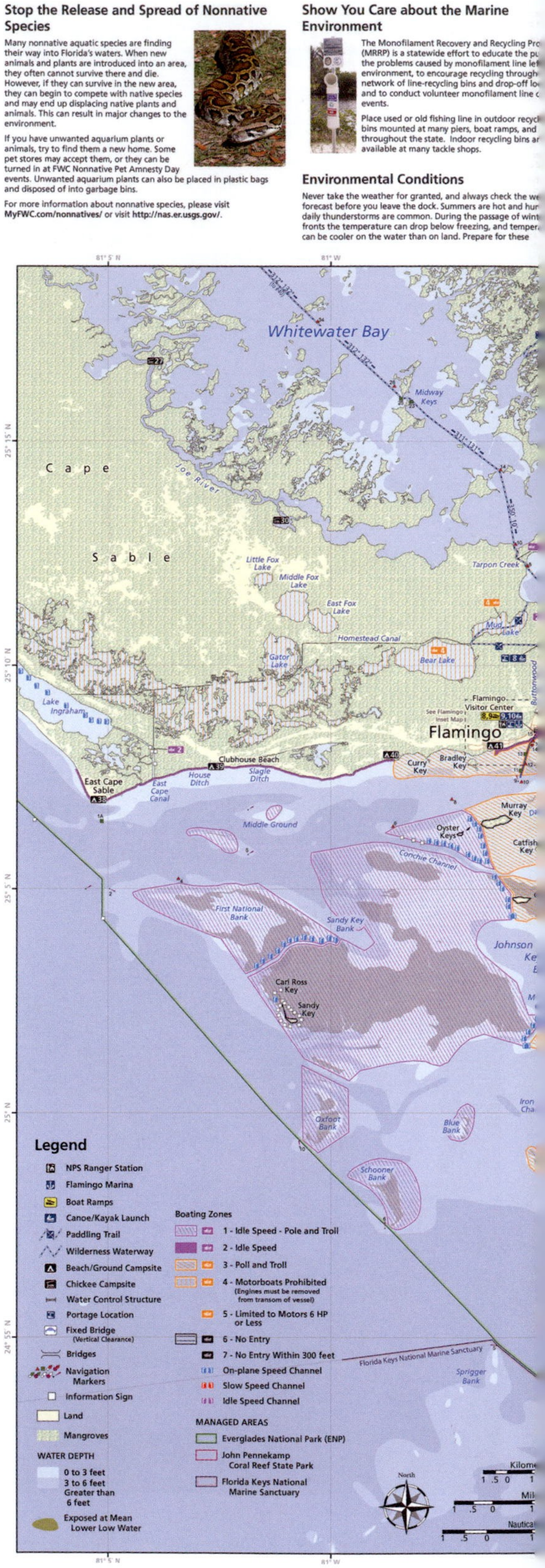

The map shows Everglades National Park pole and troll zones and slow-speed corridors. Data sources: National Park Service, Florida Fish and Wildlife Commission.

…screen, wet/cold weather gear, and …n per person per day).

…of the tidal changes within the park. …ow expanse of Florida Bay and the …ffect of tides on water level can be …amatic. It can take years of experience to …tides, wind direction, seasonal water levels, … depth in the park's waters. For this … should use only the basins and channels.

…ater depth at mean (average) low tide, but …ent than shown on charts due to the …al flows of freshwater from the mainland. … location can be a helpful tool in …er level is rising or falling. The daily tides …d or amplified by a sustained strong wind.

…hill in the air, wind can cause problems for …rs have been swamped by the choppy seas …ect the water levels throughout the … can be completely negated by a sustained …

… are PROHIBITED in all canals, ponds, …channels, and boat basins throughout the park. Elsewhere, swimming is generally discouraged due to the presence of large numbers of sharks and other predators.

Fishing Regulations

Everglades National Park regulations which can be seen at: ***https://www.nps.gov/ever/planyourvisit/fishing.htm***. Statewide regulations are managed by the Florida Fish and Wildlife Conservation Commision and can be accessed at MyFWC.com.

Catch-and-Release Information

Increasingly, anglers are practicing "catch-and-release" to do their part to preserve marine fisheries while they enjoy their outdoor fishing experiences. Here are some tips on how you can properly handle and release saltwater fish:

Know Before You Go

- Make sure you can identify the fish you are targeting.
- Always know (or have access to) the current fishing regulations. This minimizes handling time when determining whether or not you can keep the fish you caught.
- Use tackle heavy enough to bring the fish in quickly, and avoid using multi-hook rigs or lures.
- If you have a treble hook, you can remove some of the hooks and flatten the barbs.
- Make sure you have all the proper tools and gear on your vessel before heading out for the day.

Handling Fish

- Handle fish as little as possible and only with wet hands.
- Match tackle to the targeted fish to land it quickly and minimize stress on the fish.
- Never hold a fish by its jaw, gills or eyes. Hold the fish horizontally and support its weight with both hands.
- Release the fish as quickly as possible and if possible keep the fish in the water at all times.

Removing the Hook

- Using a dehooking tool is safer for the fish and for you.
- Cut the leader as close to the hook as possible if it cannot be quickly removed.
- When using natural bait (live or dead) use circle hooks to reduce internal harm and decrease dehooking time.

The Release

- Gently place the fish head first in the water, supporting its body until it swims away.
- A fish that has been stressed by the fight or handling should be revived by moving it forward in the water to promote water flow over the gills.
- If a released fish does not swim away, recover it and try again.
- For more information about catch and release fishing, please visit ***catchandrelease.org***.

Minimal Recreational Boating Equipment Checklist

Use this checklist to ensure a safer, more comfortable boating trip. Complete the Everglades National Park Boater Education course and carry your permit.

* - Coast Guard required equipment

- **Everglades National Park Boater Permit**
- Wearable Personal Floatation Devices (PFDs)*- must be available for every person on board. Children under 6 years of age are required by law to wear a life jacket on boats less than 26 feet long. Any vessel 16 feet and longer must also carry one throwable (Type IV) device. Remember, a life jacket can only save your life if you're wearing one.
- Fire Extinguisher*- is recommended on all powerboats, and mandatory for certain vessels.
- Sound Signaling Device*- Vessels under 39 feet must carry a whistle, horn or other attention getting device. Boats more than 39 feet require both a whistle and a bell.
- Visual Distress Signals* - Include flares (make sure they are stored properly and monitor their expiration dates). Mirrors, while not required, are an excellent signaling device when used on the water.
- Boat registration
- Divers-down flag
- Boating Safety Identification Card
- Extra water and food
- Polarized sunglasses
- Basic Tool Kit
- First aid kit and sunscreen
- Anchor and plenty of anchor line
- Extra fuel
- VHF radio and cell phone
- Charts and maps
- GPS, compass and chart(s)
- Pole or Paddle
- Spotlight
- Spare batteries

Note: This list is a quick reference and may not include the most up-to-date requirements. The operator of a boat whether owned, rented, or borrowed, is responsible for having ALL the USCG required equipment onboard and in good working condition. Different sized vessels may require additional safety equipment. Contact the FWC Law Enforcement (1-888-404-FWCC) or USCG (305-536-5611, or ***http://cgaux.org/vsc*** for more information.

If you see illegal or inappropriate activities, or boating accidents or incidents, report these potential violations to the Everglades National Park Dispatch at 305-242-7740 (or call 800-788-0511 in a life threatening emergency).

Send a message to: ***evergladesnavaidrepair@nps.gov*** if you see damaged or missing markers, providing as much detail as possible, including a photo of the problem observed.

Management Zones

The 2015 General Management Plan (GMP) establishes the majority of Florida Bay and other marine waters as a Boating Access Zone, which allows for safe passage on plane (at cruising speed). The following park management zones have been established to increase protection for natural and wilderness resources, and for enhanced visitor use and enjoyment.

Slow Speed Zone: There are a limited number of slow speed zones where traffic congestion or a history of impacts to natural resources occurs. Minimal wake is required in these areas.

Idle Speed Zone: There are a limited number of idle speed zones where traffic congestion or a history of impacts to natural resources occurs. No wake is required in these areas.

Pole/Troll Zones: In Everglades National Park, pole/troll zones (PTZs) are areas where the use of internal combustion motors are prohibited, in order to protect shallow water resources such as seagrass beds and mud flats. Electric trolling motors and non-motorized power sources, such as drifting, push poles, and paddles, are the ONLY allowable way to move through the area.

Why designate pole/troll zones?
These zones were created to protect sensitive aquatic vegetation, wildlife, and the submerged marine wilderness resources. They also improve the quality of flats fishing, enhance paddling and wildlife viewing opportunities, and expand education on proper shallow boating techniques.

Snake Bight and other Pole/Troll Zones
The area known as Snake Bight became Everglades National Park's first PTZ in 2011. This zone was the result of careful study by park managers and considerable input from the public.

Pole/troll zones, coupled with strong education and enforcement, have shown to be an effective management strategy for increasing resource protection and visitor enjoyment of marine areas. Everglades National Park expanded these areas in Florida Bay in its 2015 GMP and were implemented in 2017.

When entering a PTZ, shut down and raise your internal combustion motor - you may switch to an electric trolling motor, or paddle, drift, or pole through the protected zone.

Pole/Troll/Idle Zones: When the water is sufficiently deep, this zone allows the use of internal combustion motors at idle speed, in addition to poling, paddling, or using trolling motors. This zone occurs in several areas in Florida Bay with water depths that can occasionally accommodate motors operating at idle speed. This zone also protects the submerged marine wilderness.

Backcountry (Non-motorized) Zones: These zones provide opportunities for tranquil wilderness experiences, and allow paddling and other types of hand propelled vessel operation. Motor boats operating any engine or trolling motor are not permitted in these areas.

Everglades Paddling Trail: Newly established trail from Everglades City to Flamingo that includes some seasonal segments that are zoned backcountry and others that include slow or idle speed.

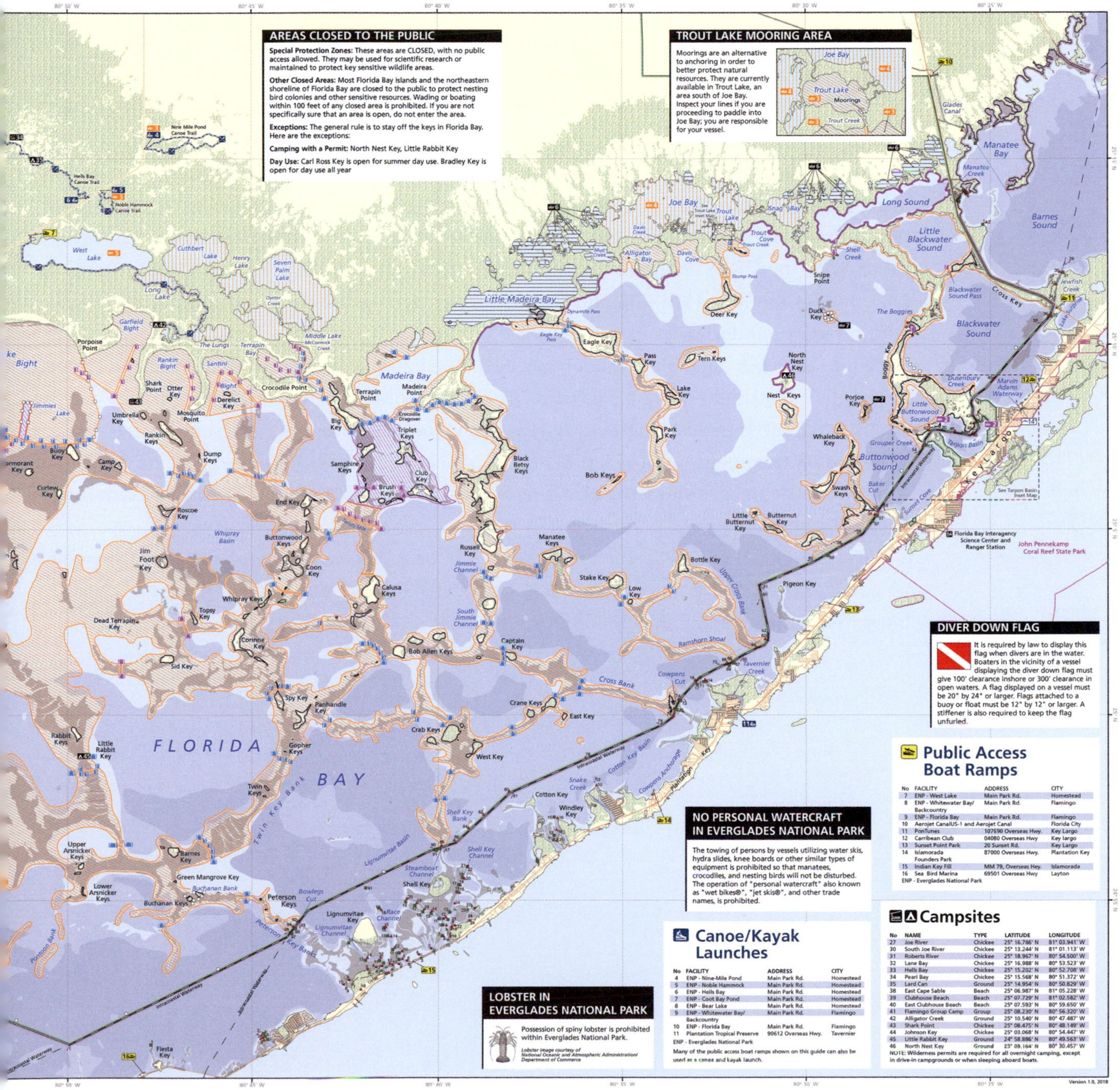

AREAS CLOSED TO THE PUBLIC

Special Protection Zones: These areas are CLOSED, with no public access allowed. They may be used for scientific research or maintained to protect key sensitive wildlife areas.

Other Closed Areas: Most Florida Bay islands and the northeastern shoreline of Florida Bay are closed to the public to protect nesting bird colonies and other sensitive resources. Wading or boating within 100 feet of any closed area is prohibited. If you are not specifically sure that an area is open, do not enter the area.

Exceptions: The general rule is to stay off the keys in Florida Bay. Here are the exceptions:

Camping with a Permit: North Nest Key, Little Rabbit Key

Day Use: Carl Ross Key is open for summer day use. Bradley Key is open for day use all year

TROUT LAKE MOORING AREA

Moorings are an alternative to anchoring in order to better protect natural resources. They are currently available in Trout Lake, an area south of Joe Bay. Inspect your lines if you are proceeding to paddle into Joe Bay; you are responsible for your vessel.

DIVER DOWN FLAG

It is required by law to display this flag when divers are in the water. Boaters in the vicinity of a vessel displaying the diver down flag must give 100' clearance inshore or 300' clearance in open waters. A flag displayed on a vessel must be 20" by 24" or larger. Flags attached to a buoy or float must be 12" by 12" or larger. A stiffener is also required to keep the flag unfurled.

NO PERSONAL WATERCRAFT IN EVERGLADES NATIONAL PARK

The towing of persons by vessels utilizing water skis, hydra slides, knee boards or other similar types of equipment is prohibited so that manatees, crocodiles, and nesting birds will not be disturbed. The operation of "personal watercraft" also known as "wet bikes®", "jet skis®", and other trade names, is prohibited.

LOBSTER IN EVERGLADES NATIONAL PARK

Possession of spiny lobster is prohibited within Everglades National Park.

Lobster image courtesy of National Oceanic and Atmospheric Administration/ Department of Commerce

Public Access Boat Ramps

No	FACILITY	ADDRESS	CITY
7	ENP - West Lake	Main Park Rd.	Homestead
8	ENP - Whitewater Bay/ Backcountry	Main Park Rd.	Flamingo
9	ENP - Florida Bay	Main Park Rd.	Flamingo
10	Aerojet CanalUS-1 and Aerojet Canal		Florida City
11	PonTunes	107690 Overseas Hwy.	Key Largo
12	Carribean Club	04080 Overseas Hwy	Key largo
13	Sunset Point Park	20 Sunset Rd.	Key Largo
14	Islamorada Founders Park	87000 Overseas Hwy.	Plantation Key
15	Indian Key Fill	MM 79, Overseas Hey.	Islamorada
16	Sea Bird Marina	69501 Overseas Hwy	Layton

ENP - Everglades National Park

Canoe/Kayak Launches

No	FACILITY	ADDRESS	CITY
4	ENP - Nine-Mile Pond	Main Park Rd.	Homestead
5	ENP - Noble Hammock	Main Park Rd.	Homestead
6	ENP - Hells Bay	Main Park Rd.	Homestead
7	ENP - Coot Bay Pond	Main Park Rd.	Homestead
8	ENP - Bear Lake	Main Park Rd.	Homestead
9	ENP - Whitewater Bay/ Backcountry	Main Park Rd.	Flamingo
10	ENP - Florida Bay	Main Park Rd.	Flamingo
11	Plantation Tropical Preserve	90612 Overseas Hwy.	Tavernier

ENP - Everglades National Park

Many of the public access boat ramps shown on this guide can also be used as a canoe and kayak launch.

Campsites

No	NAME	TYPE	LATITUDE	LONGITUDE
27	Joe River	Chickee	25° 16.786' N	81° 03.941' W
30	South Joe River	Chickee	25° 13.244' N	81° 01.113' W
31	Roberts River	Chickee	25° 18.967' N	80° 54.500' W
32	Lane Bay	Chickee	25° 16.988' N	80° 53.523' W
33	Hells Bay	Chickee	25° 15.202' N	80° 52.708' W
34	Pearl Bay	Chickee	25° 15.568' N	80° 51.372' W
35	Lard Can	Ground	25° 14.954' N	80° 50.829' W
38	East Cape Sable	Beach	25° 06.987' N	81° 05.228' W
39	Clubhouse Beach	Beach	25° 07.729' N	81° 02.582' W
40	East Clubhouse Beach	Beach	25° 07.593' N	80° 59.650' W
41	Flamingo Group Camp	Group	25° 08.230' N	80° 56.320' W
42	Alligator Creek	Ground	25° 10.540' N	80° 47.487' W
43	Shark Point	Chickee	25° 08.475' N	80° 48.149' W
44	Johnson Key	Chickee	25° 03.068' N	80° 54.447' W
45	Little Rabbit Key	Ground	24° 58.886' N	80° 49.563' W
46	North Nest Key	Ground	25° 09.164' N	80° 30.457' W

NOTE: Wilderness permits are required for all overnight camping, except in drive-in campgrounds or when sleeping aboard boats.

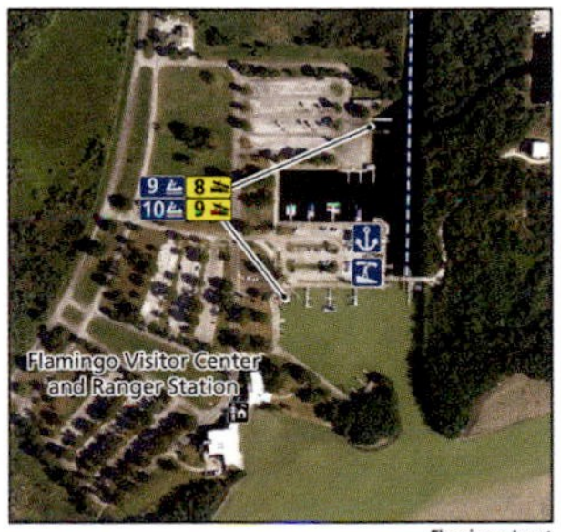

Flamingo Inset

Tarpon Basin Inset

DEVILS TOWER IN 3D: USING DRONES TO CREATE GIS DATA ON DEVILS TOWER IN A 3D MAP

THI PRUITT, YELLOWSTONE NATIONAL PARK
NEAL JANDER, NPS DENVER SERVICE CENTER PLANNING DIVISION
NELL CONTI AND KERRY SHAKARJIAN, NPS INTERMOUNTAIN REGION

Devils Tower National Monument in the Black Hills area of Wyoming contacted the NPS Intermountain Region Geographic Resources Division GIS team (IMR GIS) to help determine the best way to represent management data on the vertical aspects of Devils Tower, including exotic vegetation and climbing routes. This request proposed using a drone (a remotely piloted aircraft) to gather high-resolution imagery of the sides of Devils Tower for a 3D model that can be used to create the management data. In collaboration with the USGS National Unmanned Aircraft Systems (UAS) Project Office located in Lakewood, Colorado, IMR GIS and the monument received NPS approval for UAS flights within park boundaries.

From the UAS data, a 3D model was developed and brought into ArcGIS Pro to create exotic plant treatment polygons as well as climbing route line data on Devils Tower in a 3D environment. The exotic treatment polygons shown in the images represent different years of herbicide treatment, with green being 2015, yellow being 2013, and red being 2012. The climbing route images show the climbing routes as linear features in brown and climbing anchor locations as green points.

The 3D model of Devils Tower can currently be viewed and downloaded by the public by going to Sketchfab.com and searching for Devils Tower National Monument at https://sketchfab.com/models/f08f4bb0230f4cc899f837f2126db7b7.

The model displays a vegetation treatment scene of the top view of Devils Tower, a monolith of igneous rock rising 867 feet from its base and situated at an elevation of 5,112 feet above sea level. Data sources: NPS, Esri, DigitalGlobe, GeoEye, Earthstar Geographics, CNES/Airbus DS, USDA, USGS, AeroGRID, IGN, GIS Community.

The model displays a vegetation treatment scene of the top view of Devils Tower, a sacred site for many American Indian tribes. Data sources: NPS, Esri, DigitalGlobe, GeoEye, Earthstar Geographics, CNES/Airbus DS, USDA, USGS, AeroGRID, IGN, GIS Community.

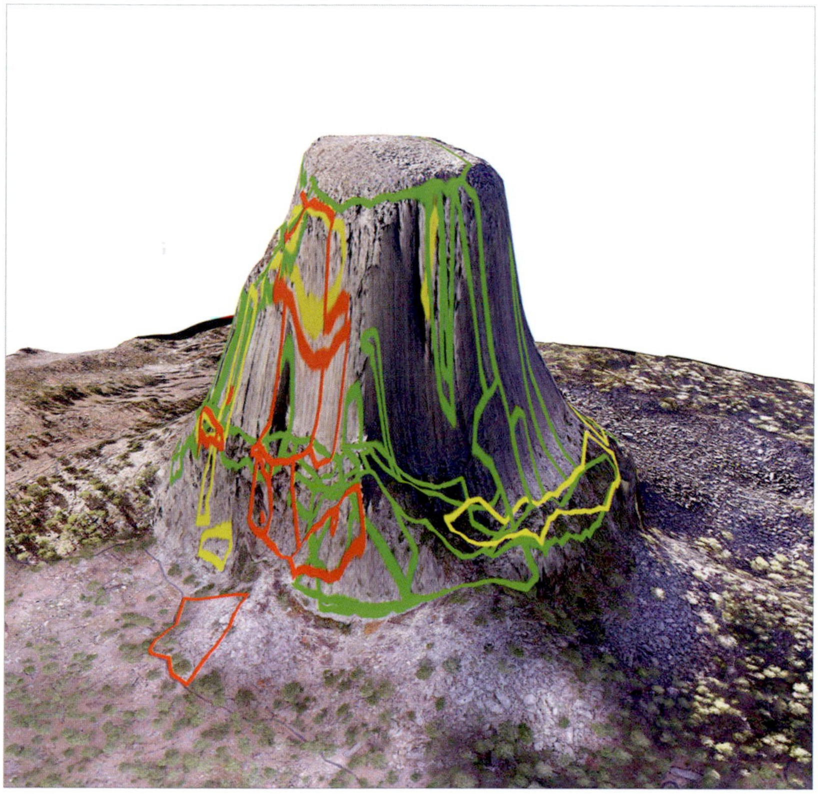

The model displays a vegetation treatment scene of the northeast face of Devils Tower. Data sources: NPS, Esri, DigitalGlobe, GeoEye, Earthstar Geographics, CNES/Airbus DS, USDA, USGS, AeroGRID, IGN, GIS Community.

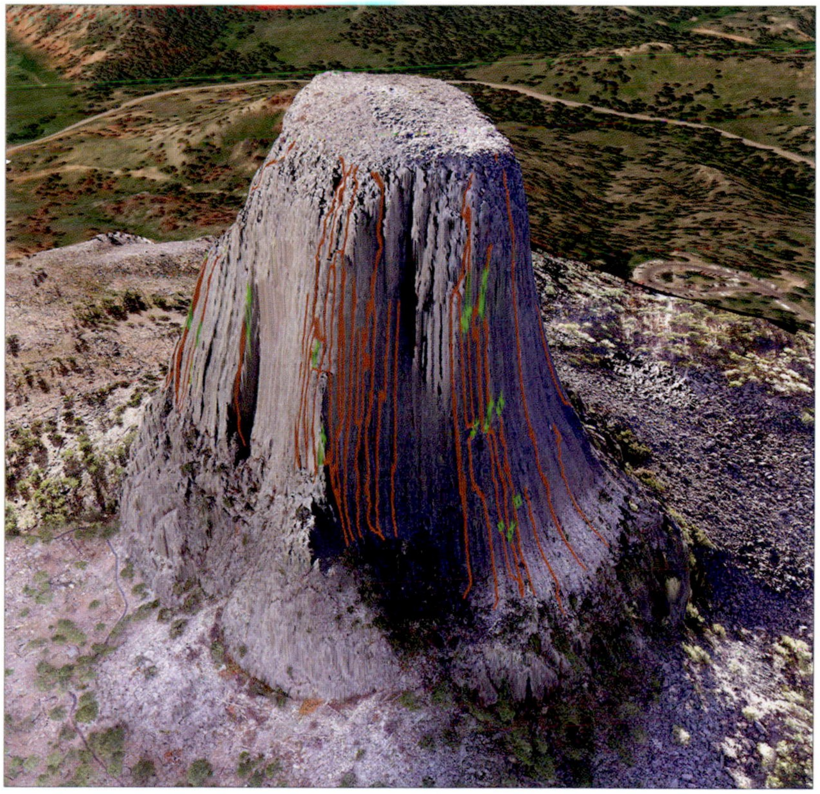

A climbing routes scene shows the north face of Devils Tower. Data sources: NPS, Esri, DigitalGlobe, GeoEye, Earthstar Geographics, CNES/Airbus DS, USDA, USGS, AeroGRID, IGN, GIS Community.

3 VISITOR AND RESOURCE PROTECTION

STEVE SHACKELTON, YOSEMITE NATIONAL PARK

PROTECTION AND EMERGENCY SERVICES FOR THE PARKS

Through the extraordinary evolution of continuous breakthroughs, digital mapping is rapidly positioning itself to become a new language of park visitor and resource protection. As travel becomes more convenient and affordable, visitors enjoying our parks are increasingly diversified by nationality, cultural filters, and language. If rangers are to do their work within the exacting mandate of providing increased visitation without damaging underlying resources and values, there must be newly efficient ways to communicate complex ideas to every visitor—no matter their origin, background, or age.

Graphic visualization through mapping is a marvelous way of communicating the why and the how behind protection strategies. These methods likely will become one of the primary ways to communicate ideas about safety and stewardship with visitors before they even leave home. After all, safeguarding visitors and the parks they visit is our top responsibility. In that light, communication and education—in all their most innovative, creative, and useful forms—are paramount. Digital mapping is part of that solution.

In search and rescue, for example, GIS provides platforms for the smart distribution of often-scarce search assets by using analytics that are as helpful in predicting where a lost visitor won't likely be found as they are in predicting where the visitor likely will be found. In a protracted search, the use of analytics is even more efficient because they can constantly update maps that convey information ranging from previous-period helicopter flight patterns to updated search team assignments. Using analytics, first responders have more information to update and comfort family members on their progress in searching for a missing loved one.

The job of protecting national park resources is vast. Rangers might be tasked with protecting a particular wildlife species or a group of peaceful protesters at a demonstration, while resource managers safeguard ecosystems. Tools being developed around GIS are revolutionizing the way rangers work with resource managers and interpretive specialists to provide layers of protection, from archeological sites to rare plant assemblages. These layers can start with strategies for education outside the park, with the result that visitors invest in stewardship. And when preventive tactics fail, other strategies support new best practices in the preparation of civil and criminal prosecutions.

Esri's new developments in *conservation intelligence*, which take tried-and-true field collection and reporting templates and couple them with rugged field applications and geo-location technologies, portend a new level of accuracy and ease in the fieldwork of scene documentation and the subsequent longitudinal work of monitoring during recovery. This intelligent case-making, when properly communicated, leads full circle to protection.

Finally, the US national parks team and their sister park systems around the world can share digital mapping, analytics, and the scientific products that support resource protection instantly with colleagues, back and forth, anywhere on the planet.

Whether it's new conditions in fire management related to climate, intelligence on wildlife trafficking, information about zoonotic disease, or protocols for search and rescue or field medicine, ideas must be shared to dominate the contravening influences and effectively take care of Earth's most precious places. This sharing strengthens the effectiveness of our community of stewardship through a common language of maps.

ASSATEAGUE ISLAND SURVEY CONTROL NETWORK

NEIL WINN, ASSATEAGUE ISLAND NATIONAL SEASHORE

Assateague Island National Seashore along the coasts of Maryland and Virginia is a barrier island with a geomorphological monitoring program that relies on accurate elevation measurements to detect changes over time. Survey control marks (benchmarks) are the backbone of such a monitoring program, and in 2015 Assateague Island National Seashore worked with partners to develop a survey network based on GPS observations. Repeat occupations of 10 to 20 hours per mark and processing baselines using the National Geodetic Survey OPUS-Projects software resulted in network accuracy to within millimeters. This accuracy provides strong consistency to our coastal monitoring efforts and assists in developing strong statistical trends in barrier island movement.

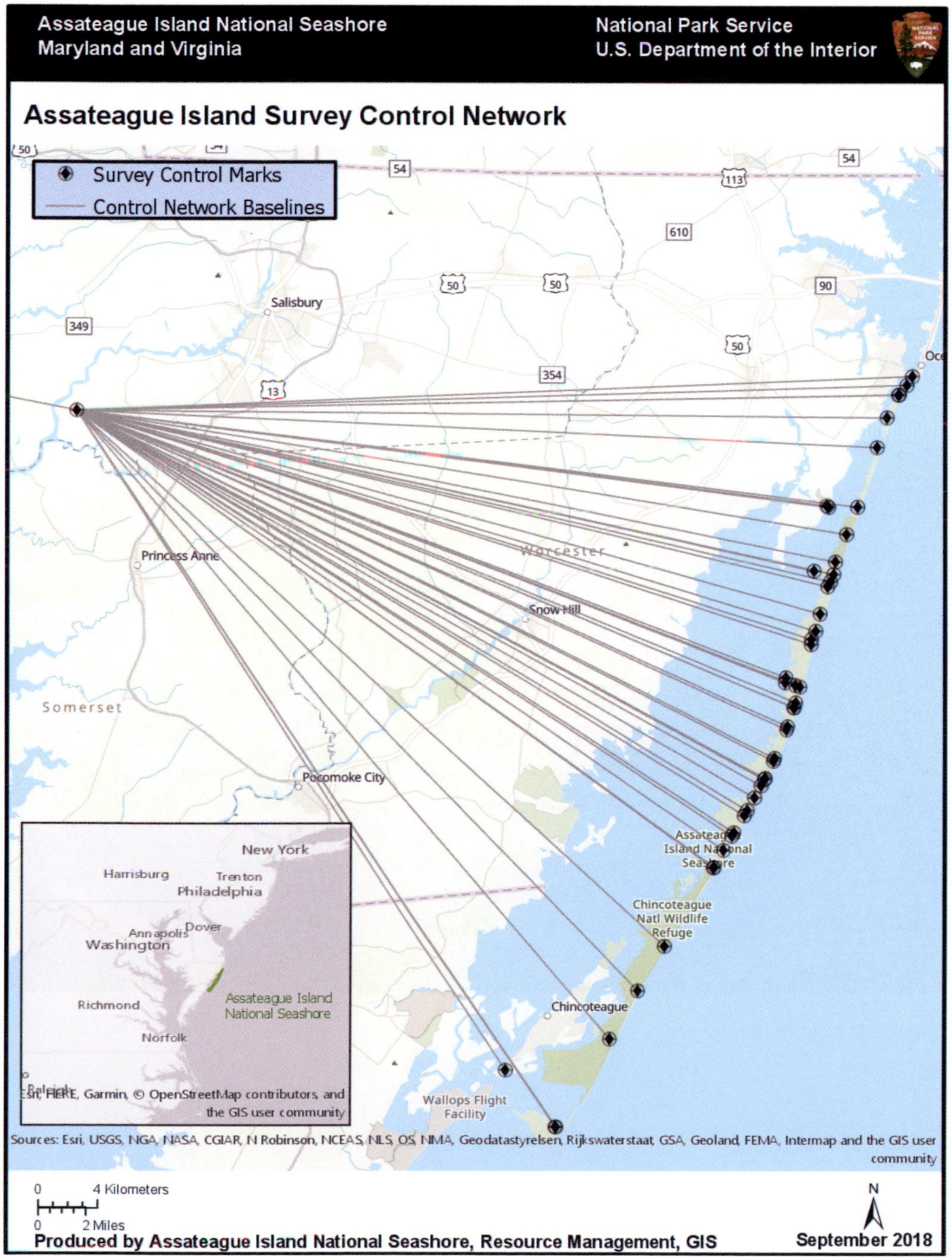

(*Top left*) A Global Navigation Satellite System (GNSS) occupied survey control mark. (*Top right*) An Assateague Island National Seashore survey control mark. (*Left*) An Assateague Island National Seashore survey control network map. Map URL: https://asis-nps.opendata.arcgis.com/datasets/asis-survey-control-net-2015. Data sources: NPS, Basemap: Esri World Light Gray Canvas Base and Esri World Topographic Map.

HURRICANE DAMAGE ASSESSMENT IN THE US VIRGIN ISLANDS

PAUL HARDWICK,
SEQUOIA AND KINGS CANYON NATIONAL PARKS, NPS WESTERN INCIDENT MANAGEMENT TEAM

In September 2017, the US Virgin Islands and Puerto Rico were hit by two powerful hurricanes, Irma and Maria, within two weeks. There was extensive damage to infrastructure and vessels across these islands. NPS incident management teams were deployed to the affected NPS units to assist staff and help with the stabilization of infrastructure, so that the parks could be reopened to the public.

Approximately 60 percent of the island of Saint John is managed as part of Virgin Islands National Park. An NPS facilities team conducted a damage assessment of the NPS infrastructure. In a parallel effort, satellite imagery was used to count the number of displaced vessels that were damaged and scattered across NPS-managed areas. This map used initial infrastructure damage and vessel assessment data to visualize concentrations of damage. Approximately 60 percent of the park buildings on Saint John were damaged, 14 of which were destroyed. Approximately 200 displaced vessels were identified within park boundaries. The incident management team then utilized this map to help prioritize the placement of team members and recovery efforts.

Displaced vessels float in a bay.

Damage to buildings.

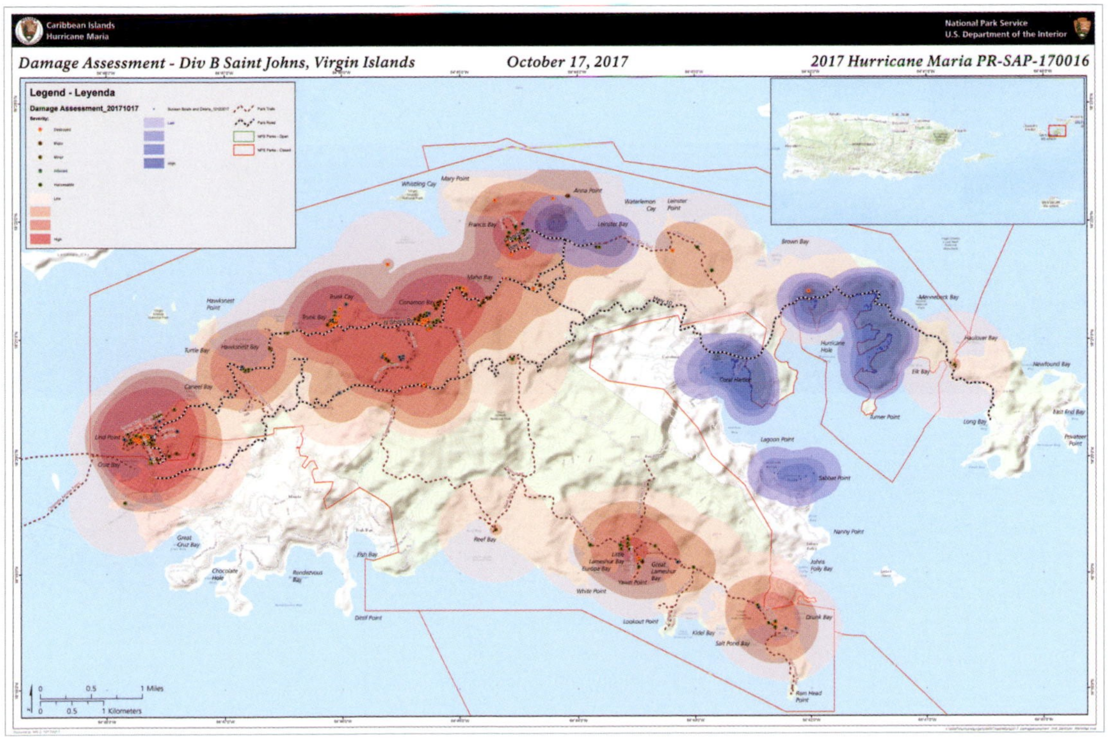

A damage assessment hot spot map of *Hurricane Maria* in 2017. Data sources: NPS.

2017 ANNUAL SAR REPORT DASHBOARD

HILARY MELTON CROWLEY, NPS LAW ENFORCEMENT, SECURITY AND EMERGENCY SERVICES

NPS units submit annual search and rescue (SAR) reports to the Washington Office (WASO) Emergency Services division. This dashboard represents the first visualization of this data on a national level, as well as furthering NPS Emergency Services' vision of collecting SAR data that can be used for visual presentation and analysis. WASO and park units can use this interactive dashboard to analyze the annual SAR data by region, state, or park unit, and then view the number and types of incidents experienced by each unit. In previous years, this data had only been presented by way of a large spreadsheet. This dashboard allows NPS Emergency Services to see what areas are experiencing large numbers of SAR incidents and to see breakdowns of incident data quickly and easily. This type of platform is also more easily consumable by WASO senior leadership, the regions, individual park units, and the public.

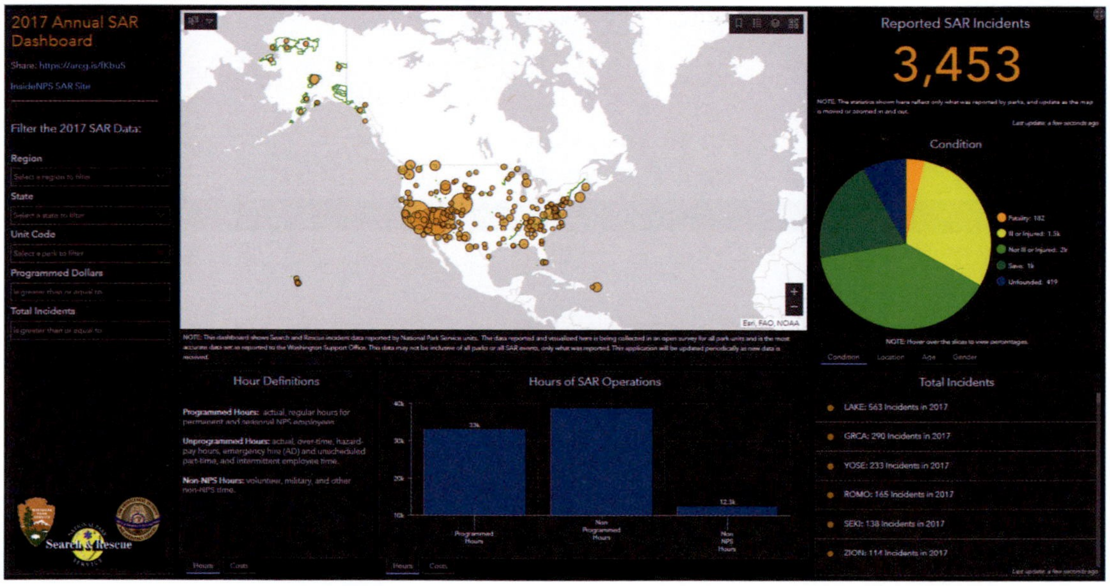

The 2017 Annual Search and Rescue (SAR) Report Dashboard shows US data. Map URL: https://arcg.is/fKbuS. Data sources: NPS Emergency Services, NPS Land Resources Division, Basemap: Esri ArcGIS Online Light Gray Canvas Basemap.

YOSEMITE MOBILE STREET MAP BOOK IN CARRYMAP

ANSLEY SINGER, JEREME CHANDLER, AND VANESSA GLYNN-LINARIS, YOSEMITE NATIONAL PARK

Yosemite National Park's Visitor and Resource Protection Division needed a way for emergency responders to locate addresses, place names, building numbers, fire hydrants, pay phones, and other front-country facilities to ensure a quick response. Beginning in 2009, the Protection Division started a large-scale mapping project updating GIS data throughout the park. In 2011, Yosemite Dispatch used the updated GIS data to produce a park-wide paper map book for use by emergency responders. The map book was so popular that rangers, firefighters, and park staff from all divisions requested copies.

Having a paper map book is important; however, digital maps with better search capabilities were the next logical step. In 2014, the Protection Division requested that the GIS specialist convert the printed map book into a format that the rangers and firefighters could utilize on their mobile devices. The tool would need to work offline, have a simple interface, have search capabilities, be easily accessible, and allow for cartographic control. After researching multiple products, the CarryMap Builder extension was chosen because it met all the requirements. For example, it allows users to turn layers on and off, display different symbology at various scales, and create an expiration date on maps so field users know when they need to obtain updated maps. Using CarryMap, the GIS specialist can create maps of the entire park in relatively small file sizes to email or transfer them via Bluetooth USB to park rangers in the field. Assistant fire module captain Jasper Peach provided critical feedback after field testing CarryMap during the early testing phase.

Using CarryMap on their mobile devices, emergency responders are now able to query an address or location and zoom to that area to determine the location of the incident. The map is updated frequently and distributed to field staff via email.

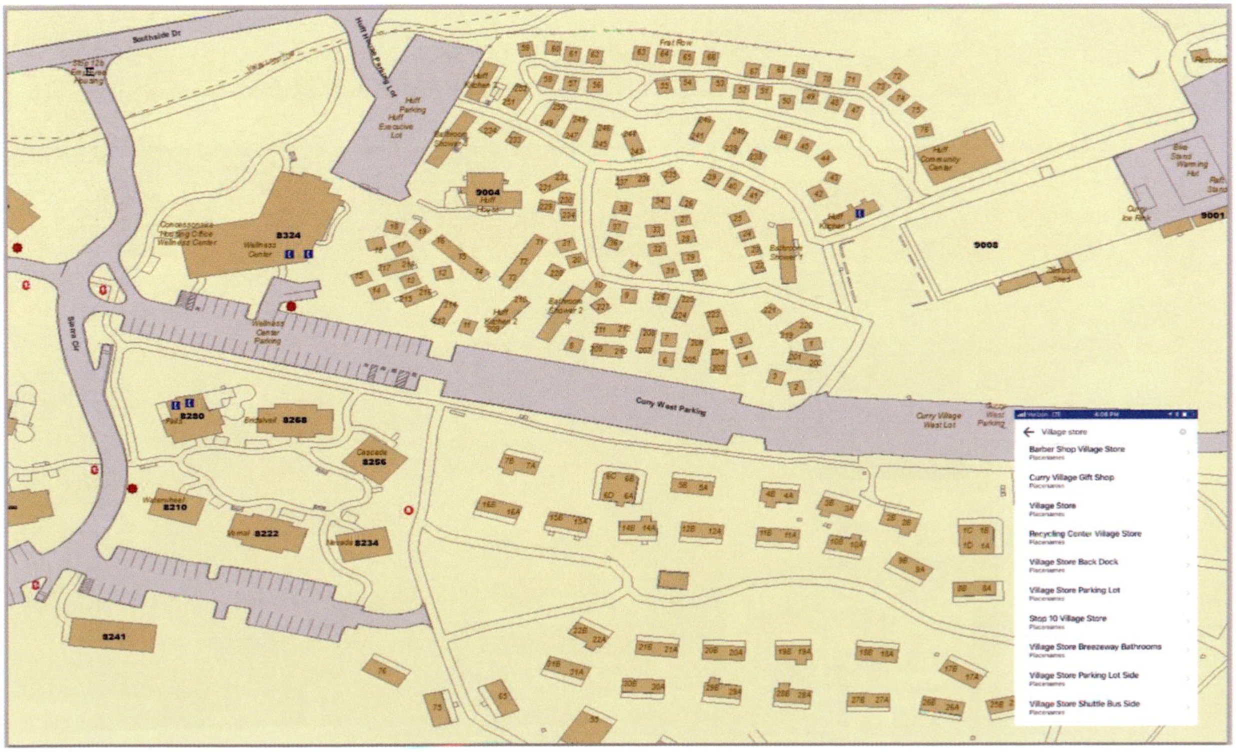

The map shows areas of Yosemite Valley as they would appear on a mobile device using the Yosemite map book in the CarryMap application. Specifically, this screenshot shows the zoom capabilities of the CarryMap application. Data sources: NPS, USGS.

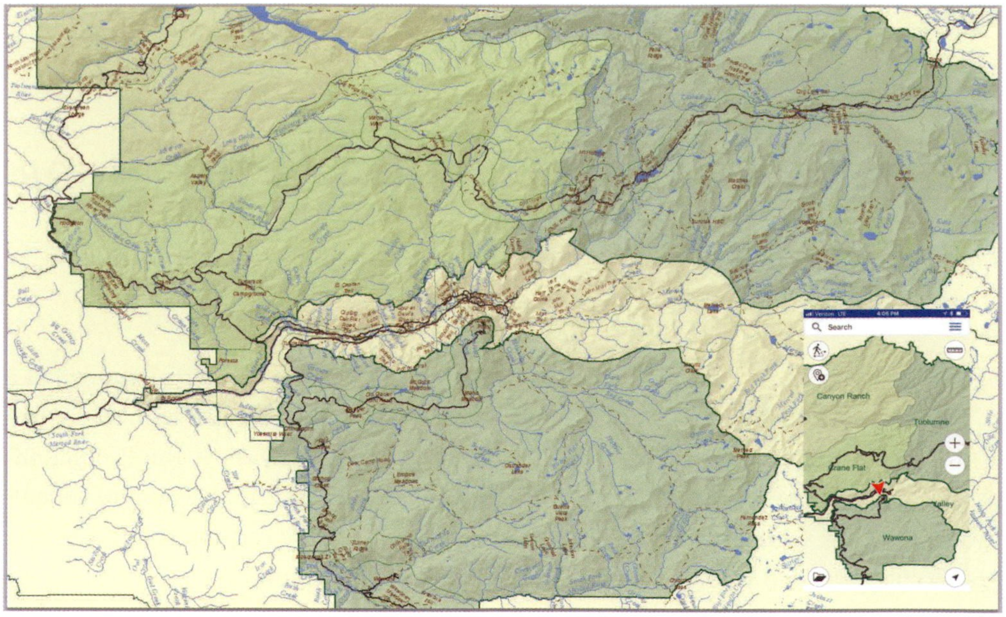

A screenshot shows the search functionality of the Yosemite map book in the CarryMap application.

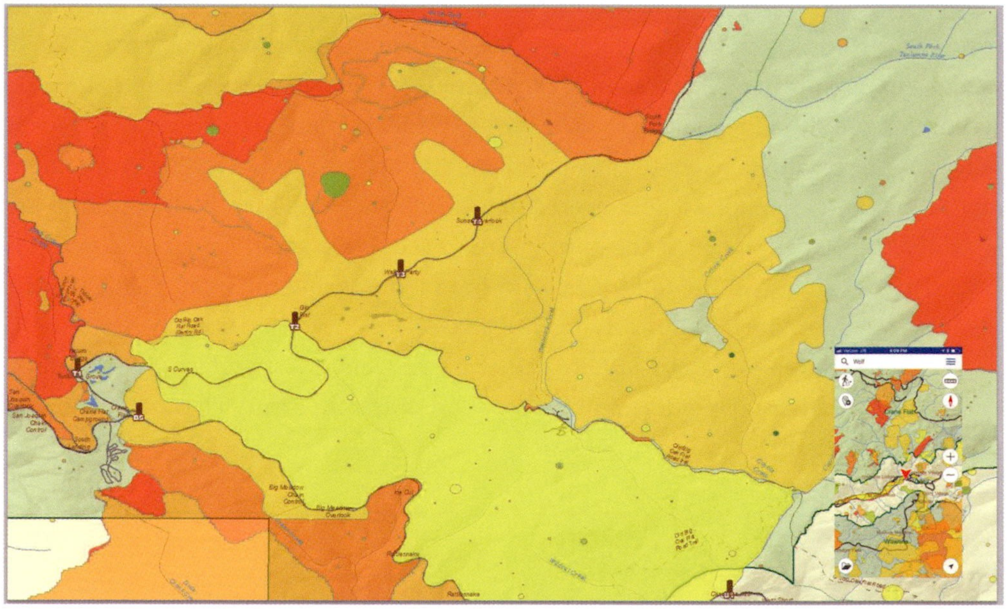

A screenshot shows fire history data in Yosemite National Park with road mileage markers in the CarryMap application.

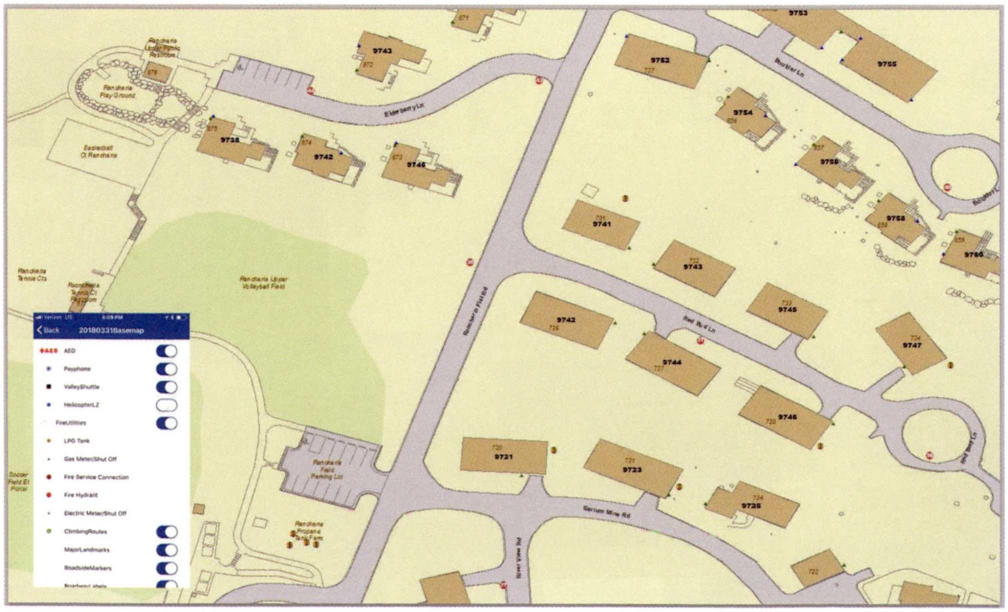

A screenshot shows layers and their visibility options in the CarryMap application. Data sources on this page: NPS, USGS.

2018 HAWAI'I VOLCANOES NATIONAL PARK INCREASED VOLCANIC ACTIVITY INCIDENT MAPS

MARK WASSER, HAWAI'I VOLCANOES NATIONAL PARK

A new, active phase of the eruption of Kīlauea Volcano on Hawai'i Island (The Big Island) began in May 2018. Twenty-four fissures opened in the Lower East Rift Zone, and the associated lava flows covered 8,782 acres in the Puna District. Within Hawai'i Volcanoes National Park, the continued withdraw of magma stored below the summit of Kīlauea led to regular earthquakes and significant subsidence (sinking) of the summit caldera—almost 500 meters in some places. These daily earthquakes (8,000 in total) produced significant damage in the summit area and led to the nearly full closure of the park for 134 days. The National Park Service (NPS) Incident Management Team worked with the USGS Hawaiian Volcano Observatory (HVO) to use data collected by HVO and a Department of Interior drone team to produce maps illustrating the latest changes and current conditions at the Kīlauea Summit area. This data included lava flow extents and updated digital elevation models and orthomosaics, and it was instrumental in detailing the dynamic changes happening at the summit during the three months of the eruption, as shown in these selected maps:

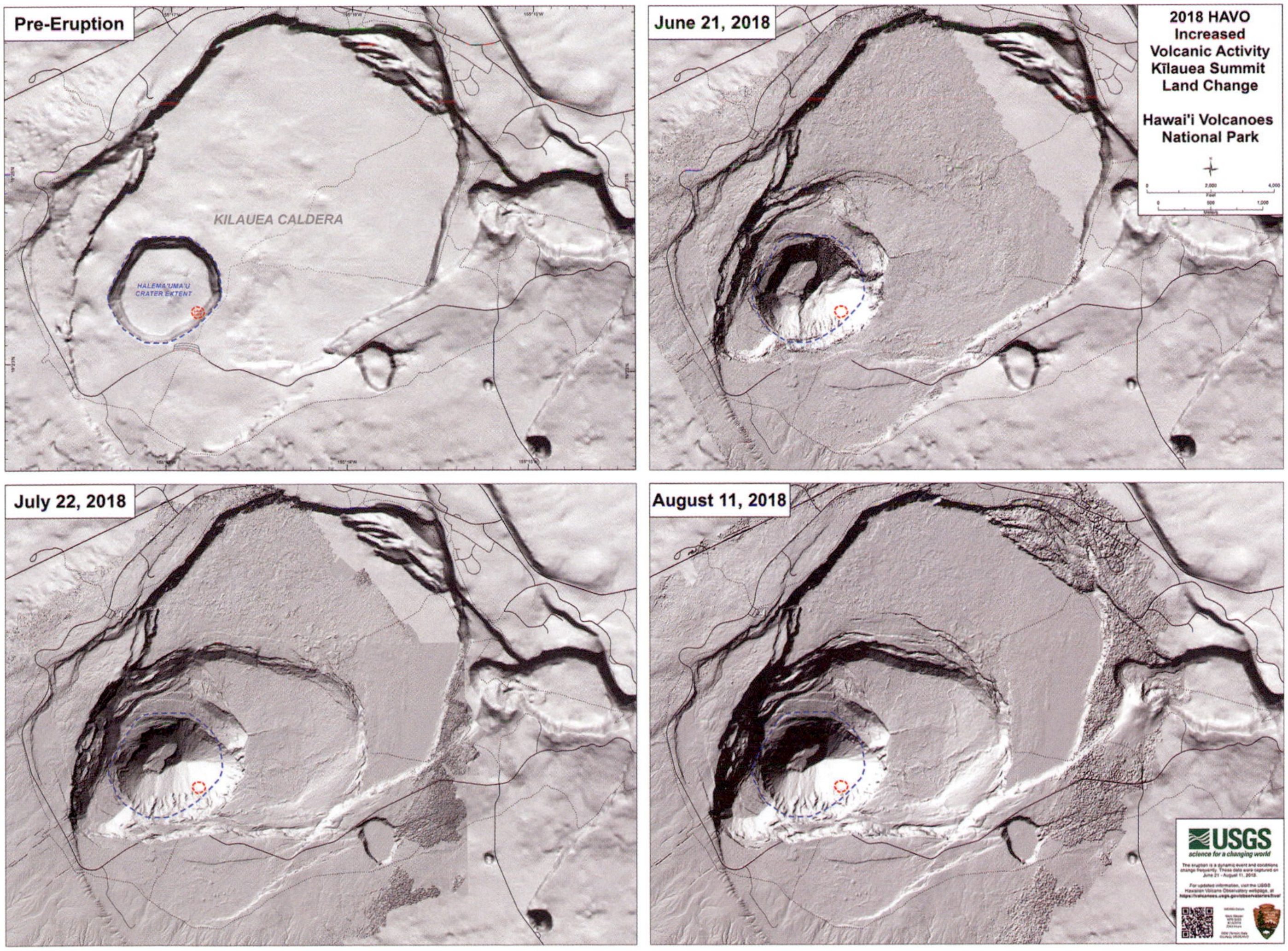

1. A comparison of four digital elevation models generated at pre-eruption and three stages of the eruption.

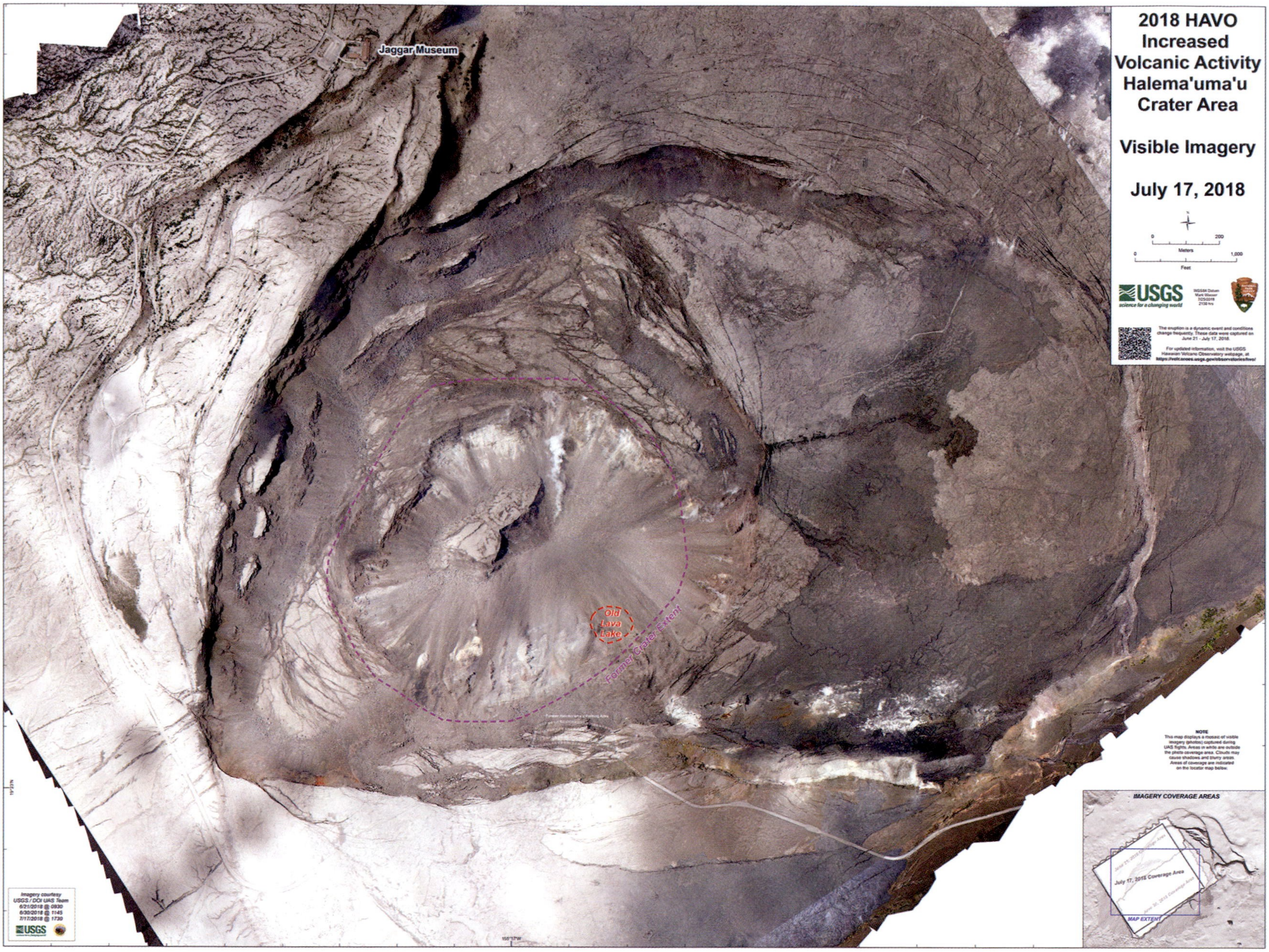

2. An orthomosaic (composite of visible imagery) of the summit area taken during the eruption.

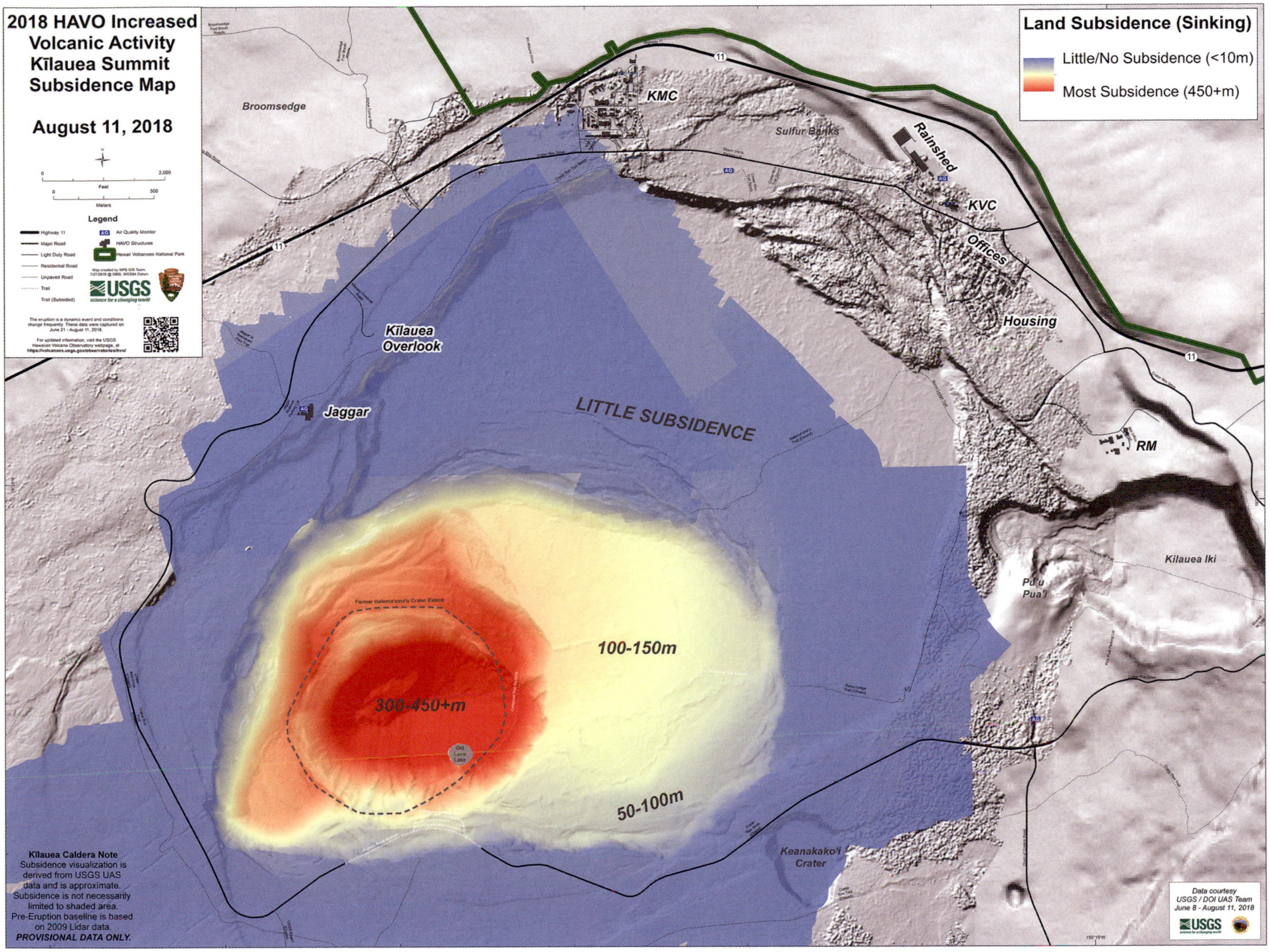

3. A map details the overall amount of subsidence around the Kīlauea Summit area.

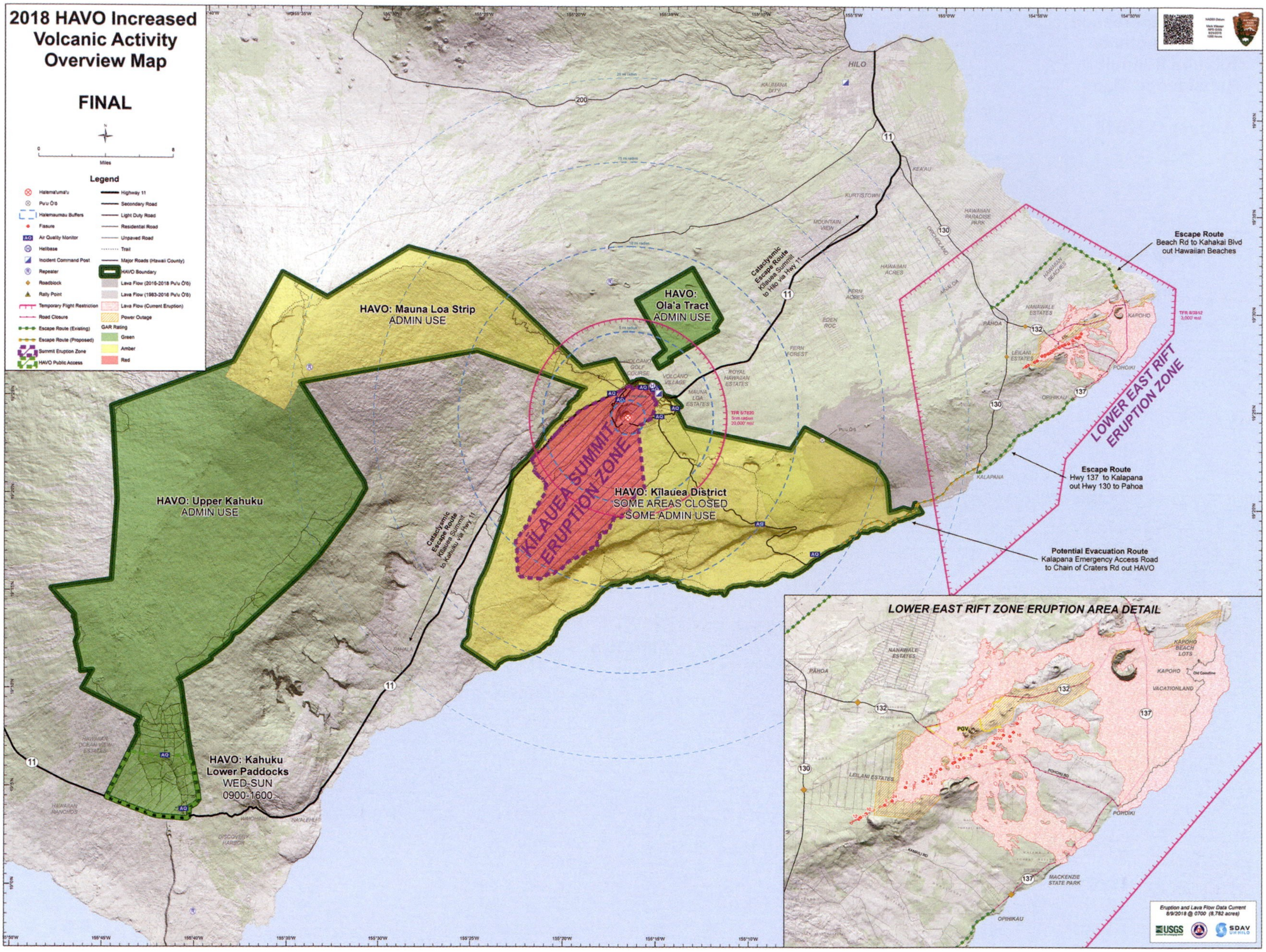

4. An overview of the eruption activity and incident status, both inside and outside the park, is shown. Data sources on pages 45–48: Lava flow outlines courtesy of University of Hawai'i at Hilo, Spatial Data Analysis and Visualization Labs. Data sources: US Department of Interior, NPS, USGS, Hawai'ian Volcano Observatory, University of Hawai'i at Hilo Spatial Data Analysis and Visualization Lab.

NPS HURRICANE SANDY COASTAL RESILIENCY RESEARCH STORY

THOM CURDTS, NPS RESOURCE INFORMATION SERVICES DIVISION
SARA STEVENS, NPS INVENTORY AND MONITORING PROGRAM
KRISTEN HYCHKA, UNIVERSITY OF RHODE ISLAND
DENNIS SKIDDS, NPS INVENTORY AND MONITORING PROGRAM

Hurricane Sandy made landfall in New Jersey on October 29, 2012, devastating local towns and communities along the Atlantic Coast. National Park units of the Northeast Coastal and Barrier Inventory and Monitoring Network—including Assateague Island, Cape Cod, and Fire Island National Seashores and Gateway National Recreation Area—were dramatically affected by the storm. Natural systems, cultural resources, and infrastructure were all severely impacted. Following the storm, Congress appropriated funding for response, recovery, and mitigation projects for DOI to repair and rebuild parks and refuges, but also to invest in scientific data and studies to support recovery in the region.

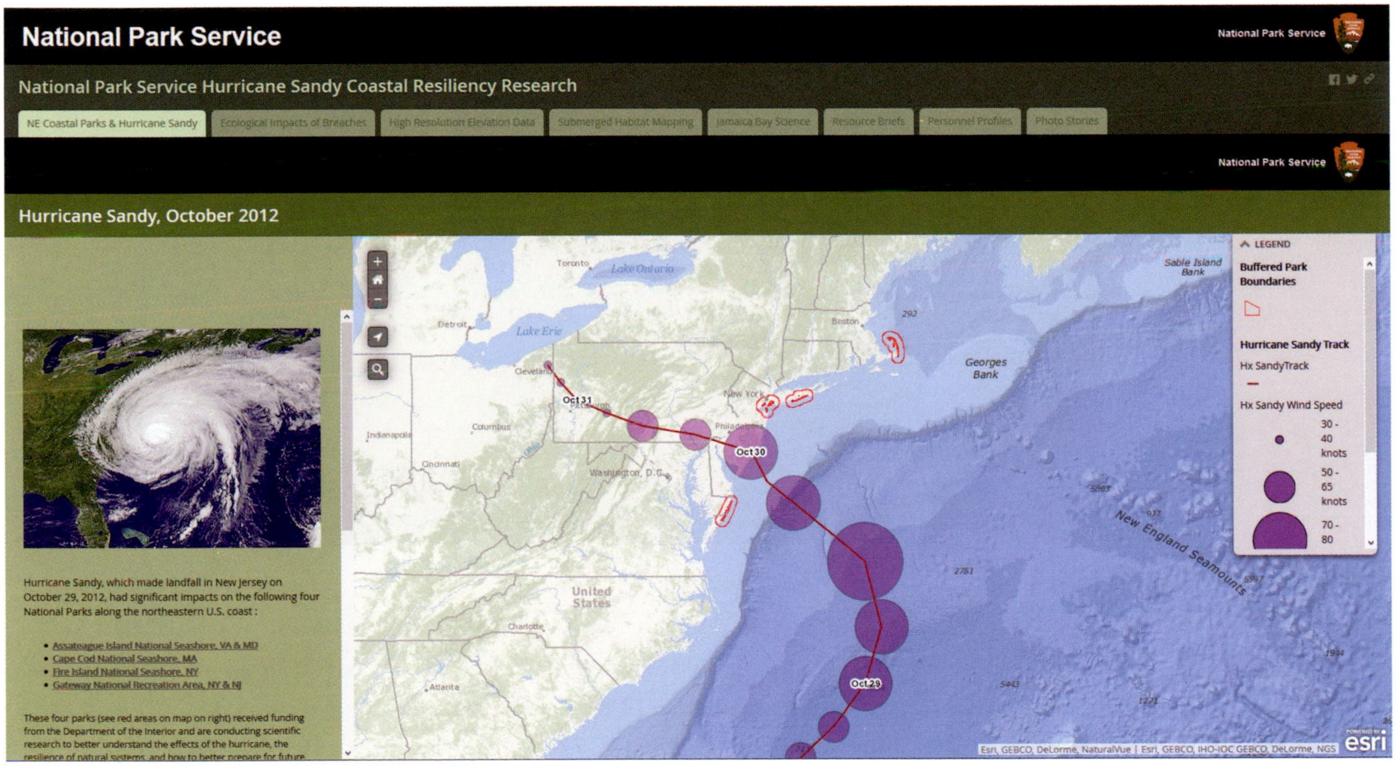
National Park Service
National Park Service Hurricane Sandy Coastal Resiliency Research
NE Coastal Parks & Hurricane Sandy
Hurricane Sandy, October 2012
Hurricane Sandy, which made landfall in New Jersey on October 29, 2012, had significant impacts on the following four National Parks along the northeastern U.S. coast :
Assateague Island National Seashore, VA & MD
Cape Cod National Seashore, MA
Fire Island National Seashore, NY
Gateway National Recreation Area, NY & NJ
These four parks (see red areas on map on right) received funding from the Department of the Interior and are conducting scientific research to better understand the effects of the hurricane, the
LEGEND
Buffered Park Boundaries
Hurricane Sandy Track
Hx SandyTrack
Hx Sandy Wind Speed
30 - 40 knots
50 - 65 knots
70 - 80
Oct 31
Oct 30
Oct 29
Georges Bank
New England Seamounts
United States
esri

A screenshot of a story tab displays an interactive topographic map of the Northeast Coastal Parks and Hurricane Sandy in October 2012.

URL: http://go.nps.gov/sandy_resiliency. Data sources : NPS, Esri World Ocean, Esri World Imagery, New York State 2010 Orthoimagery.

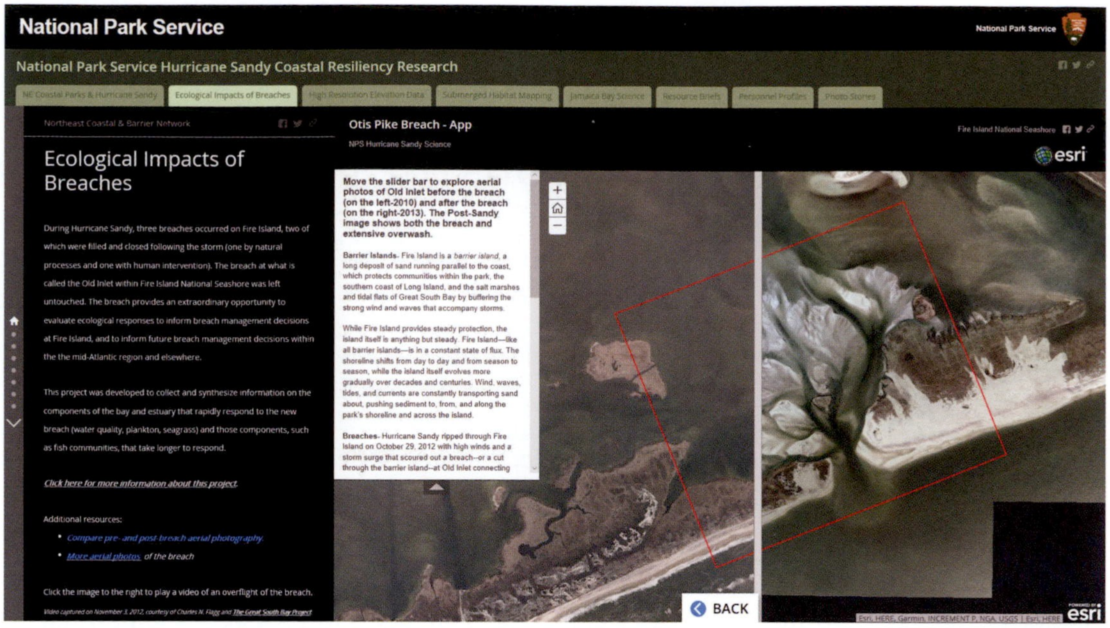
National Park Service
National Park Service Hurricane Sandy Coastal Resiliency Research
Ecological Impacts of Breaches
Northeast Coastal & Barrier Network
Ecological Impacts of Breaches
During Hurricane Sandy, three breaches occurred on Fire Island, two of which were filled and closed following the storm (one by natural processes and one with human intervention). The breach at what is called the Old Inlet within Fire Island National Seashore was left untouched. The breach provides an extraordinary opportunity to evaluate ecological responses to inform breach management decisions at Fire Island, and to inform future breach management decisions within the the mid-Atlantic region and elsewhere.
This project was developed to collect and synthesize information on the components of the bay and estuary that rapidly respond to the new breach (water quality, plankton, seagrass) and those components, such as fish communities, that take longer to respond.
Click here for more information about this project.
Additional resources:
Compare pre- and post-breach aerial photography.
More aerial photos of the breach
Click the image to the right to play a video of an overflight of the breach.
Otis Pike Breach - App
NPS Hurricane Sandy Science
Move the slider bar to explore aerial photos of Old Inlet before the breach (on the left-2010) and after the breach (on the right-2013). The Post-Sandy image shows both the breach and extensive overwash.
Barrier Islands- Fire Island is a barrier island, a long deposit of sand running parallel to the coast, which protects communities within the park, the southern coast of Long Island, and the salt marshes and tidal flats of Great South Bay by buffering the strong wind and waves that accompany storms.
While Fire Island provides steady protection, the island itself is anything but steady. Fire Island—like all barrier islands—is in a constant state of flux. The shoreline shifts from day to day and from season to season, while the island itself evolves more gradually over decades and centuries. Wind, waves, tides, and currents are constantly transporting sand about, pushing sediment to, from, and along the park's shoreline and across the island.
Breaches- Hurricane Sandy ripped through Fire Island on October 29, 2012 with high winds and a storm surge that scoured out a breach--or a cut through the barrier island--at Old Inlet connecting
esri
BACK

A screenshot of a story tab displays an interactive slider bar of aerial photos depicting ecological impacts of breaches.

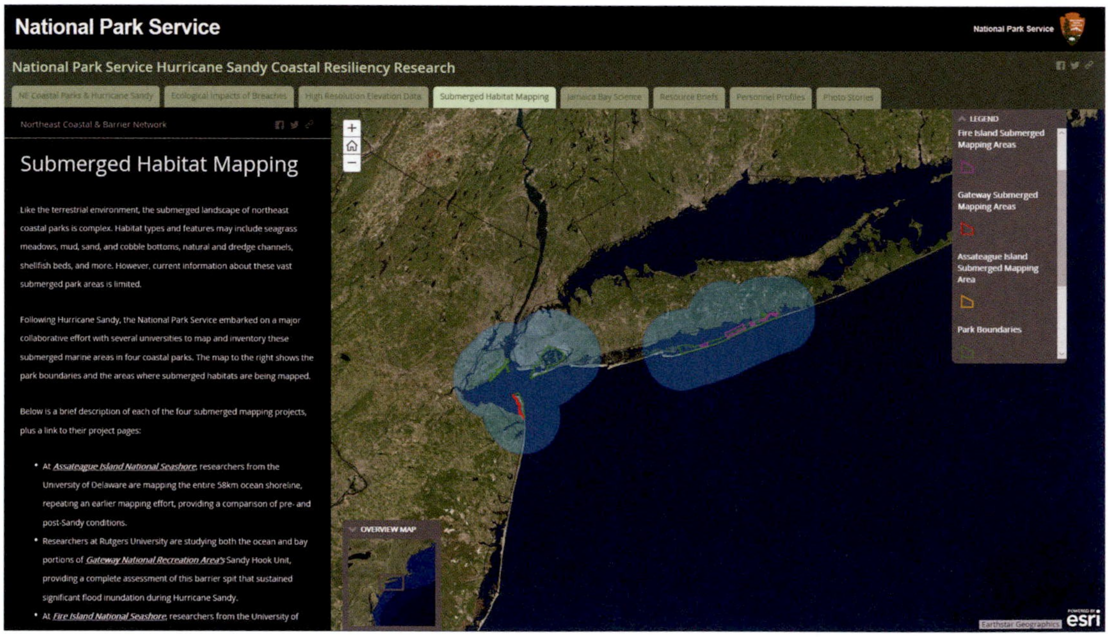
National Park Service
National Park Service Hurricane Sandy Coastal Resiliency Research
Submerged Habitat Mapping
Northeast Coastal & Barrier Network
Submerged Habitat Mapping
Like the terrestrial environment, the submerged landscape of northeast coastal parks is complex. Habitat types and features may include seagrass meadows, mud, sand, and cobble bottoms, natural and dredge channels, shellfish beds, and more. However, current information about these vast submerged park areas is limited.
Following Hurricane Sandy, the National Park Service embarked on a major collaborative effort with several universities to map and inventory these submerged marine areas in four coastal parks. The map to the right shows the park boundaries and the areas where submerged habitats are being mapped.
Below is a brief description of each of the four submerged mapping projects, plus a link to their project pages:
At Assateague Island National Seashore, researchers from the University of Delaware are mapping the entire 58km ocean shoreline, repeating an earlier mapping effort, providing a comparison of pre- and post-Sandy conditions.
Researchers at Rutgers University are studying both the ocean and bay portions of Gateway National Recreation Area's Sandy Hook Unit, providing a complete assessment of this barrier spit that sustained significant flood inundation during Hurricane Sandy.
At Fire Island National Seashore, researchers from the University of
LEGEND
Fire Island Submerged Mapping Areas
Gateway Submerged Mapping Areas
Assateague Island Submerged Mapping Area
Park Boundaries
OVERVIEW MAP
esri

A screenshot of a story tab displays a map with satellite imagery describing submerged habitat.

The NPS Northeast Region Science Team submitted proposals and was awarded $21 million for projects developed to assess the condition of park natural resources following Hurricane Sandy, enhance knowledge of the long-term resiliency of park resources, and help increase coastal resiliency and capacity to withstand future storm damage in these areas.

The NPS created a site to provide information about the diverse coastal resiliency research efforts that were funded in these four coastal parks. The site is designed around a series of tabs devoted to different topics concerning Hurricane Sandy science in these parks, including the ecological impacts of barrier island breaches caused by the storm, the mapping of changes in submerged habitats, and the collection of detailed elevation data for the modeling of coastal flood risks. Each tab contains links to dozens of related maps, documents, websites, photos, and videos.

National Park Service
National Park Service
National Park Service Hurricane Sandy Coastal Resiliency Research
NE Coastal Parks & Hurricane Sandy
Ecological Impacts of Breaches
High Resolution Elevation Data
Submerged Habitat Mapping
Jamaica Bay Science
Resource Briefs
Personnel Profiles
Photo Stories
Northeast Coastal & Barrier Network
Jamaica Bay Science
Salt Marsh Health
Salt marshes serve important ecosystem roles as habitat, nurseries for marine organisms, filters for contaminant inputs to coastal waters, and buffers against the impact of storms on the coastal environment. However, salt marsh loss is a worldwide phenomenon and this is especially evident in Jamaica Bay, with losses in acreage of ~63% between 1951 and 2003. Marsh loss may be due to numerous factors; stresses on Jamaica Bay marshes include rising sea levels, changing configuration of the bay through activities such as dredging of the Bay bottom, and eutrophication of the Bay system.
This project, being conducted in collaboration with the National Park Service and scientists from the School of Marine & Atmospheric Sciences at Stony Brook University, will characterize both the sediment dynamical and geochemical constraints on salt marsh resiliency at selected marsh sites in Jamaica Bay. Resiliency is defined here as the ability of the Jamaica Bay marshes to cope with the increased rates of sea level rise and continued loadings of nutrients and resultant organic matter production in the Bay.
The proposed work offers a unique opportunity to re-sample sites that were sampled in 2005-2007. In particular, the new data coupled with prior data will enable the assessment of the health of restored marshes approximately 5 to10 years after restoration.
Click here for more information on this project.
Photo at right: Clapper Rail (Stephen P. Broker, All rights reserved)

A screenshot of a tab displays information about the science of Jamaica Bay salt marsh health.

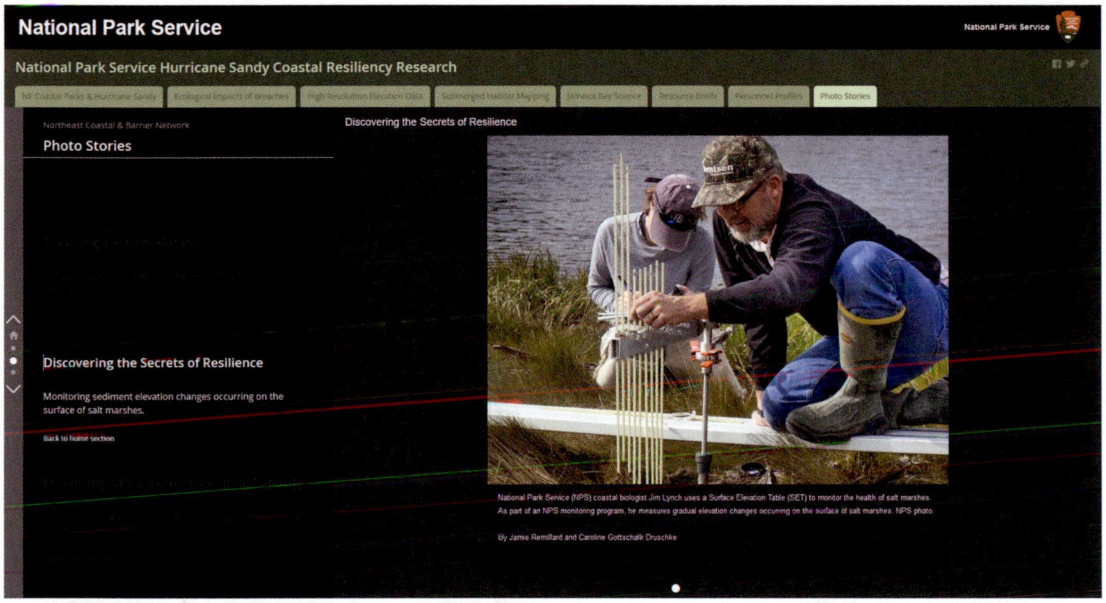
National Park Service
National Park Service
National Park Service Hurricane Sandy Coastal Resiliency Research
NE Coastal Parks & Hurricane Sandy
Ecological Impacts of Breaches
High Resolution Elevation Data
Submerged Habitat Mapping
Jamaica Bay Science
Resource Briefs
Personnel Profiles
Photo Stories
Northeast Coastal & Barrier Network
Photo Stories
Discovering the Secrets of Resilience
Monitoring sediment elevation changes occurring on the surface of salt marshes.
Back to home section
Discovering the Secrets of Resilience
National Park Service (NPS) coastal biologist Jim Lynch uses a Surface Elevation Table (SET) to monitor the health of salt marshes. As part of an NPS monitoring program, he measures gradual elevation changes occurring on the surface of salt marshes. NPS photo
By Jamie Remillard and Caroline Gottschalk Druschke

A screenshot of a tab displays a photo of a park scientist monitoring sediment elevation changes occurring on the surface of a salt marsh.

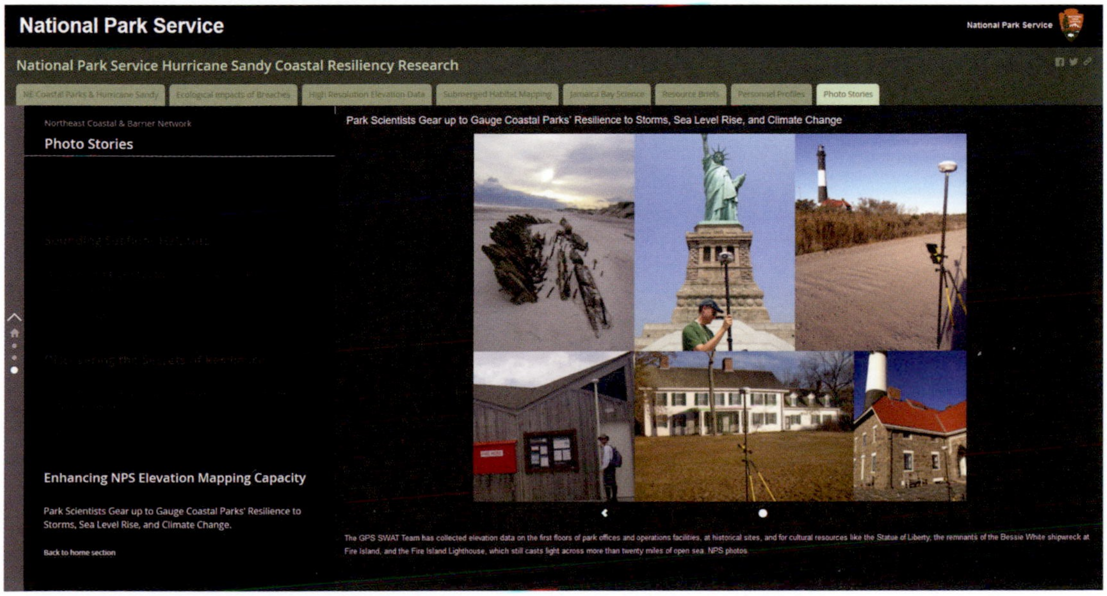
National Park Service
National Park Service
National Park Service Hurricane Sandy Coastal Resiliency Research
NE Coastal Parks & Hurricane Sandy
Ecological Impacts of Breaches
High Resolution Elevation Data
Submerged Habitat Mapping
Jamaica Bay Science
Resource Briefs
Personnel Profiles
Photo Stories
Northeast Coastal & Barrier Network
Photo Stories
Enhancing NPS Elevation Mapping Capacity
Park Scientists Gear up to Gauge Coastal Parks' Resilience to Storms, Sea Level Rise, and Climate Change.
Back to home section
Park Scientists Gear up to Gauge Coastal Parks' Resilience to Storms, Sea Level Rise, and Climate Change
The GPS SWAT Team has collected elevation data on the first floors of park offices and operations facilities, at historical sites, and for cultural resources like the Statue of Liberty, the remnants of the Bessie White shipwreck at Fire Island, and the Fire Island Lighthouse, which still casts light across more than twenty miles of open sea. NPS photos

A screenshot of a story tab displays a photo compilation of park scientists collecting elevation data in a variety of NPS locations. URL and data sources for pages 49-51: http://go.nps.gov/sandy_resiliency; NPS, Esri World Ocean, Esri World Imagery, New York State 2010 Orthoimagery.

ICE AGE COMPLEX PREFERRED ALTERNATIVE WITH HUNTING AREAS

MATT COLWIN, NPS MIDWEST REGION GEOSPATIAL RESOURCES

This map was developed as part of a series created during the planning process for the Ice Age Complex at Cross Plains, Wisconsin, a piece of land acquired by the Ice Age National Scenic Trail. This map depicts the preliminary preferred alternative that was developed by trail staff for managing the land, as well as areas where hunting is restricted. Park trail staff noted that there were issues with local residents being unaware of hunting restrictions, so this map was developed to better visualize those areas. Other information on the map includes important viewing areas, camping and picnic areas, and the historic Wilke homestead.

The map depicts the preliminary preferred alternative for managing the land and areas where hunting is restricted. Ice Age Trail line data courtesy of the Ice Age Trail Alliance. Data sources: Ice Age Complex data: NPS, Ice Age National Scenic Trail; Ice Age Trail Line: Ice Age Trail Alliance; Roads, Railroad, Cities, State/County and other boundaries data: Esri; Esri World Street Map: Esri, HERE, Garmin, USGS, Intermap, INCREMENT P, NRCAN, Esri Japan, METI, Esri China (Hong Kong), NOSTRA, © OpenStreetMap contributors, and the GIS user community.

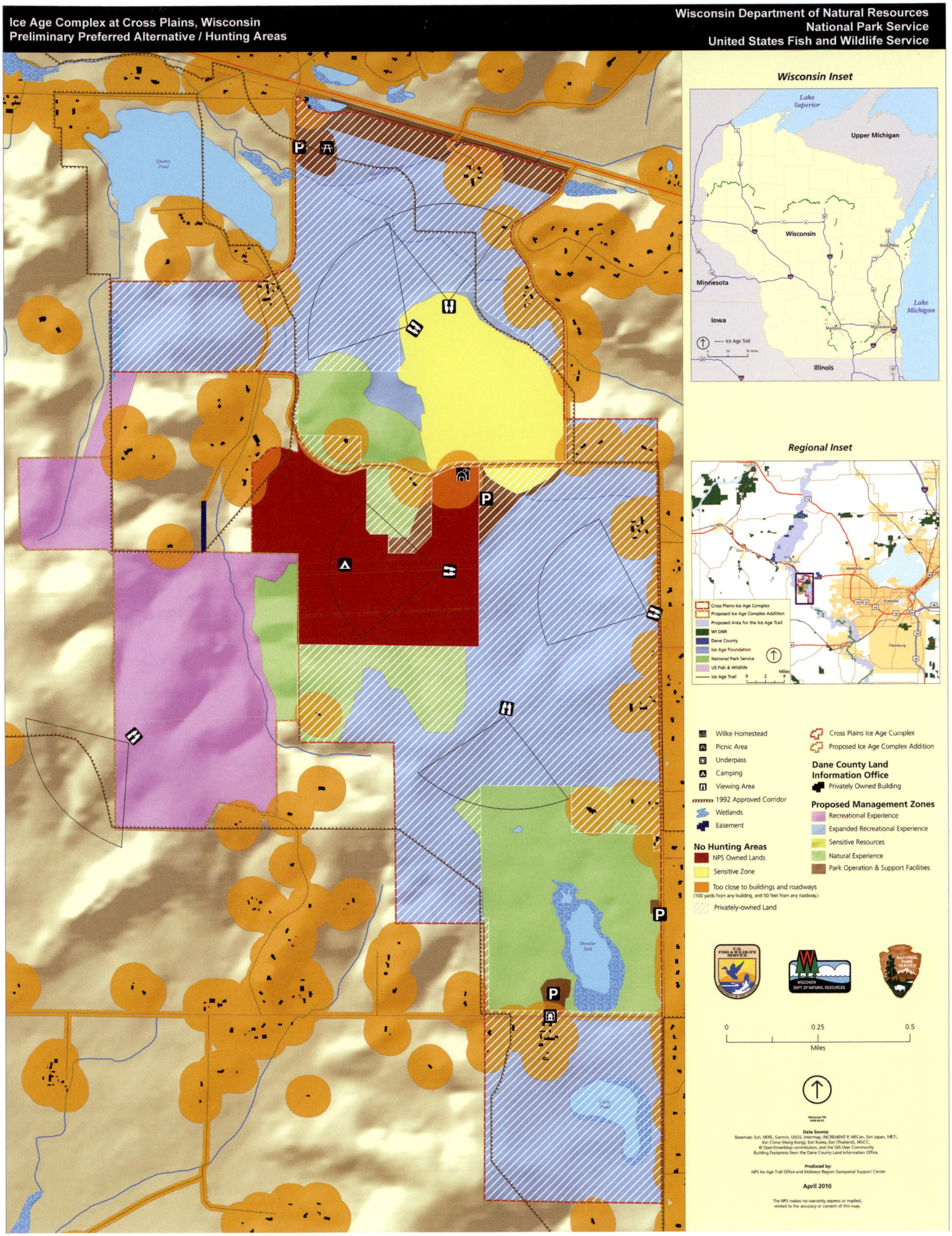
Ice Age Complex at Cross Plains, Wisconsin
Preliminary Preferred Alternative / Hunting Areas
Wisconsin Department of Natural Resources
National Park Service
United States Fish and Wildlife Service
Wisconsin Inset
Lake Superior
Upper Michigan
Wisconsin
Minnesota
Iowa
Lake Michigan
Illinois
Ice Age Trail
Regional Inset
Cross Plains Ice Age Complex
Proposed Ice Age Complex Addition
Proposed Area for the Ice Age Trail
WI DNR
Dane County
Ice Age Foundation
National Park Service
US Fish & Wildlife
Ice Age Trail
Wilke Homestead
Picnic Area
Underpass
Camping
Viewing Area
1992 Approved Corridor
Wetlands
Easement
Cross Plains Ice Age Complex
Proposed Ice Age Complex Addition
Dane County Land Information Office
Privately Owned Building
Proposed Management Zones
Recreational Experience
Expanded Recreational Experience
Sensitive Resources
Natural Experience
Park Operation & Support Facilities
No Hunting Areas
NPS Owned Lands
Sensitive Zone
Too close to buildings and roadways
(100 yards from any building, and 50 feet from any roadway.)
Privately-owned Land
0
0.25
0.5
Miles
Data Source:
Basemap: Esri, HERE, Garmin, USGS, Intermap, INCREMENT P, NRCan, Esri Japan, METI, Esri China (Hong Kong), Esri Korea, Esri (Thailand), NGCC, © OpenStreetMap contributors, and the GIS User Community
Building Footprints from the Dane County Land Information Office.
Produced by:
NPS Ice Age Trail Office and Midwest Region Geospatial Support Center
April 2010
The NPS makes no warranty, express or implied, related to the accuracy or content of this map.

4 MANAGING FIRE

KAREN FOLGER, SEQUOIA AND KINGS CANYON NATIONAL PARKS
SKIP EDEL, NPS DIVISION OF FIRE AND AVIATION MANAGEMENT, BRANCH OF WILDLAND FIRE

GIS FOR PLANNING—DURING AND AFTER FIRE RESPONSE

Wildland fires burn across hundreds of thousands of acres of national parklands each year. These fires dramatically alter the vegetation and landscape and are a powerful force of change in national parks. The National Park Service (NPS) aggressively fights and suppresses fires that occur near populated areas, visitor facilities, and valuable natural and cultural resources. However, in many cases, wildland fires are allowed to burn to fulfill their natural role upon the landscapes of the national parks. Wildland fires benefit nature by rejuvenating soils, opening wildlife habitat, and clearing excess vegetation below the canopy of the forest. Additionally, many parks use prescribed fire to treat areas of the landscape during conditions that are more favorable than those that may occur during unplanned wildland fires.

Fire management in national parks today reflects a commitment to public safety and a realization that fire plays an important role in the natural cycle of renewal. When fires burn in our national parks, fire managers turn to an array of tools—from the venerable Pulaski ax dating to the early 1900s to modern-day laptops, smartphones, tablets, and satellite imagery. The park service increasingly relies on geographic information systems (GIS) as a high-tech fire management tool, which has spread through all levels and geographic areas of our vast national park system. Fire managers use GIS to protect invaluable natural, historical, and cultural resources in national parklands. The technology also plays a key role in strategic planning and budgeting for fire management.

Mapping has long been a part of fire management in the park service. Maps from nearly a century ago showed the extent of fires within Yellowstone and other national parks. Park service fire managers began to envision the potential of GIS mapping as the technology developed in the early 1980s. By the mid-1990s, GIS proved it belonged in the national arsenal of fire management tools with the development of software programs that predict fire behavior. Recent advents in technology, including web mapping and applications such as ArcGIS StoryMaps, have taken mapping to a whole new level.

Using the latest GIS technology, fire managers today can combine an almost unlimited number of layers of information onto a single map or web map that displays a fire's relationship to a national park and surrounding area. These individual layers, or themes, can represent vegetation and fuels, historic fire origins and burn patterns, and the location of cabins, campsites, and important habitat and archeological sites within park boundaries. Firefighters equipped with smartphones and tablets can call up these maps and quickly respond while on the fire lines. Using GIS and related technologies such as global positioning systems, they can send data on fire perimeters and burn rates to the news media and public via the internet and to fire managers at distant command centers for near-real-time analysis.

Fire managers use GIS to help predict a fire's intensity, rate of spread, and potential maximum size. Displaying this data on maps and websites helps fire managers determine the placement of firefighters and the need for evacuations. Maps of fire histories also show where fire has not occurred at normal or natural intervals. This provides invaluable information about the buildup of vegetation and fuel, and thus the potential for a catastrophic fire and the need for prescribed burns, fuel breaks, and long-term prevention strategies. Today, the challenge is not so much to discover new applications of GIS to fire management, but rather for fire management to keep up with the continuously evolving field of GIS and the new tools that put this technology into the hands of everyone who needs it.

GLACIER NATIONAL PARK STAND-AGE REPLACEMENT IN THE FIRES OF 2003

GINA MAZZA, GLACIER NATIONAL PARK

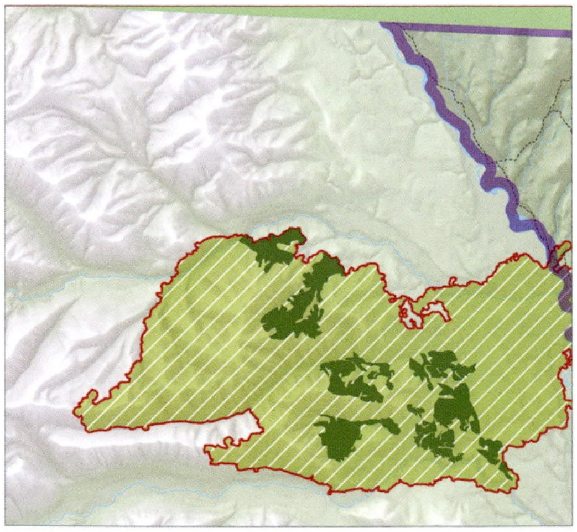

The map shows Glacier National Park stand-age replacement in the fires of 2003. Data sources: USGS, State of Montana, NPS.

This map was created as a service project for Glacier National Park Fire Management through the University of Montana. The map shows the 2003 fire perimeters for fires that burned more than 100 acres. Using the 2015 NAIP imagery, areas were digitized where fires left stands of trees. The map shows that most of the forest within the fire perimeters was entirely burned, or replaced, with few stands of old-growth trees remaining. This data provides an insight into how the fires behaved. This project will be expanded by assessing stand-age replacement for fire seasons both past and present in order to assemble an accurate and up-to-date stand-age map. Future fire managers can refer to this data, or they can use it to assess how fire severity changes over time. Large fires in Glacier National Park in northern Montana seem to occur with a higher frequency than in the past. Data showing locations of old-growth stands, where potential fire breaks occur, or whatever a creative spatial analyst can draw from this assessment can aid managers and ecologists alike in understanding this dynamic force known as wildfire.

2018 FERGUSON FIRE INFORMATION AND PROGRESSION MAPS

KENT VAN WAGTENDONK, YOSEMITE NATIONAL PARK

The Ferguson Fire started auspiciously on Friday the 13th, July 2018, just seven miles downstream from Yosemite National Park in the Merced River drainage. The fire spread quickly, forcing evacuations and closing the highway along the Merced River into Yosemite National Park, thus impacting employees, residents, tourists, and gateway communities. Within two weeks, the fire had closed the remaining two roads into Yosemite Valley, prompting the park to close the valley further and extending impacts on the local region.

To manage such a complex fire, an Incident Management Team (IMT) was ordered. Their public information officer provided fire updates on a regular basis, but the associated map lacked the needed detail to assist park staff with operations. To alleviate this problem, the park created two custom maps that complemented the IMT updates and emailed them to park, regional, and national NPS staff and park partners each morning.

The first map incorporated incident-specific data such as fire perimeter, hand-dozer line, drop points, helispots, and division breaks, as well as infrared (IR) data. The IMT and IR interpreter posted their respective data to the fire incident data repository. These publicly available datasets were downloaded to create a map showing where certain control actions were being taken in addition to where the IR sensors were detecting heat intensity of the fire. The second map showed daily progression data that tracked fire growth over the course of the fire. These products provided readers a visual link to the IMT's fire updates.

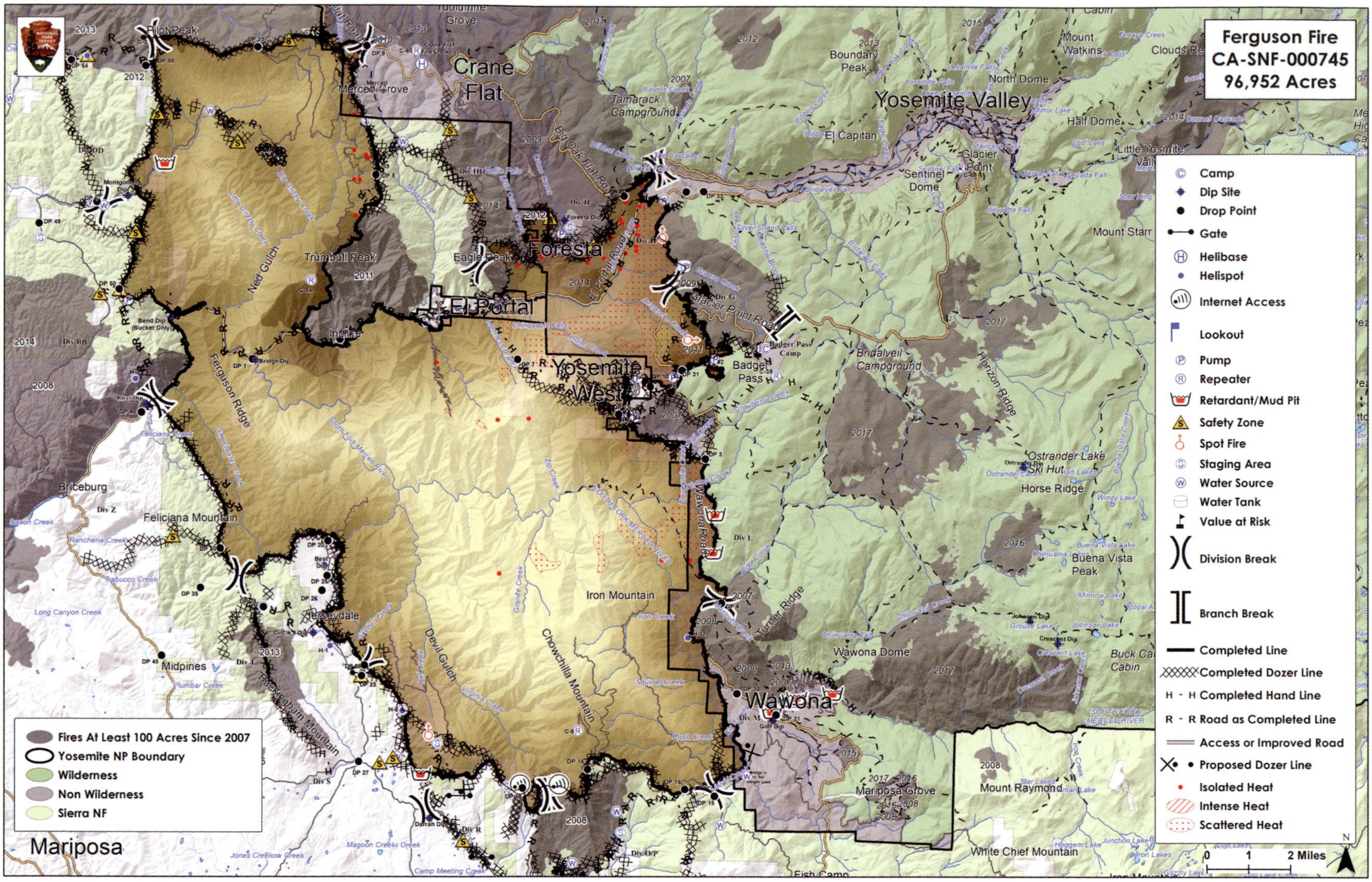

The Ferguson Fire Information Map shows the boundary and fire area. Data sources: USGS, National Interagency Fire Center, CalFire, NPS.

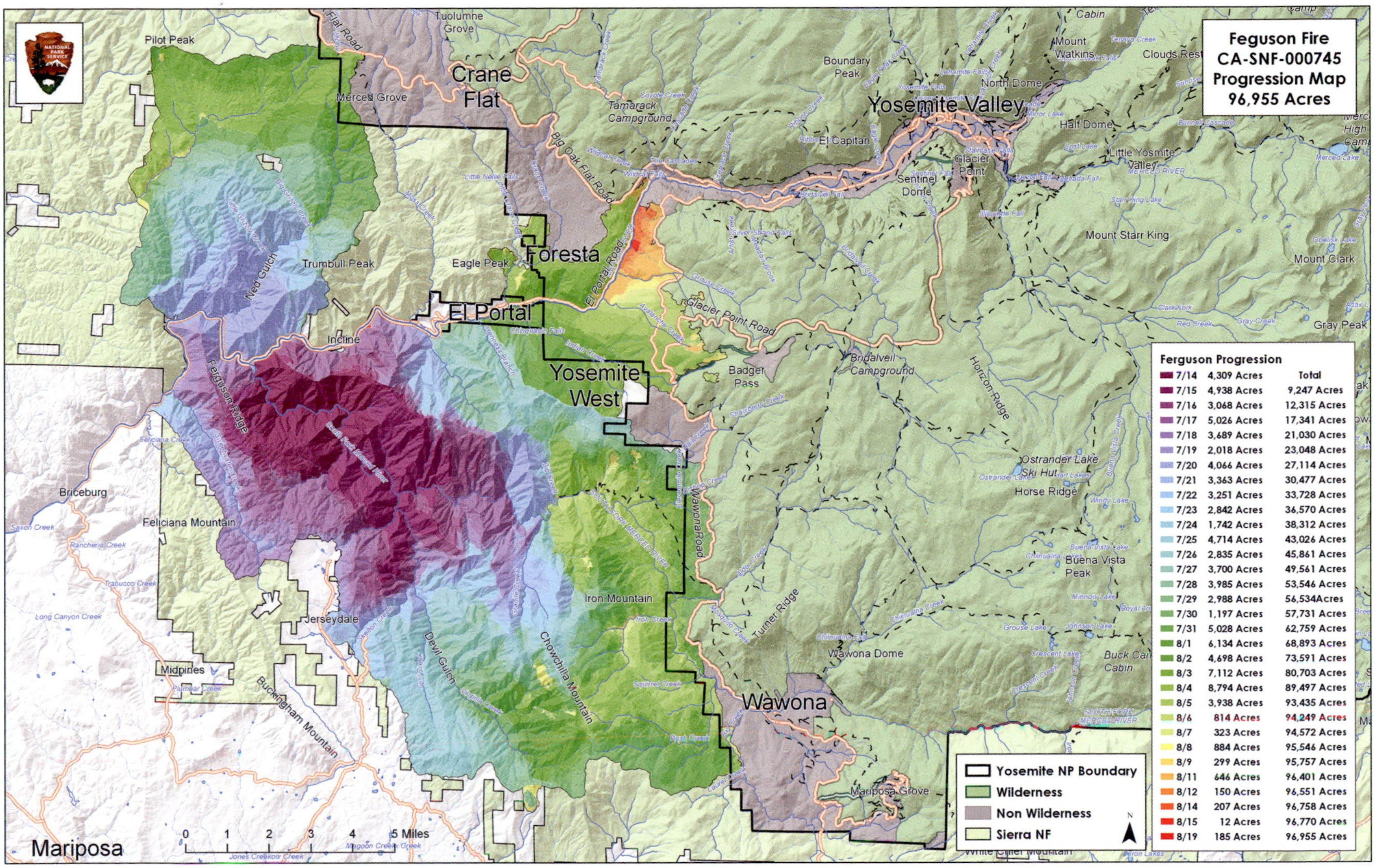
Feguson Fire
CA-SNF-000745
Progression Map
96,955 Acres
Ferguson Progression
7/14 4,309 Acres Total
7/15 4,938 Acres 9,247 Acres
7/16 3,068 Acres 12,315 Acres
7/17 5,026 Acres 17,341 Acres
7/18 3,689 Acres 21,030 Acres
7/19 2,018 Acres 23,048 Acres
7/20 4,066 Acres 27,114 Acres
7/21 3,363 Acres 30,477 Acres
7/22 3,251 Acres 33,728 Acres
7/23 2,842 Acres 36,570 Acres
7/24 1,742 Acres 38,312 Acres
7/25 4,714 Acres 43,026 Acres
7/26 2,835 Acres 45,861 Acres
7/27 3,700 Acres 49,561 Acres
7/28 3,985 Acres 53,546 Acres
7/29 2,988 Acres 56,534Acres
7/30 1,197 Acres 57,731 Acres
7/31 5,028 Acres 62,759 Acres
8/1 6,134 Acres 68,893 Acres
8/2 4,698 Acres 73,591 Acres
8/3 7,112 Acres 80,703 Acres
8/4 8,794 Acres 89,497 Acres
8/5 3,938 Acres 93,435 Acres
8/6 814 Acres 94,249 Acres
8/7 323 Acres 94,572 Acres
8/8 884 Acres 95,546 Acres
8/9 299 Acres 95,757 Acres
8/11 646 Acres 96,401 Acres
8/12 150 Acres 96,551 Acres
8/14 207 Acres 96,758 Acres
8/15 12 Acres 96,770 Acres
8/19 185 Acres 96,955 Acres
Yosemite NP Boundary
Wilderness
Non Wilderness
Sierra NF
N
0 1 2 3 4 5 Miles
Crane Flat
Foresta
El Portal
Yosemite West
Yosemite Valley
Wawona
Mariposa
Pilot Peak
Merced Grove
Tuolumne Grove
Trumbull Peak
Eagle Peak
Incline
Briceburg
Feliciana Mountain
Jerseydale
Midpines
Iron Mountain
Badger Pass
Bridalveil Campground
El Capitan
Sentinel Dome
Glacier Point
Half Dome
North Dome
Boundary Peak
Mount Watkins
Clouds Rest
Little Yosemite Valley
Mount Starr King
Mount Clark
Gray Peak
Ostrander Lake Ski Hut
Horse Ridge
Buena Vista Peak
Wawona Dome
Buck Camp Cabin
Mariposa Grove
Tamarack Campground
Ned Gulch
Ferguson Ridge
Devil Gulch
Chowchilla Mountain
Buckingham Mountain
Turner Ridge
Horizon Ridge
Big Oak Flat Road
El Portal Road
Glacier Point Road
Wawona Road

The Ferguson Fire Progression Map shows the spread of fire. Data sources: USGS, National Interagency Fire Center, CalFire, NPS.

ENTERPRISE GIS FOR FIRE SUPPRESSION REPAIR AT YOSEMITE NATIONAL PARK

HEIDI OGLE, NPS INTERMOUNTAIN REGION, GEOGRAPHIC RESOURCES DIVISION
ELIZABETH HALE, YOSEMITE NATIONAL PARK

As mega-fires become more common and wildfires burn practically year-round, enterprise GIS technology is being called upon to provide a common operating picture for personnel working on all aspects of the fire. The Ferguson Fire burned into Yosemite National Park in the summer of 2018. For resource advisors working on the fire lines to minimize impacts to natural and cultural resources, it was critical to see where sensitive resources were located relative to fire management activities. GIS specialists set up ArcGIS Collector as a mobile app for viewing different map layers and collecting data, and, after the fire was contained, for mapping and communicating about needed and accomplished suppression repair work. The dozer lines, retardant drops, and other locations they mapped in Collector went directly into the National Incident Feature Service (NIFS). For plans, operations, and incident command staff tasked with rapid personnel and equipment response decisions, having up-to-date information was paramount. GIS specialists set up a web map to dynamically display the map layers in the NIFS. With an operations dashboard, suppression repair statistics automatically calculated and updated when resource advisors made edits in Collector.

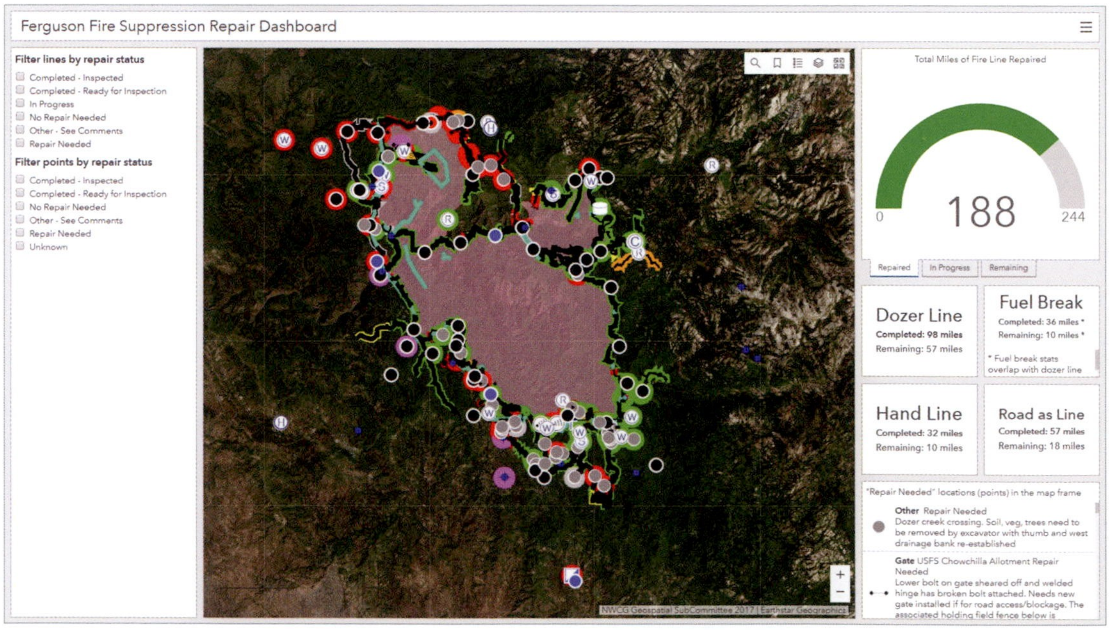

The Ferguson Fire suppression repair operations dashboard shows a legend and a fire map.

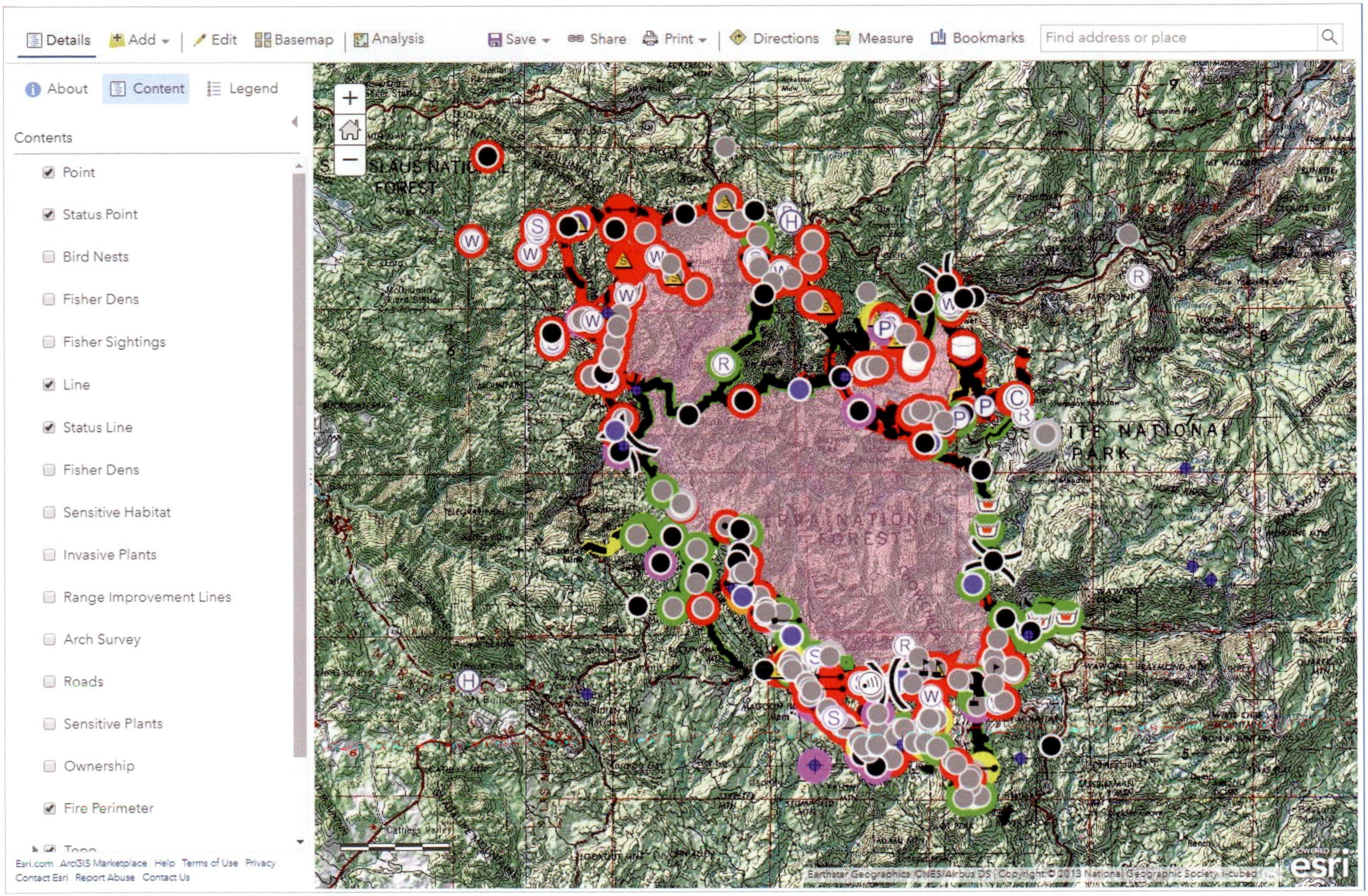

The Ferguson Fire suppression repair web map shows the fire perimeter and repair status of contingency lines.

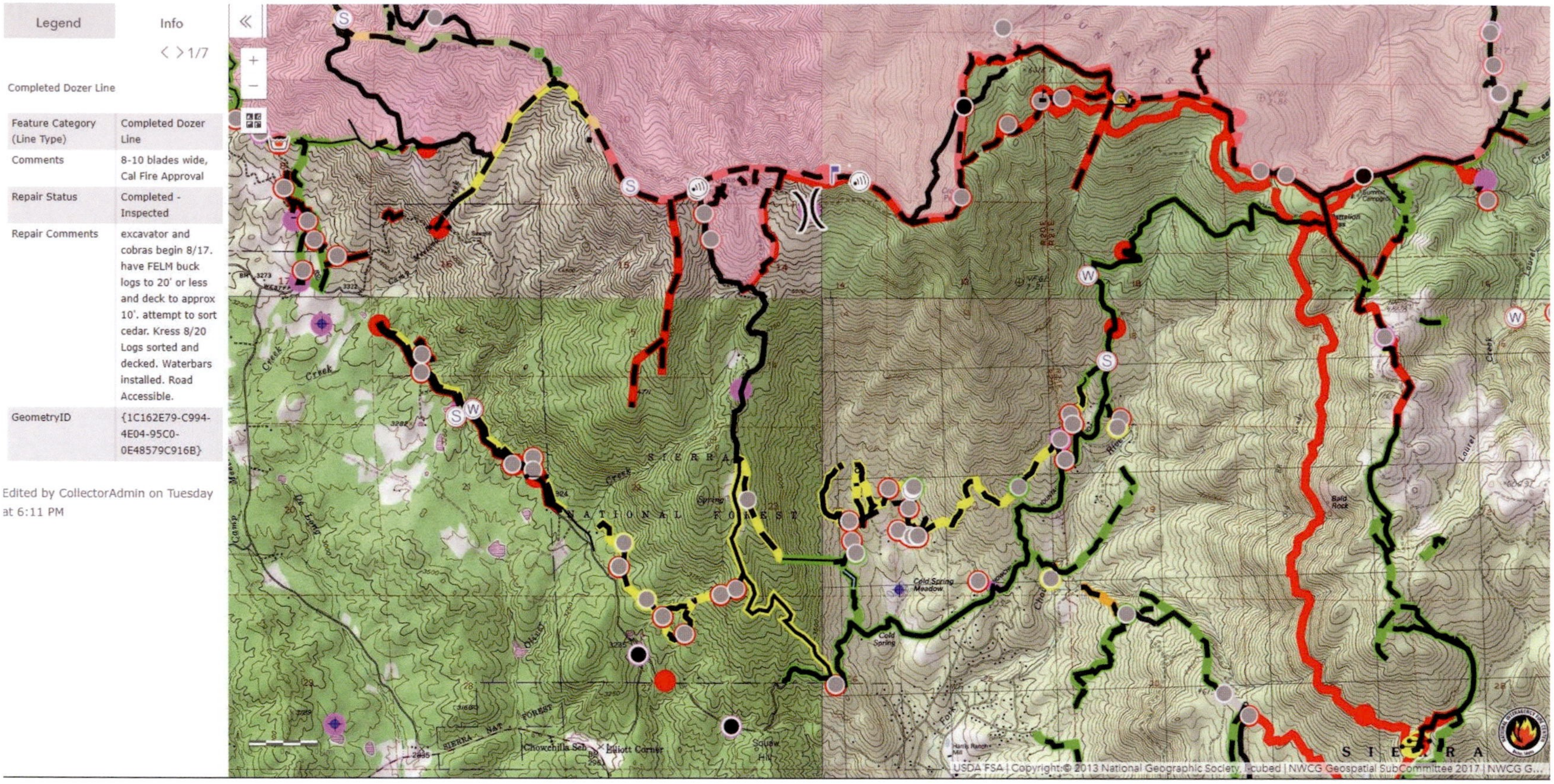

The Ferguson Fire suppression repair web app shows layers, data, and a legend.

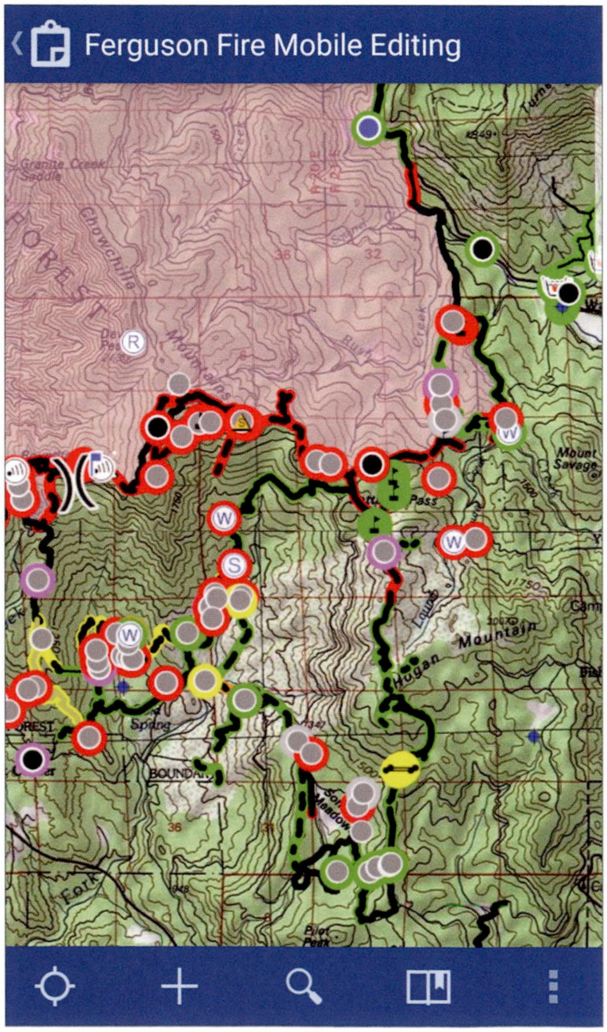

Ferguson Fire suppression repair Collector application provides mobile editing capability.

Data sources, pages 60–62: National Interagency Fire Center, NPS.

A resource advisor captures and edits fire data with Collector. NPS photo/Erin Dickman.

An NPS GIS specialist uses Collector on the fire line. NPS photo/Oliver Anderson.

2015 JOURNEY OF THE ROUGH FIRE

KAREN FOLGER, SEQUOIA AND KINGS CANYON NATIONAL PARKS

The 2015 Rough Fire burned into Sequoia and Kings Canyon National Parks, ultimately becoming the second-largest wildfire ever recorded in the Sierra Nevada at the time. Previous prescribed fires and managed wildland fires that were allowed to burn helped keep the fire from destroying homes and park structures. While areas of the parks were evacuated, nearly $400 million in facility assets were saved due to these previous treatments. Since 1968, these parks have been a leader in implementing prescribed fire to protect communities and resources. Fire is a natural process and forests in the Sierra Nevada will continue to burn. These parks used a story map to take visitors on the journey of this fire and highlight major events. It also provides context to the complex job of mapping large, fast-moving wildfires. It shows where these previous fires occurred and highlights the interaction with the Rough Fire.

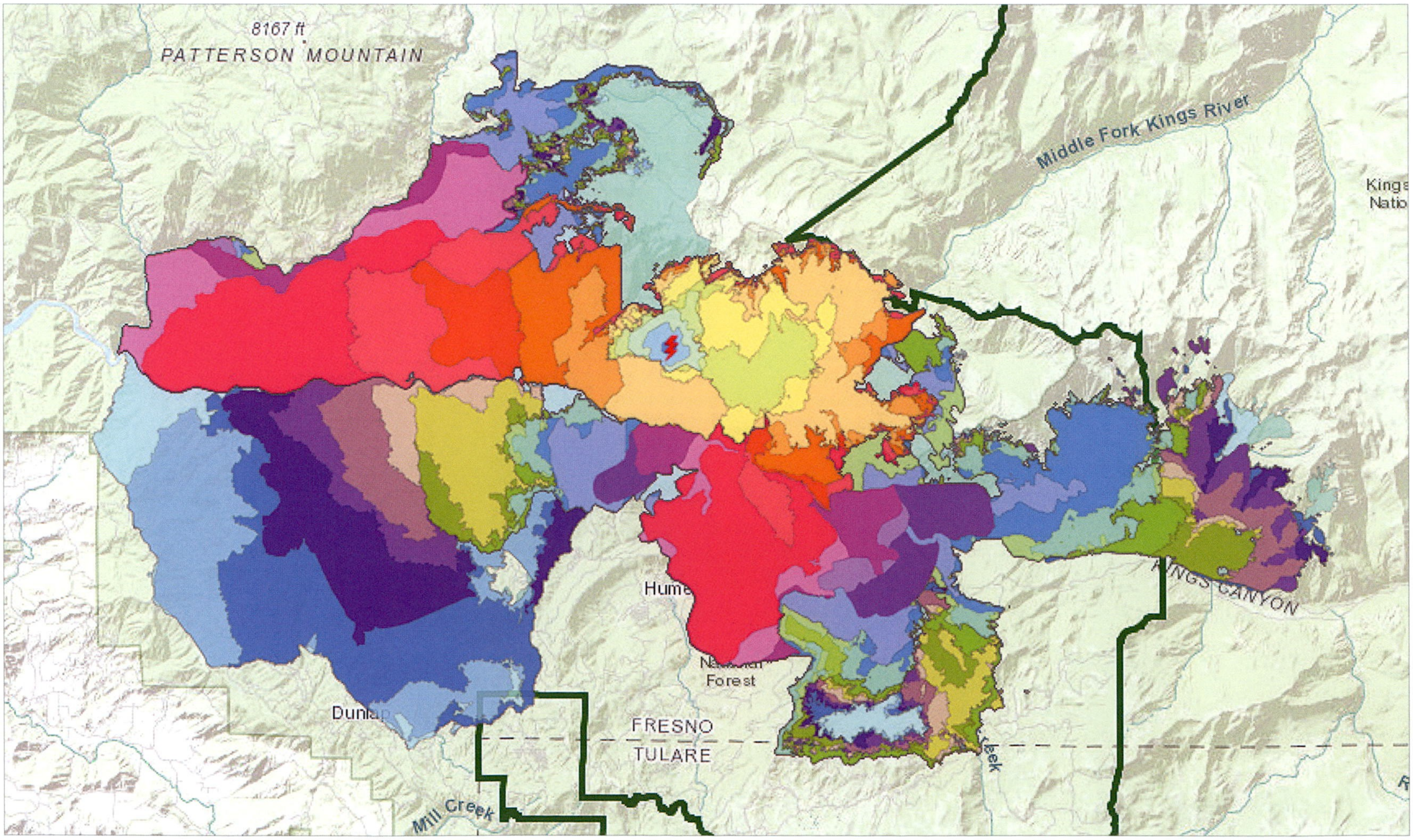

Rough Fire progression map shows the Sierra Nevada. Data sources: Map by K. Folger, NPS; data from NPS archive; basemap credit(s), Esri.

This photo shows the Rough Fire ignition, July 31, 2015. NPS photo/ M. Donnelly.

This aerial photo shows the Rough Fire on August 9, 2015. NPS photo/ C. Vernon.

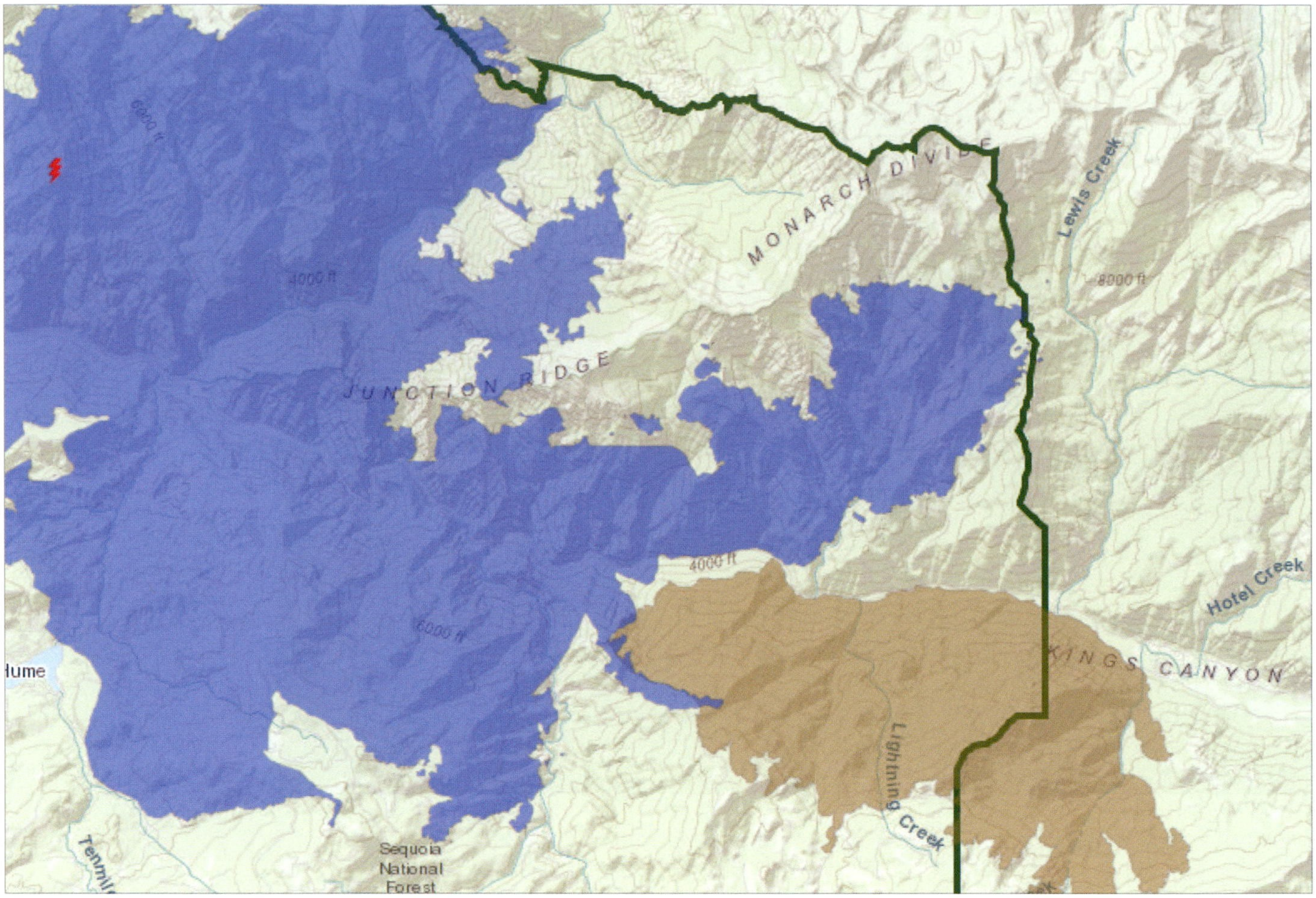

The map shows the Rough Fire on August 30, 2015. Data sources: Map by K. Folger, NPS; data from NPS archive; basemap credit(s), Esri.

The photo shows the Rough Fire September 10, 2015, north of Pam Point. NPS photo/ T. Caprio.

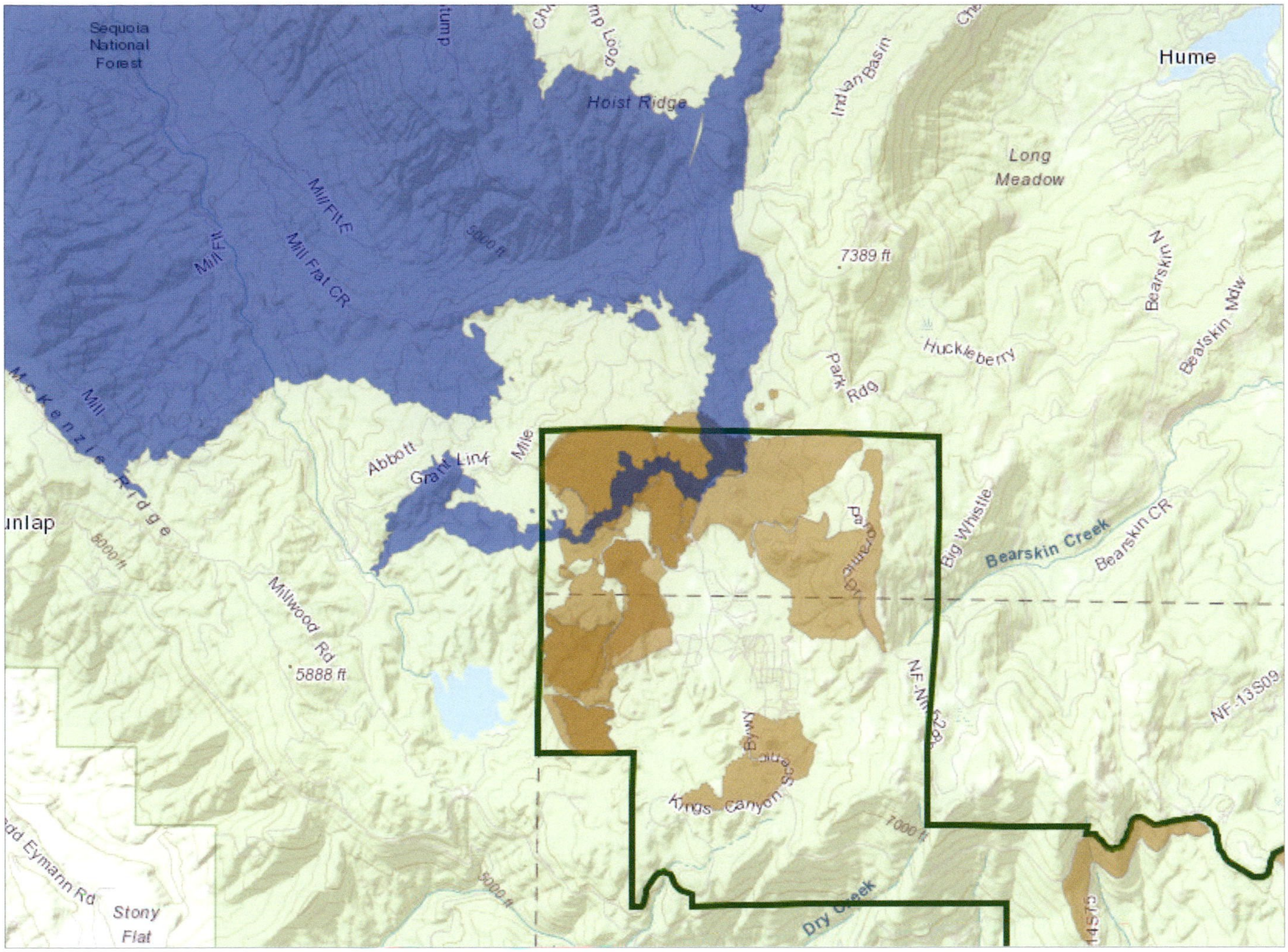

The map shows the Rough Fire boundary from September 12, 2015.

BANDELIER NATIONAL MONUMENT FIRE MANAGEMENT PLAN

SKY SKACH, COLORADO STATE UNIVERSITY
NELL CONTI, NPS INTERMOUNTAIN REGION, GEOGRAPHIC RESOURCES DIVISION

Fire is considered a natural process in maintaining structural and functional integrity of Bandelier National Monument's vegetation communities in New Mexico. Most of the vegetation communities and wildlife that have persisted through time are now fire dependent. A high concentration of lightning strikes, climatic conditions, and topography make fire one of the dominant natural disturbance processes at Bandelier where fires frequently occur.

These conceptual maps depict potential alternatives for an updated fire management plan and were developed as part of a data gathering effort. The project area is the entire legislated boundary of the monument, including the noncontiguous 799-acre Tsankawi unit, which is 12 miles from the main park and includes archeological sites and the ancestral Pueblo village of Tsankawi. Bandelier National Monument protects roughly 33,000 acres of rugged canyon and mesa country, where evidence of human presence dates 10,000 years.

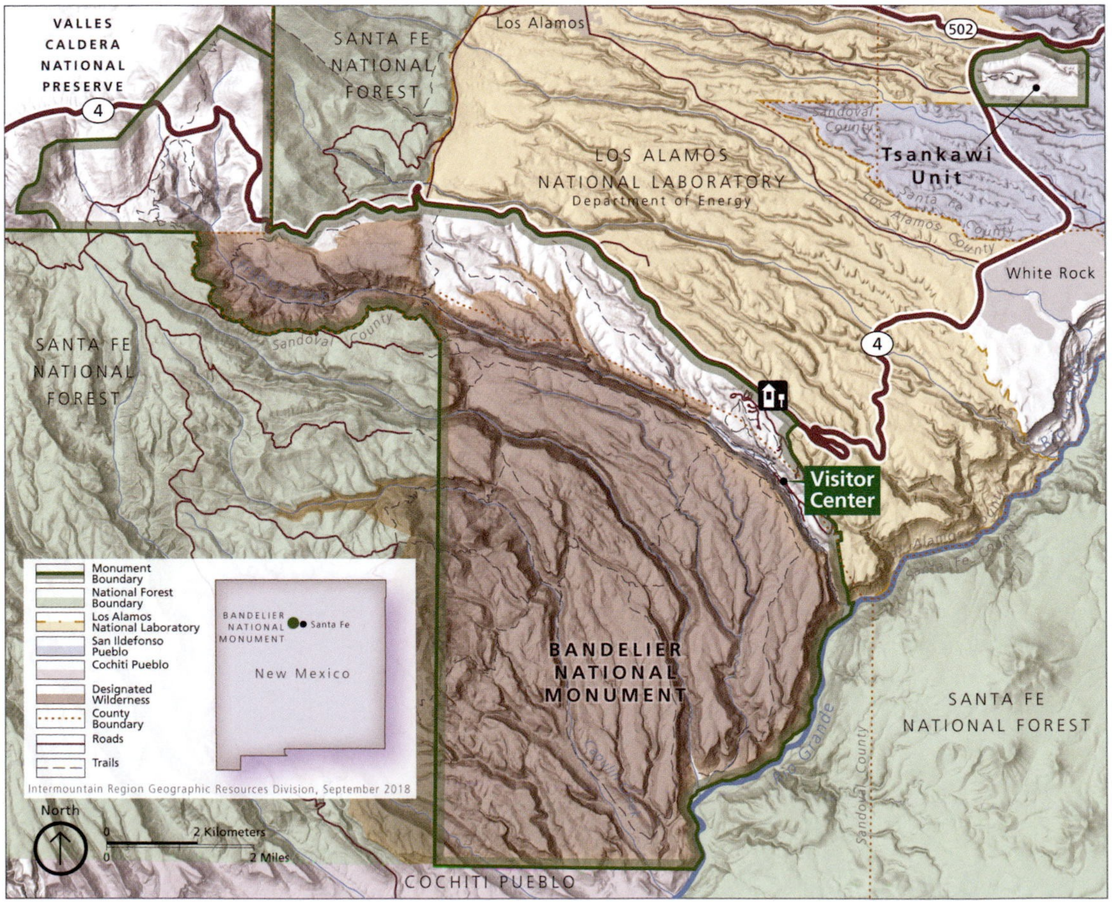

The context map shows the Bandelier National Monument Fire Management Plan. Data sources: NPS, Esri.

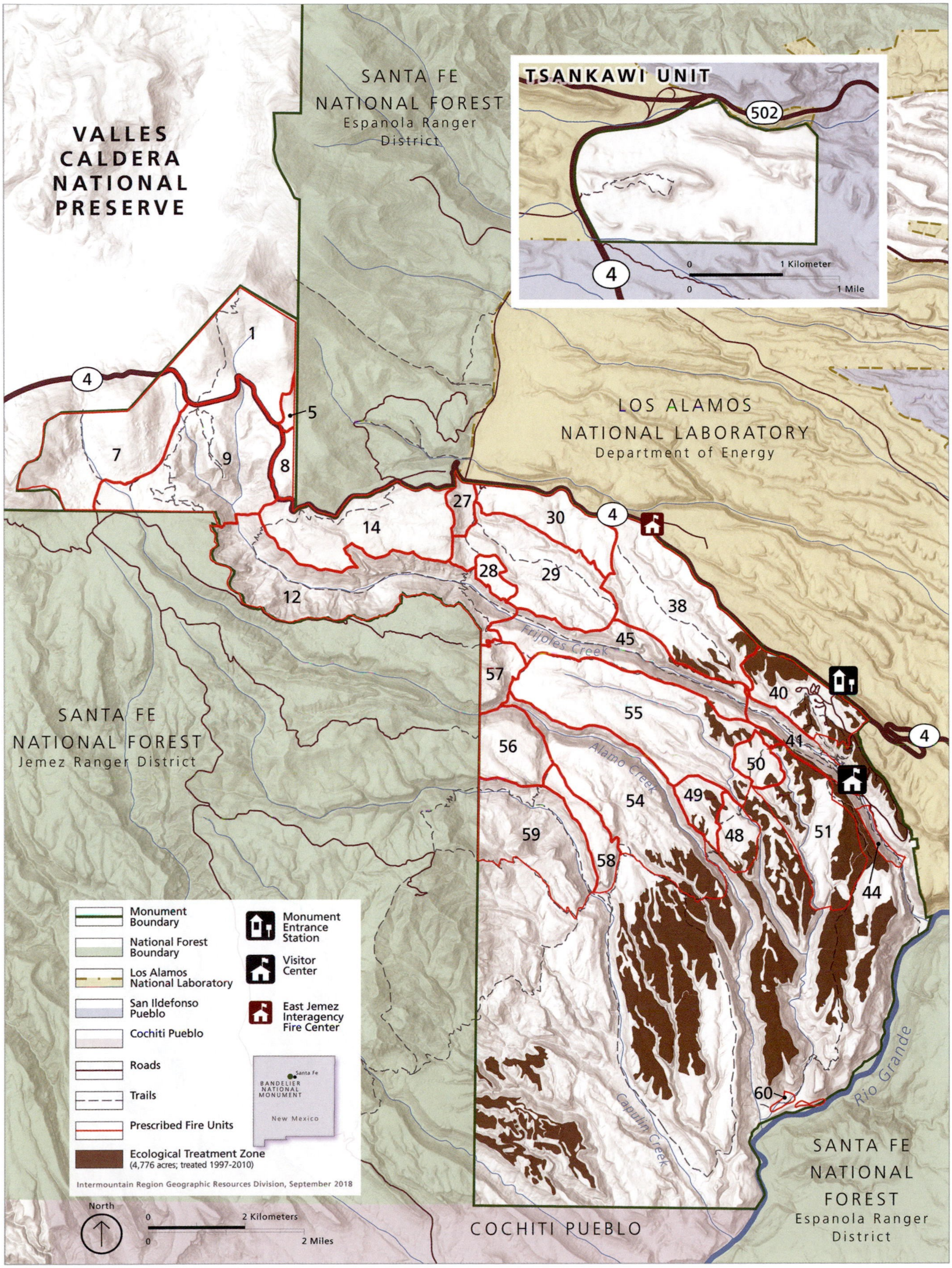

The context map shows the Bandelier National Monument Fire Management Plan prescribed fire units. Data sources: NPS, Esri.

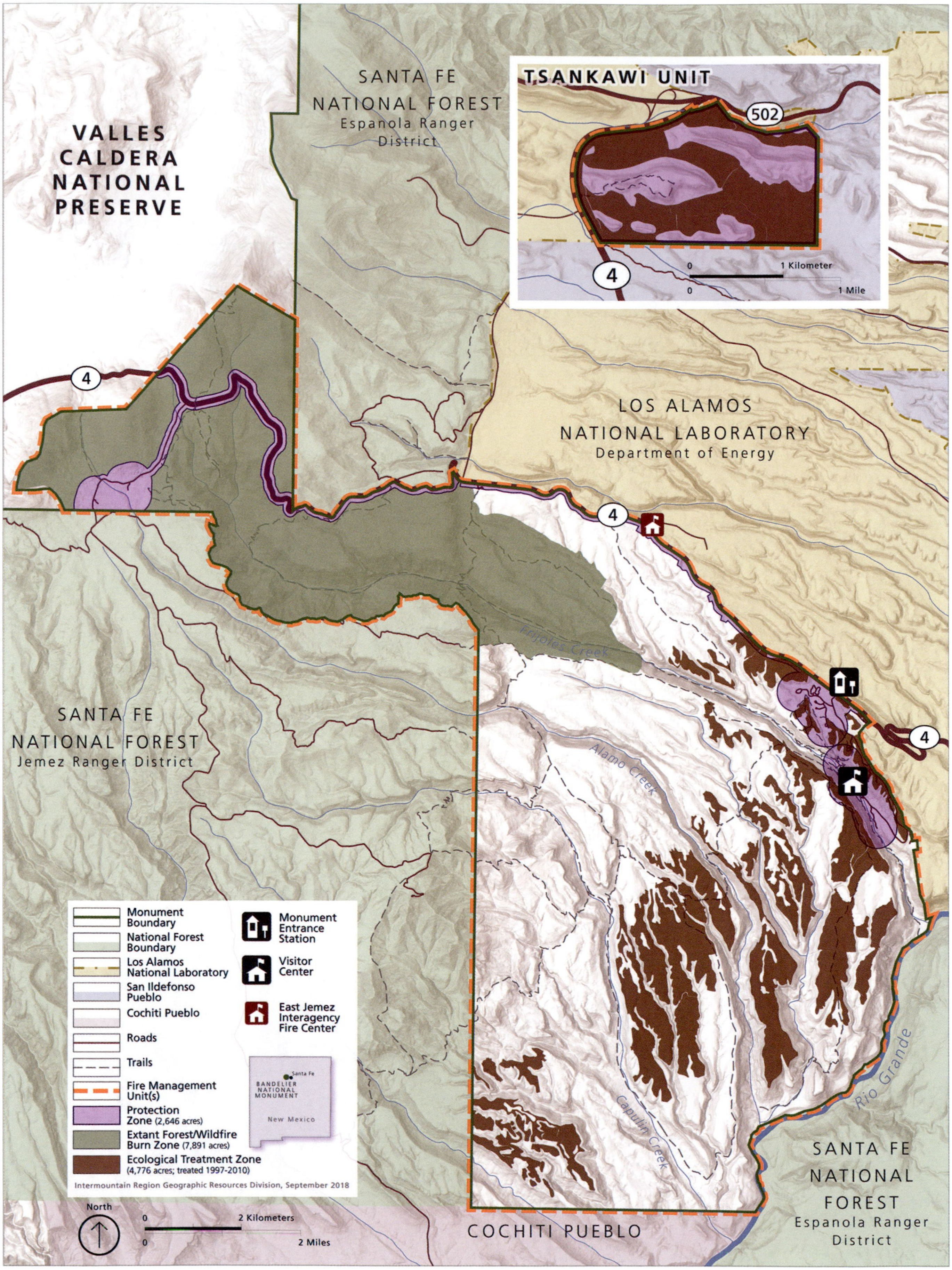

The map shows the Bandelier National Monument Fire Management Plan, Alternative 1. Data sources: NPS, Esri.

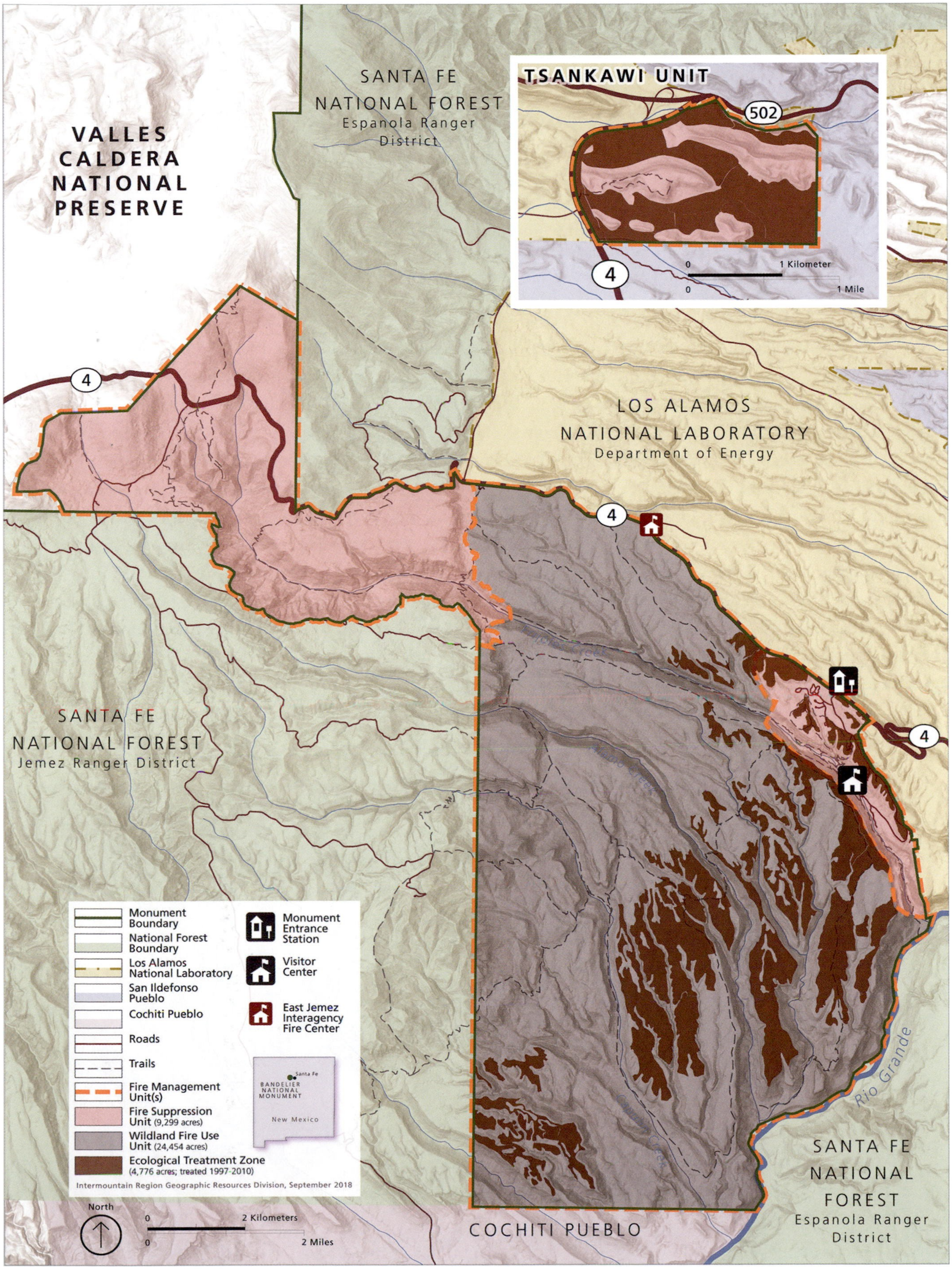

The map shows the Bandelier National Monument Fire Management Plan, Alternative 2. Data sources: NPS, Esri.

EQUIPPING FIREFIGHTERS WITH THE MAPS THEY NEED TO PRESERVE HISTORY

JUSTIN SHEDD, CENTER FOR GEOSPATIAL ANALYTICS, NORTH CAROLINA STATE UNIVERSITY
DR. MEGAN SKRIP, CENTER FOR GEOSPATIAL ANALYTICS, NORTH CAROLINA STATE UNIVERSITY
DR. JELENA VUKOMANOVIC, DEPARTMENT OF PARKS, RECREATION, AND TOURISM MANAGEMENT, NORTH CAROLINA STATE UNIVERSITY, AND CENTER FOR GEOSPATIAL ANALYTICS, NORTH CAROLINA STATE UNIVERSITY

When you step onto the hallowed grounds of Kings Mountain National Military Park in South Carolina, you are also stepping back in time. Here, a battlefield from America's Revolutionary War is preserved as it was in the late 1700s, when colonial armies won a tide-turning victory in their fight for independence. History buffs and recreationists alike cherish the park for its time capsule views, and to maintain its unique cultural landscape, the park regularly resets the landscape's clock with controlled burns. These prescribed fires also reduce risks from wildland fires by reducing the amount of material available to burn. In the event of a wildland fire, coordinating with local fire departments presents a logistical challenge, so park staff partnered with researchers from the Center for Geospatial Analytics at North Carolina State University to compile GIS data in large, high-quality printed maps to accompany the park's official fire management plan.

These maps display an array of valuable information about the park and surrounding area, and their great size allows large numbers of people to gather around them to plan. Each map combines information about buildings and monuments, topography, management objectives, vegetation, and more onto a 3-foot by 4-foot paper sheet. Other national parks have adopted this map-sheet approach as well, and project partners tailor the maps to the parks' specific needs. As communal sources of easy-to-read GIS data, map sheets help ensure that everyone involved is always on the same page.

Stakeholders gather around a map sheet to determine the information that should be placed on the map. NPS photo/Justin Shedd.

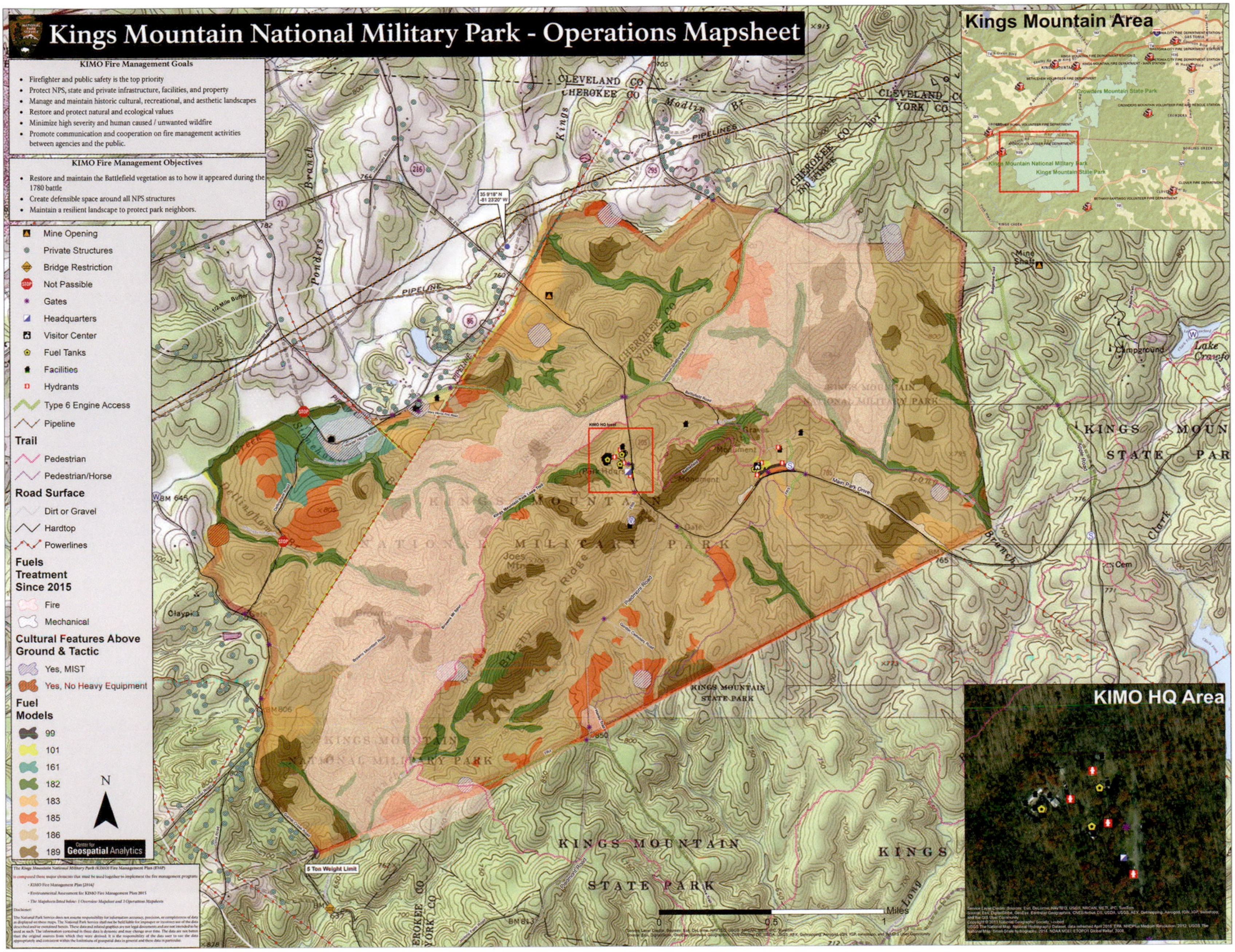

This map sheet was developed for park managers at Kings Mountain National Military Park. "Fuel Model" represents the flammability of vegetation on ground.

Data sources: NPS (layer data sources), Esri Street Map World 2D, Esri World Imagery, Esri USA Topo.

Reenactors in period clothing commemorate the 1780 battle of Kings Mountain. NPS photo/Leah Taber.

Controlled burns are also used at Gettysburg National Military Park in Pennsylvania to re-create historical landscapes present at the time of the Civil War. NPS photo/Grace Crawford.

New growth sprouts from the ashes in front of a Civil War monument after a controlled burn at Gettysburg National Military Park in Pennsylvania. NPS photo/Grace Crawford.

FIRE HISTORY AT CRATER LAKE NATIONAL PARK

CHRIS WAYNE, CRATER LAKE NATIONAL PARK

The size of fires at Crater Lake National Park in Oregon have dramatically increased since 2015, as illustrated in these maps showing historical fire perimeters. From 1933 (when fires were first recorded) through 2014, approximately 20,000 acres were burned by wildland fire and prescribed burns. However, from 2015 through 2017, nearly 42,000 acres burned in wildland fire, which is more than double the acreage of the previous 81 years combined. This seemingly simple pair of maps tells a powerful story of how the largest fires in the park's recorded history have occurred in the last four years.

This change can be attributed to many factors:

- More fires across the West lead to a scarcity of firefighting resources.
- Earlier onset of summer contributes to a longer fire season.
- Heavy snowpack in some years leads to increased fuel loads.
- Lower snowpack in other years leads to drought conditions.
- Historic policies of aggressive fire suppression have resulted in the buildup of downed dead wood and a subsequent increase in ladder fuels.

The drivers behind these changes are complex and the subject of much research and debate. Both natural cycles and anthropogenic actions play a role. Whether this cycle is short-term or long-term remains to be seen. For the foreseeable future, however, resource management goals and strategies will need to adapt to deal with these changes.

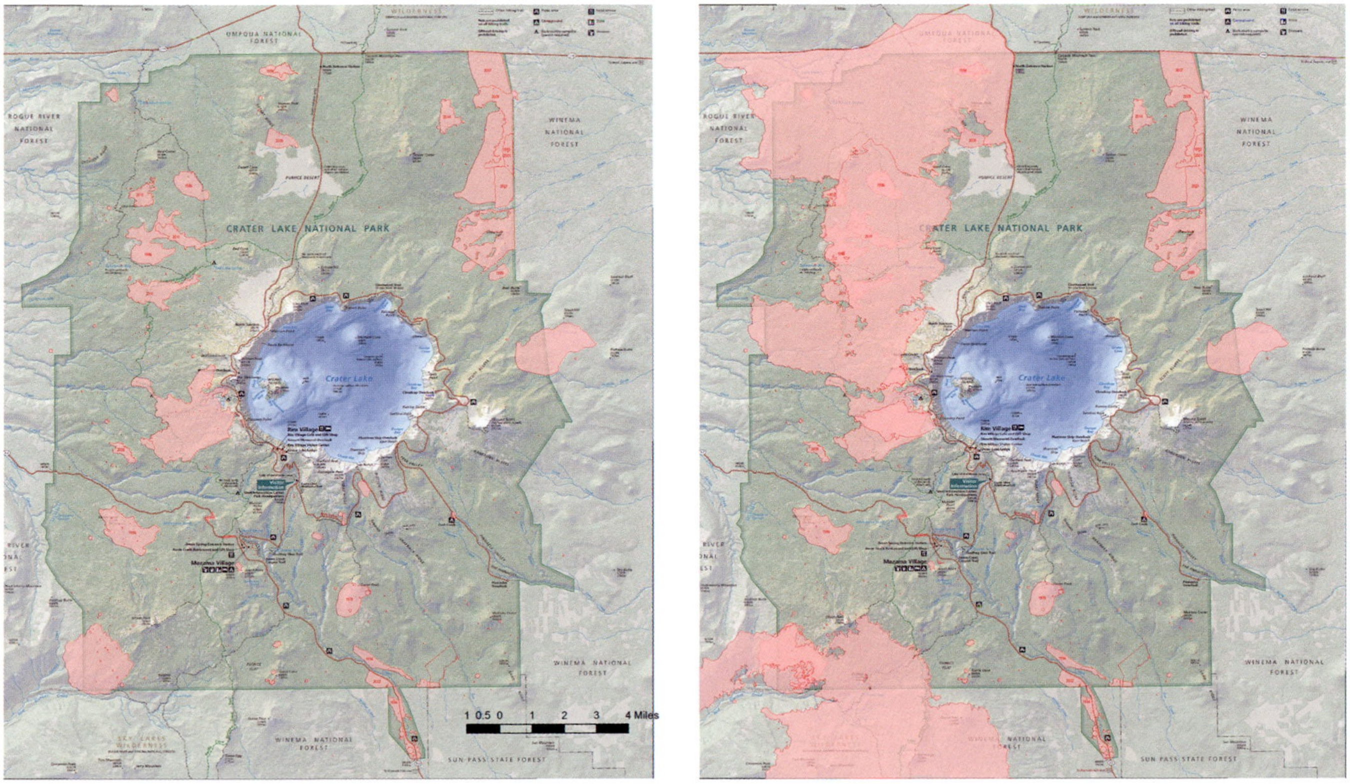

This comparison map shows Crater Lake National Park wildland fire perimeters from 1933–2014 (*left*) and 1933–2017. Data sources: NPS.

5 NATURAL RESOURCES

RAYMOND M. SAUVAJOT, NPS NATURAL RESOURCE STEWARDSHIP AND SCIENCE

PROTECTING AND SUSTAINING NATURAL RESOURCES

Maps have always been indispensable to the National Park Service (NPS) in its role as the preeminent caretaker of some of the world's most significant places. Geographic information is critical for those who manage these superlative resources and a necessity to visitors who play in and enjoy the values of our national parks. Computer-based mapping and increasingly sophisticated techniques for collecting and analyzing geographic information have created a whole new world of mapped information that has revolutionized how the NPS understands, visualizes, and communicates about the resources under its care. The benefit of using this technology for stewardship has been especially true for natural resources—from diverse habitats and ecosystems to geologic features, scenic views, clean water, and dark night skies—where the use of GIS is now common and necessary for the full range of NPS natural resource stewardship and science activities.

GIS capability and highly sophisticated remote sensing tools provide NPS natural resource managers with unprecedented ability to monitor and evaluate ecological change in dynamic park ecosystems. For example, the NPS Inventory and Monitoring Program has identified key ecosystem indicators that provide insights into how changing climates, shifting fire regimes, nonnative invasive species, and other factors affect park landscapes, inform resource management actions, and influence how visitors experience parks. Through the use of mapped information, remotely sensed data, and spatial statistics, park managers can ask questions about how, when, and where change is occurring; predict potential hazard areas prone to flooding or landslides; and identify where invasive species may next take hold.

In areas where ecosystem damage has already occurred, the NPS uses GIS to prioritize restoration activities and monitor the effectiveness of its restoration investments. For example, from the vast wetlands of the Everglades to coastal ecosystems of the Pacific Northwest, geographic analysis is used to effectively restore vulnerable and sensitive areas. GIS techniques are also extremely valuable for identifying wildlife corridors and habitat connectivity, pointing scientists and resource managers to areas that are most important for ensuring wildlife access to winter and summer ranges, migratory pathways, and other habitats necessary for long-term viability.

In addition to the scientific and management applications of maps and GIS, natural resource data can be portrayed through compelling geographic visualizations that literally open new windows for appreciating national park resources. For example, park vegetation maps, when coupled with on-the-ground experiences in eastern forests, southwestern deserts, alpine peaks, tropical islands, and many other ecosystems protected in national parks, offer complementary information that enhances our understanding of the diversity and wonder of our natural heritage. Advanced benthic mapping techniques can provide similar opportunities under the water by illuminating unseen or unknown resources offshore in coastal parks and the Great Lakes. As this volume demonstrates, the combination of analytical capability with incredible visualization techniques can create new insights and inspiration by park managers and visitors to national parks.

The capabilities and applications of GIS, digital cartography, remote sensing, and spatial analysis are now indispensable and standard tools for NPS natural resource management. As the examples in this volume attest, NPS resource managers and scientists continue to find new and innovative ways to map America's national parks for better understanding and more effective natural resource conservation. Indeed, maps and GIS are critical to the NPS mission, helping ensure that natural resources found in parks are well understood, effectively managed and conserved, and, when necessary, properly restored so that they may be enjoyed by future generations.

HOLE-IN-THE-DONUT WETLAND RESTORATION

JONATHAN TAYLOR, MICHELLE TONGUE, AND JOHN DOLLARD, EVERGLADES NATIONAL PARK

The Hole-in-the-Donut (HID) is located on the southern end of Long Pine Key in Everglades National Park in Florida. Long Pine Key is a large pine-dominated ridge surrounded by native Everglades habitats including marl prairie and freshwater sloughs (see HID project map).

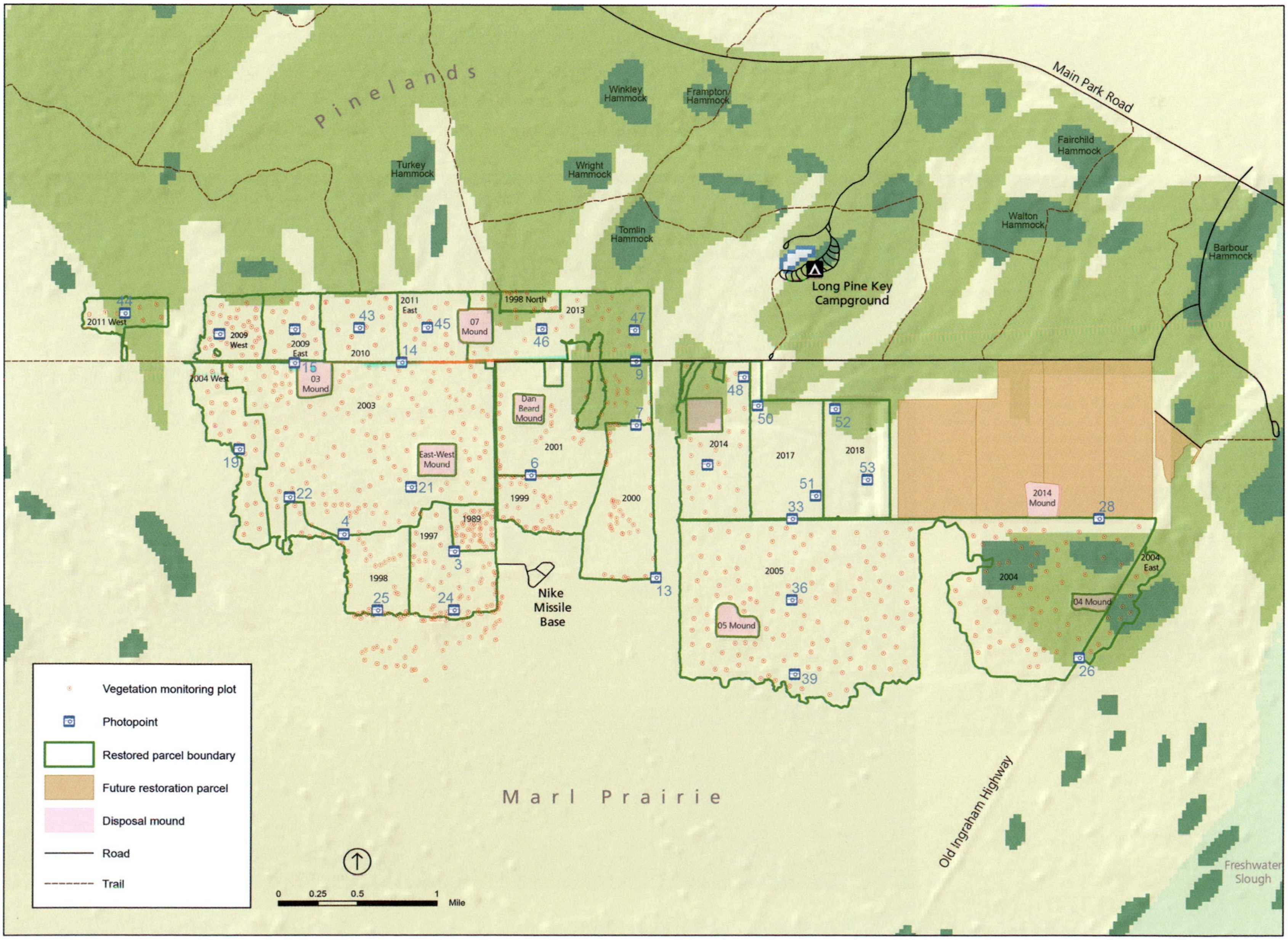

Vegetation monitoring plots, photo points, restored parcel boundaries, and future restoration parcels are shown within the Hole-in-the-Donut.

Data sources: NPS, USGS, National Aeronautics and Space Administration, State of Florida, USDA FSA, Digital Globe, GeoEye, CNES/Airbus DS, Esri.

The HID wetland restoration project is an *in-lieu fee mitigation bank* managed in cooperation with Miami-Dade County, the Florida Department of Environmental Protection, Florida Water Management Districts, and the US Corps of Engineers. These mitigation banks are typically a natural resource such as a wetland that is restored or created and preserved to compensate for loss of other wetlands permitted by federal or state environmental regulations. This one is the only wetland mitigation bank in the NPS.

HID program staff use GIS to manage and track the number of mitigation acres available in the mitigation bank for developers to purchase credits, mapping acres restored, vegetation response monitoring, prescribed fire management, and exotic plant treatment. The project is implemented in distinct parcels (see the 1989–2018 Restoration Parcels section in the accompanying Hole-in-the-Donut story) per restoration-year and involves using heavy equipment to mechanically remove the exotic and invasive Brazilian pepper-dominated vegetation and soils down to the native limestone that revegetates from adjacent native seed sources (see photo point images).

An NPS story (http://go.nps.gov/HID_storymap) about Hole-in-the-Donut has increased awareness and understanding of the project.

An NPS story map
Hole-in-the-Donut
1989 - 2018 Restored Areas
The time series web map at the right progressively displays the areas of restoration by year.
Click on the Time Slider icon at the top left corner of the web map to play the time series. Adjust the speed of the timer, or stop and play using the date tab at the bottom.
More than 5,328 acres have been restored to date. During the dry season of 2019 (January through May) a larger than normal restoration is planned. HID managers hope to scrape and dispose of approximately 500 acres of Brazilian pepper.
Click here for future restoration areas.
Hole In Donut Restoration Project (16), NPSPhoto.jpg
Basemap Image: Landsat 8 OLI, 15m multitemporal PanSharpened Natural Color images; Esri, USGS, AWS, NASA.
Hole-in-the-Donut Restoration Project Time Series
Everglades National Park
2011
2009
2009
2010
2011
1998
2013
2001
2004
2003
1999
2000
1989
1998
1997
2014
2017
2018
2005
2004
2004
esri

This web map of restored parcels within the Hole-in-the-Donut shows a time series between 1989 and 2018.
Landsat 8 OLI, 15m multitemporal PanSharpened Natural Color Images; Esri, USGS, AWS, NASA.

This photo point #06 image within the Hole-in-the-Donut was taken looking north depicting a recently scraped area in February 2001.

This photo point #06 image within the Hole-in-the-Donut was taken looking north depicting a restored area in June 2005.

An NPS story map
Hole-in-the-Donut
The Hole-in-the-Donut is located on the southern end of Long Pine Key, which is a large pine-dominated ridge surrounded by native Everglades habitats including sawgrass prairie, glades marsh, and slough marsh. The HID wetland restoration project in Everglades National Park is the only wetland mitigation bank in the National Park Service.
The project is implemented in distinct parcels per restoration-year, and involves using heavy equipment to physically remove the Brazilian Pepper-dominated vegetation and soils down to the native limestone that re-vegetates from adjacent native seed sources.

This aerial view shows the geographic location of the Hole-in-the-Donut.

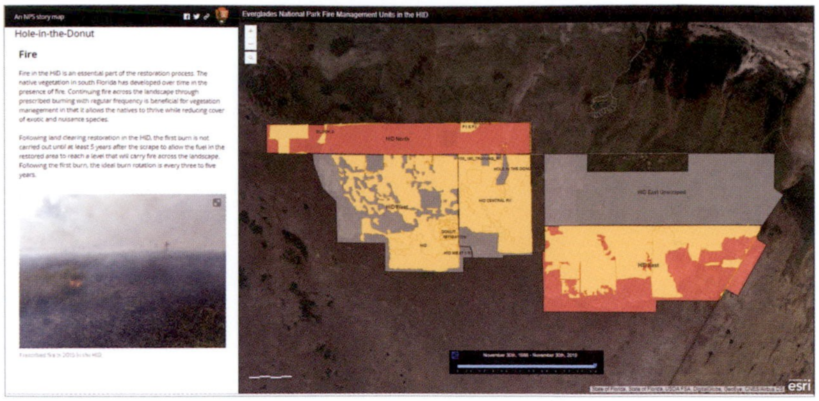
An NPS story map
Hole-in-the-Donut
Fire
Fire in the HID is an essential part of the restoration process. The native vegetation in south Florida has developed over time in the presence of fire. Continuing fire across the landscape through prescribed burning with regular frequency is beneficial for vegetation management in that it allows the natives to thrive while reducing cover of exotic and nuisance species.
Following land clearing restoration in the HID, the first burn is not carried out until at least 5 years after the scrape to allow the fuel in the restored area to reach a level that will carry fire across the landscape. Following the first burn, the ideal burn rotation is every three to five years.
Everglades National Park Fire Management Units in the HID
esri

This Hole-in-the-Donut fire map displays a timeline within Everglades National Park. State of Florida, State of Florida, USDA FSA, Digital Globe, GeoEye, CNES/Airbus DS.

An NPS story map
2017 Continuous Average Hydroperiod in HID
Everglades National Park
Hole-in-the-Donut
Hydrology
The DEM GIS layer in the map at the right displays the continuous average hydroperiod for year 2017. The points on the map display the percent burned vegetation observed after a prescribed fire conducted in 2018.
Higher water levels in Restoration Area 2004 East (lower right corner in map) prevented vegetation from burning. Cattails and other invasive species begin to dominate in areas with higher water levels and no fire.
Vegetation
Exotics treatments
0.6mi
State of Florida, State of Florida, USDA FSA, DigitalGlobe, GeoEye, CNES/Airbus DS | NPS
esri

The hydroperiod (the average time a wetland is covered in water) and percentage of vegetation burned in a prescribed fire conducted in 2018. State of Florida, State of Florida, USDA FSA, Digital Globe, GeoEye, CNES/Airbus DS. Data sources: NPS, USGS, National Aeronautics and Space Administration, State of Florida, State of Florida, USDA FSA, Digital Globe, GeoEye, CNES/Airbus DS, Esri.

VEGETATION COMMUNITIES IN THE STONY MAN MOUNTAIN AREA OF SHENANDOAH NATIONAL PARK

WENDY CASS, SHENANDOAH NATIONAL PARK
GARY FLEMING, VIRGINIA DEPARTMENT OF CONSERVATION AND RECREATION NATURAL HERITAGE PROGRAM

Shenandoah National Park lies atop the Blue Ridge Mountains 75 miles west of Washington, DC. Despite its relative proximity to the city, the park provides habitat for plants and animals found nowhere else in the state of Virginia. The park's native vegetation communities are particularly rich and varied. Past vegetation mapping efforts used a combination of multi- and hyperspectral imagery and field vegetation data collection to document 35 natural vegetation communities in the park. These communities span everything from the common oak hickory forest of the mountain slopes to the rare wetlands of Big Meadows.

Vegetation mapping has been central to enhancing the park's ability to effectively plan for disturbances such as wildfire, to protect economically important species such as American ginseng, and to protect rare plant communities. Vegetation mapping has shown that globally rare plant communities inhabit certain high elevation outcrop areas of the park. Hawksbill Mountain and Stony Man Mountain support several examples of these sensitive outcrop plant communities and are also some of the most popular hiking destinations in the park.

This map illustrates the vegetation communities of Stony Man, clearly demonstrating the highly restricted occurrence of the globally rare and park-endemic high-elevation Greenstone Barren plant community on the north slope of Stony Man Mountain. Mapping images such as this educate visitors on the extreme rarity of these sensitive areas and encourage better stewardship. Rare community vegetation mapping has been integral in identifying the locations of pristine examples of this community and in supporting visitor education and use restrictions to ensure their protection.

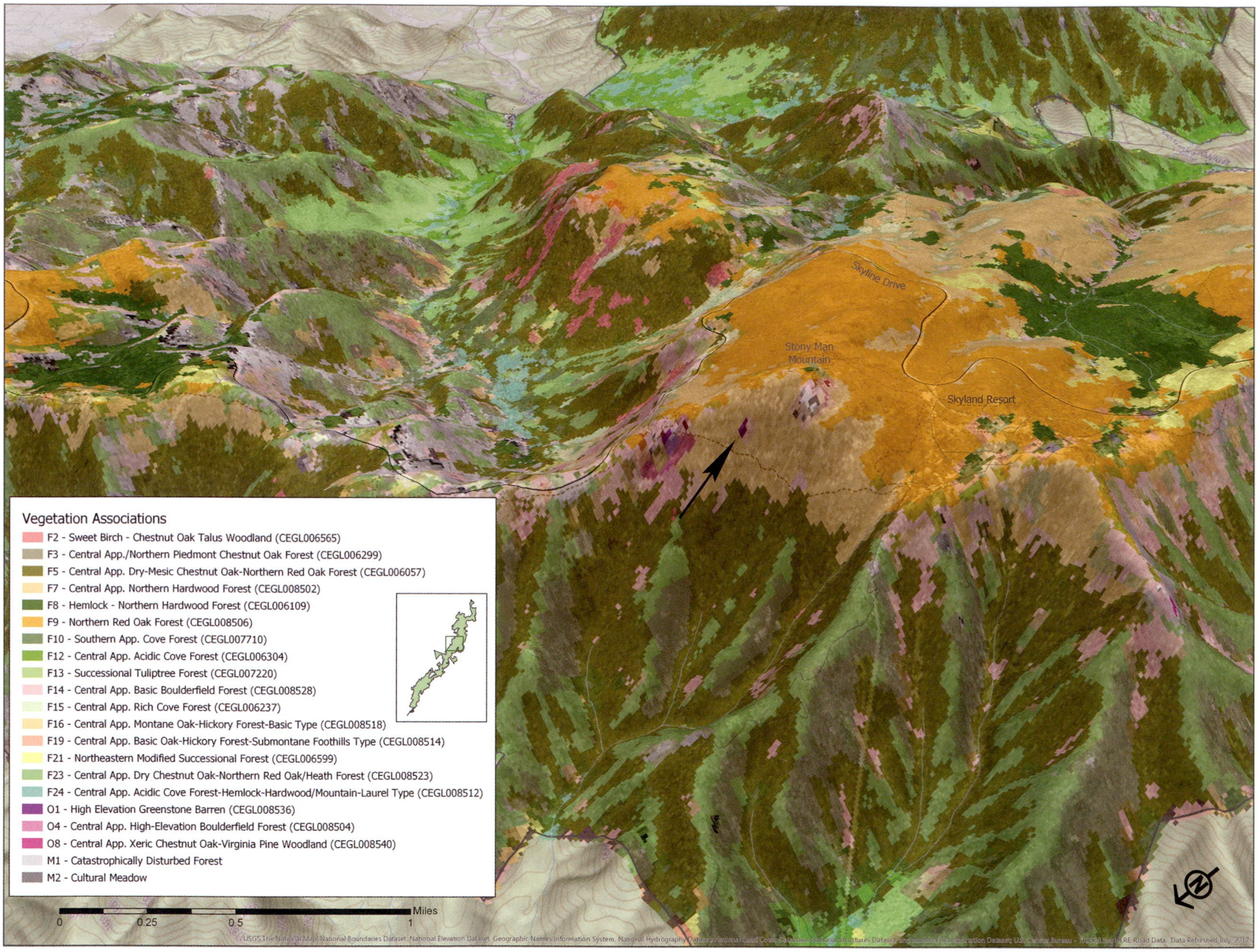

This map shows vegetation communities in the Stony Man Mountain area of Shenandoah National Park. Young, J., et al. 2009. Data sources: SHEN Vegetation Community Types: Young, J., G. Fleming, W. Cass, and C. Lea. 2009. *Vegetation of Shenandoah National Park in Relation to Environmental Gradients, Version 2.0.* Technical Report NPS/NER/NRTR 2009/142. NPS. Philadelphia, Pennsylvania. USGS The National Map: National Boundaries Dataset, National Elevation Dataset, Geographic Names Information System, National Hydrography Dataset, National Land Cover Database, National Structures Dataset, and National Transportation Dataset; US Census Bureau - TIGER/Line; HERE Road Data. Data Refreshed July 2017.

A high-elevation Greenstone Barren plant community is shown at Hawksbill Mountain area of Shenandoah National Park. ©VA CDR-DNH, Gary P. Fleming.

A park-endemic high-elevation Greenstone Barren plant community is shown on the Stony Man rock outcrop. ©VA CDR-DNH, Gary P. Fleming.

UAS-COLLECTED LIDAR FOR FLOOD HAZARD ASSESSMENT

NELL CONTI AND KERRY SHAKARJIAN, NPS INTERMOUNTAIN REGION, GEOGRAPHIC RESOURCES DIVISION

Fort Laramie National Historic Site experienced back-to-back 100-year flood events in 2015 and 2016 that significantly impacted the Laramie River channel within the fort's boundary. The flood events also impacted cultural and archeological resources at the park, including the artifact-rich Quartermaster's Dump.

Fort Laramie National Historic Site and the NPS Intermountain Region Resource Stewardship and Science Directorate teamed with the US Geological Survey (USGS) National Unmanned Aircraft Systems (UAS) Project Office to collect high-resolution lidar data along the river corridor to identify previous flood impacts and assess future flood hazard to the historic site. Ground control was installed and surveyed during a pre-flight field visit in late August 2017. The UAS flights were conducted in October 2017 during leaf-off conditions with the USGS Pulse Aerospace Vapor 55 helicopter carrying a YellowScan lidar payload. Photogrammetry was also collected during separate flights. The team placed targets over the ground control points to aid in photogrammetry orthorectification. Eleven separate flights were needed to cover an estimated 1.2 square mile area. The lidar flights were flown at 150 feet above ground level (AGL) and yielded 100 points per square meter (PPSM) point cloud and the photogrammetry flights were flown at 400 feet AGL, which produced orthophotos with a one-inch resolution. The lidar was classified to differentiate ground and non-ground points and a bare earth digital terrain model (DTM) was produced.

The data collected will be a fundamental component of floodplain and hydrological modeling at the site and will help identify past and potential future flood impacts to sensitive cultural resources that are integral to the site's purpose and significance as a National Park unit. This work is part of a pre-river corridor management planning process that will ultimately inform management decision-making regarding the river channel and the important cultural and natural resources within this changing environment.

This digital terrain model (DTM) was derived from the UAS-collected lidar point cloud of the Laramie River within Fort Laramie National Historic Site. Data sources: NPS, USGS.

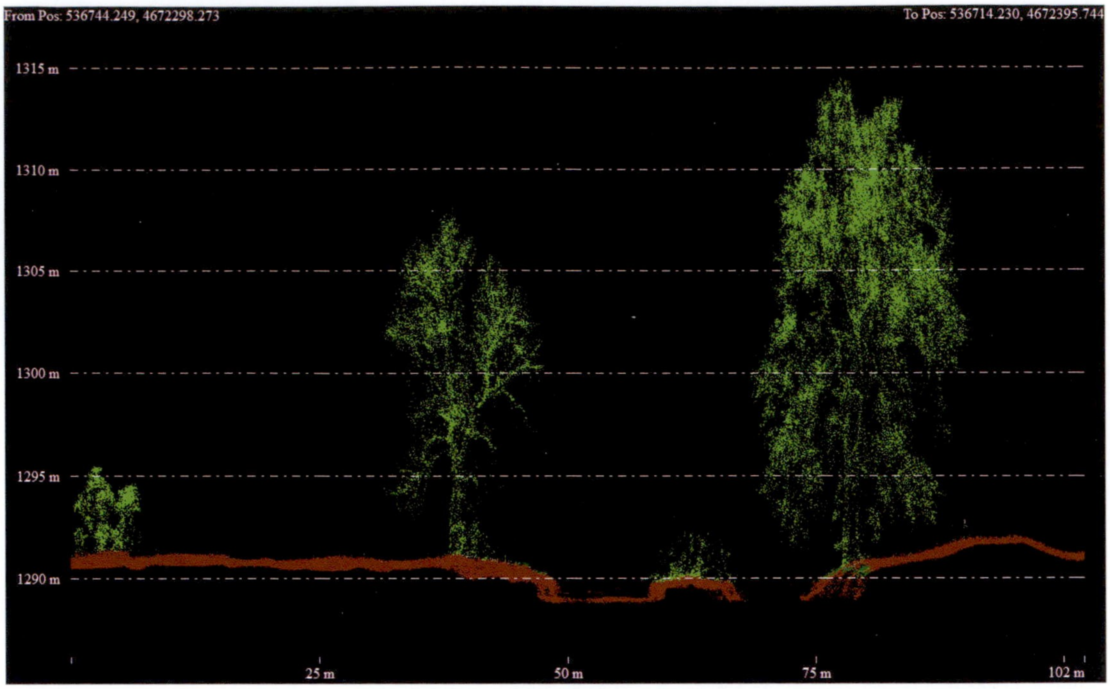

A path profile of a UAS-collected lidar point cloud of the Laramie River within Fort Laramie National Historic Site. Data sources: NPS, USGS.

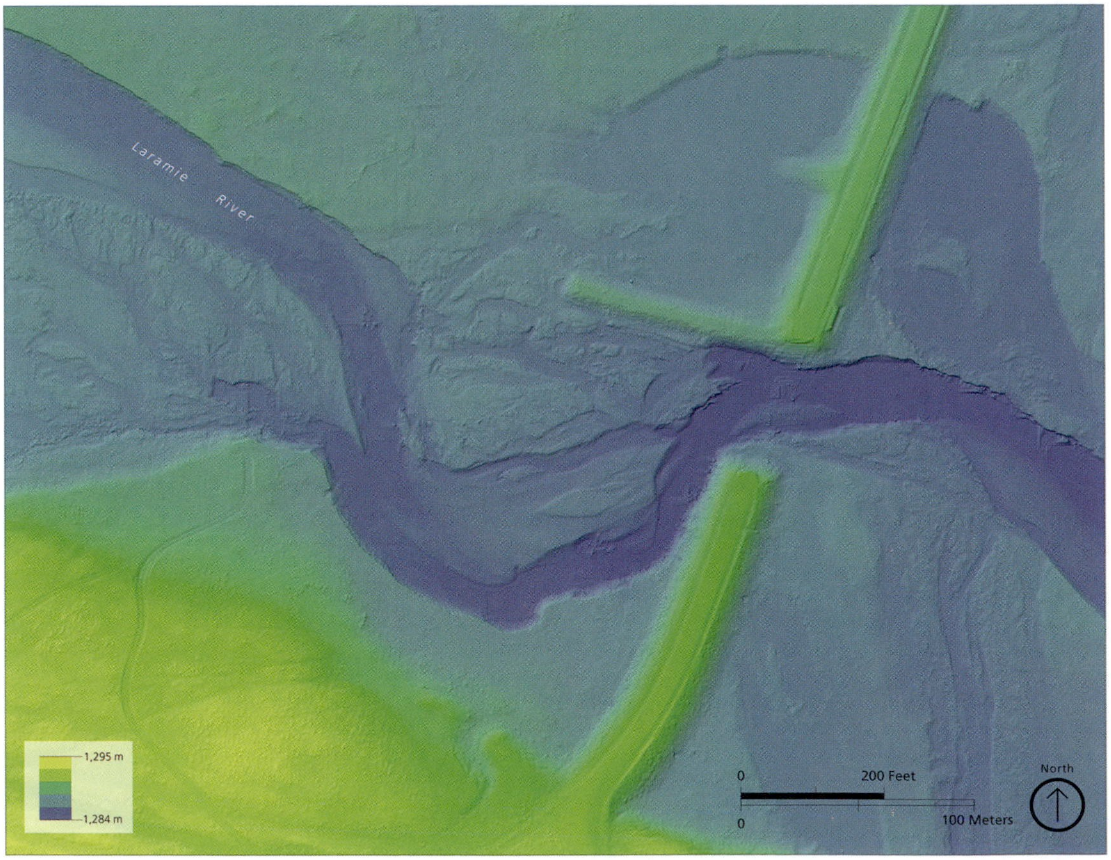

A DTM shows detail of the Laramie River bridge area. Data sources: NPS, USGS.

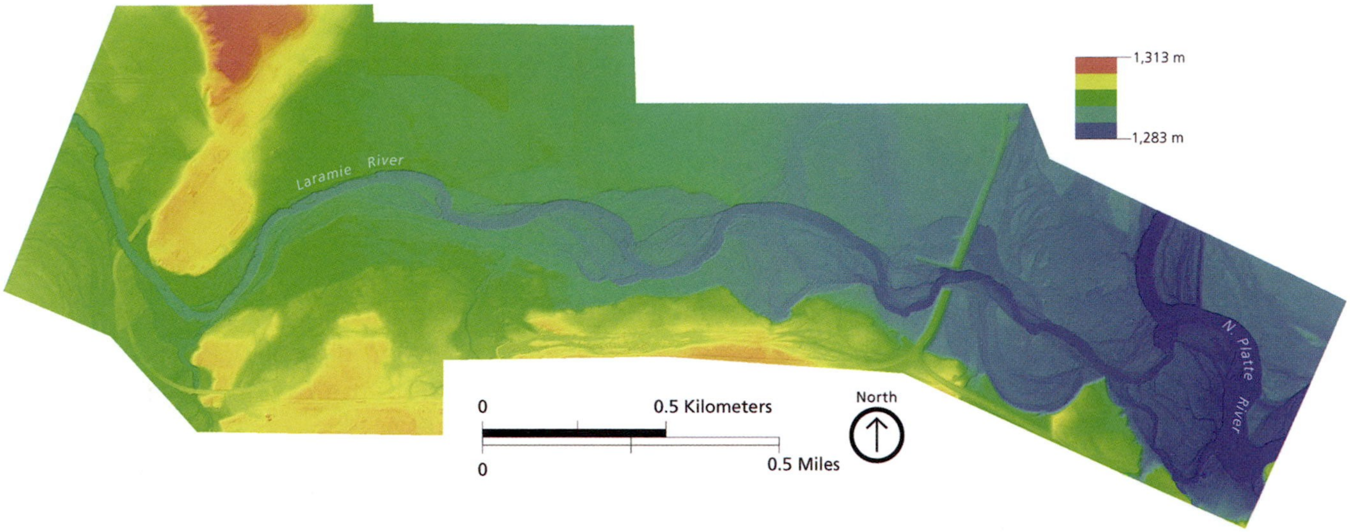

UAS-collected orthoimagery of the Laramie River within Fort Laramie National Historic Site is shown with detail of the Laramie River bridge area.

Data sources: NPS, USGS.

This photograph is of the Pulse Vapor 55 in flight at Fort Laramie National Historic Site.

TRACKING CHANGES IN MOUNTAIN LANDSCAPES

ALEX EDDY, SYLVIA HAULTAIN, ANDI HEARD, JONATHAN NESMITH, AND LINDA MUTCH,
NPS SIERRA NEVADA INVENTORY AND MONITORING NETWORK
PAUL HARDWICK, SEQUOIA AND KINGS CANYON NATIONAL PARKS
ELIZABETH HALE, YOSEMITE NATIONAL PARK

National parks in the Sierra Nevada protect varied and rugged landscapes, with elevations that range from 1,370 feet in the foothills to 14,494 feet at the peak of Mount Whitney. These parks include Yosemite National Park, Sequoia and Kings Canyon National Parks, and Devils Postpile National Monument. Millions of people visit the parks every year to experience the spectacular terrain and variety of wild plants and animals. Each park is predominately designated wilderness, which is the highest level of protection for public lands. The Sierra Nevada Network is one of 32 NPS Inventory and Monitoring Program networks that collaborate with parks, share information, and track the health of changing landscapes. The Sierra Nevada Network measures trends in birds, climate, forests, lakes, rivers, and wetlands.

Vital signs are key physical, chemical, and biological elements that represent the health of natural landscapes. The Sierra Nevada Network works with park scientists and managers to identify vital signs based on their ecological significance, sensitivity to change, and management importance. Long-term data is collected according to six peer-reviewed protocols designed to track changes over time across the landscape. Sample locations are selected based on spatially randomized survey designs that allow scientists to make inferences about vital signs across large areas. This data gives park stewards years of perspective and context to make the best, scientifically informed decisions for the Sierra Nevada parks.

Half Dome is an iconic example of the outstanding glacially shaped features found in Yosemite National Park. NPS photo/Greg Stock, Yosemite National Park.

The Devils Postpile columnar basalt formation is spectacular for its 60-foot high columns and glacially polished surfaces. NPS photo.

Sierra Nevada Network field scientists travel to high mountain lakes to sample water chemistry, lake level, and amphibians. NPS photo/Mandy Holmgren.

Sequoia and Kings Canyon National Parks are home to giant sequoias, the world's largest tree by volume. NPS photo/Tony Caprio, Sequoia and Kings Canyon National Parks.

The program measures changes in plant and macroinvertebrate communities that live in vibrant mountain wetlands.

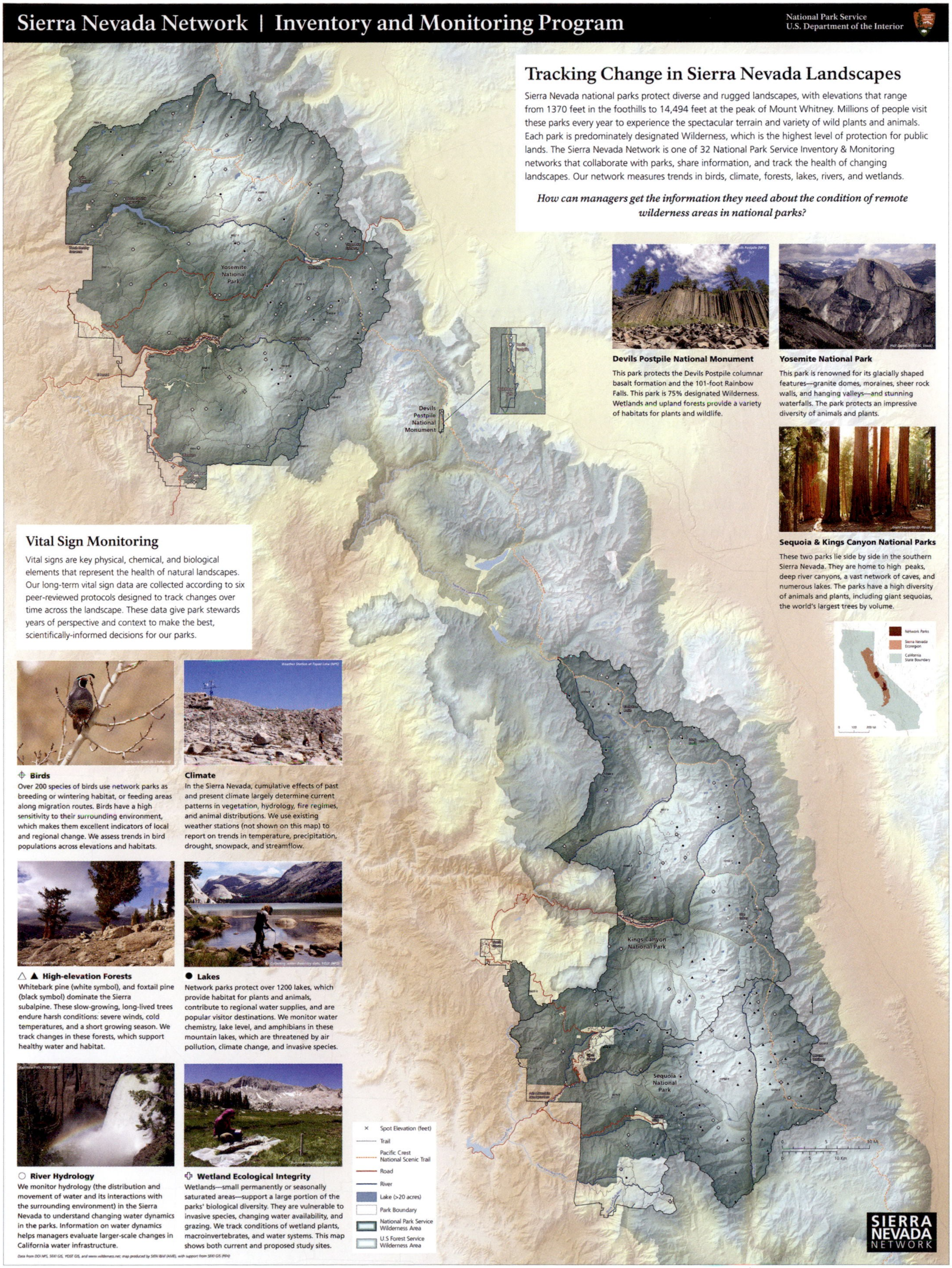
Sierra Nevada Network | Inventory and Monitoring Program
National Park Service
U.S. Department of the Interior
Tracking Change in Sierra Nevada Landscapes
Sierra Nevada national parks protect diverse and rugged landscapes, with elevations that range from 1370 feet in the foothills to 14,494 feet at the peak of Mount Whitney. Millions of people visit these parks every year to experience the spectacular terrain and variety of wild plants and animals. Each park is predominately designated Wilderness, which is the highest level of protection for public lands. The Sierra Nevada Network is one of 32 National Park Service Inventory & Monitoring networks that collaborate with parks, share information, and track the health of changing landscapes. Our network measures trends in birds, climate, forests, lakes, rivers, and wetlands.
How can managers get the information they need about the condition of remote wilderness areas in national parks?
Devils Postpile National Monument
This park protects the Devils Postpile columnar basalt formation and the 101-foot Rainbow Falls. This park is 75% designated Wilderness. Wetlands and upland forests provide a variety of habitats for plants and wildlife.
Yosemite National Park
This park is renowned for its glacially shaped features—granite domes, moraines, sheer rock walls, and hanging valleys—and stunning waterfalls. The park protects an impressive diversity of animals and plants.
Sequoia & Kings Canyon National Parks
These two parks lie side by side in the southern Sierra Nevada. They are home to high peaks, deep river canyons, a vast network of caves, and numerous lakes. The parks have a high diversity of animals and plants, including giant sequoias, the world's largest trees by volume.
Yosemite National Park
Devils Postpile National Monument
Kings Canyon National Park
Sequoia National Park
Vital Sign Monitoring
Vital signs are key physical, chemical, and biological elements that represent the health of natural landscapes. Our long-term vital sign data are collected according to six peer-reviewed protocols designed to track changes over time across the landscape. These data give park stewards years of perspective and context to make the best, scientifically-informed decisions for our parks.
Birds
Over 200 species of birds use network parks as breeding or wintering habitat, or feeding areas along migration routes. Birds have a high sensitivity to their surrounding environment, which makes them excellent indicators of local and regional change. We assess trends in bird populations across elevations and habitats.
Climate
In the Sierra Nevada, cumulative effects of past and present climate largely determine current patterns in vegetation, hydrology, fire regimes, and animal distributions. We use existing weather stations (not shown on this map) to report on trends in temperature, precipitation, drought, snowpack, and streamflow.
High-elevation Forests
Whitebark pine (white symbol), and foxtail pine (black symbol) dominate the Sierra subalpine. These slow-growing, long-lived trees endure harsh conditions: severe winds, cold temperatures, and a short growing season. We track changes in these forests, which support healthy water and habitat.
Lakes
Network parks protect over 1200 lakes, which provide habitat for plants and animals, contribute to regional water supplies, and are popular visitor destinations. We monitor water chemistry, lake level, and amphibians in these mountain lakes, which are threatened by air pollution, climate change, and invasive species.
River Hydrology
We monitor hydrology (the distribution and movement of water and its interactions with the surrounding environment) in the Sierra Nevada to understand changing water dynamics in the parks. Information on water dynamics helps managers evaluate larger-scale changes in California water infrastructure.
Wetland Ecological Integrity
Wetlands—small permanently or seasonally saturated areas—support a large portion of the parks' biological diversity. They are vulnerable to invasive species, changing water availability, and grazing. We track conditions of wetland plants, macroinvertebrates, and water systems. This map shows both current and proposed study sites.
Spot Elevation (feet)
Trail
Pacific Crest National Scenic Trail
Road
River
Lake (>20 acres)
Park Boundary
National Park Service Wilderness Area
U.S Forest Service Wilderness Area
SIERRA NEVADA NETWORK

The poster showcases the monitoring projects, which are designed to track changes in four national parks in the Sierra Nevada. Data sources: NPS, USGS.

TIDAL WETLAND RESTORATION AT THE FORT CLATSOP UNIT OF LEWIS AND CLARK NATIONAL HISTORICAL PARK

KAYLA FERMIN, LEWIS AND CLARK NATIONAL HISTORICAL PARK

Using ArcGIS, the NPS created this story illustrating Tidal Wetland Restoration at the Fort Clatsop unit of Lewis and Clark National Historical Park in Oregon for educational, outreach, and informative purposes and is featured on the park's website. The story features information about the ongoing tidal wetland restoration projects in the park. Since 2006, the NPS has been actively restoring areas along the Lewis and Clark River to more closely resemble the landscape encountered by the Corps of Discovery in 1805 and allow for natural processes to return to the system. Through passive and active restoration of tidal wetlands, the park is contributing to the recovery of the Youngs Bay watershed and endangered salmon stocks. The park conducts vegetation monitoring to measure each project's success. The story guides the viewer through the restoration process with interactive maps displaying what restoration actions have been implemented and when. The monitoring protocols are explained through maps and pictures taken in the field. The story culminates with information explaining what visitors can do to participate in similar projects through the NPS Volunteers-In-Parks (VIP) program and citizen science events such as BioBlitz. Online mapping helps connect visitors to the park's foundation and understanding of park science.

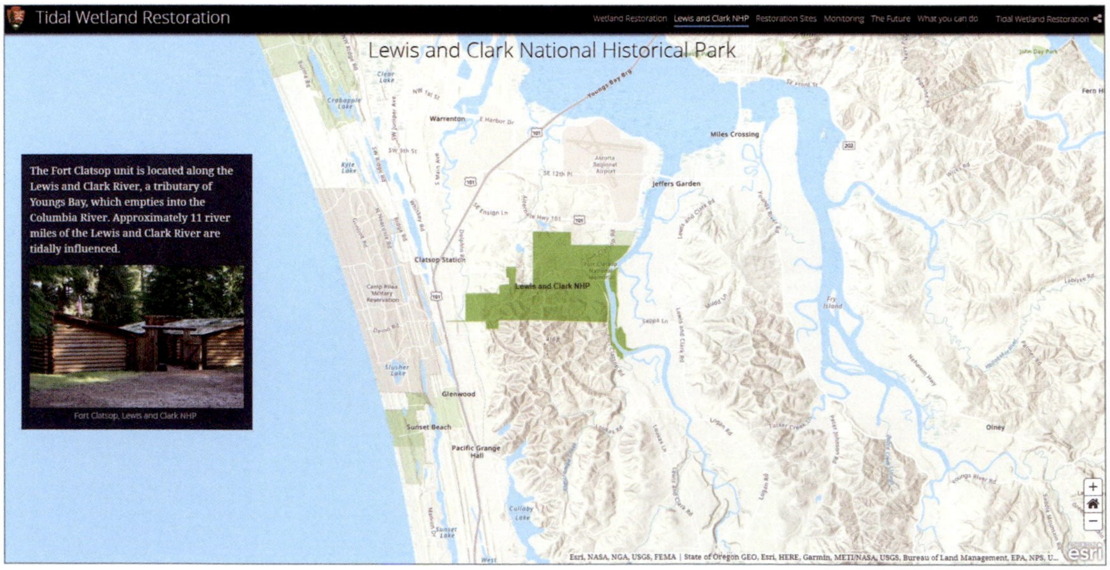

The Fort Clatsop unit is located along the Lewis and Clark River in Astoria, Oregon.

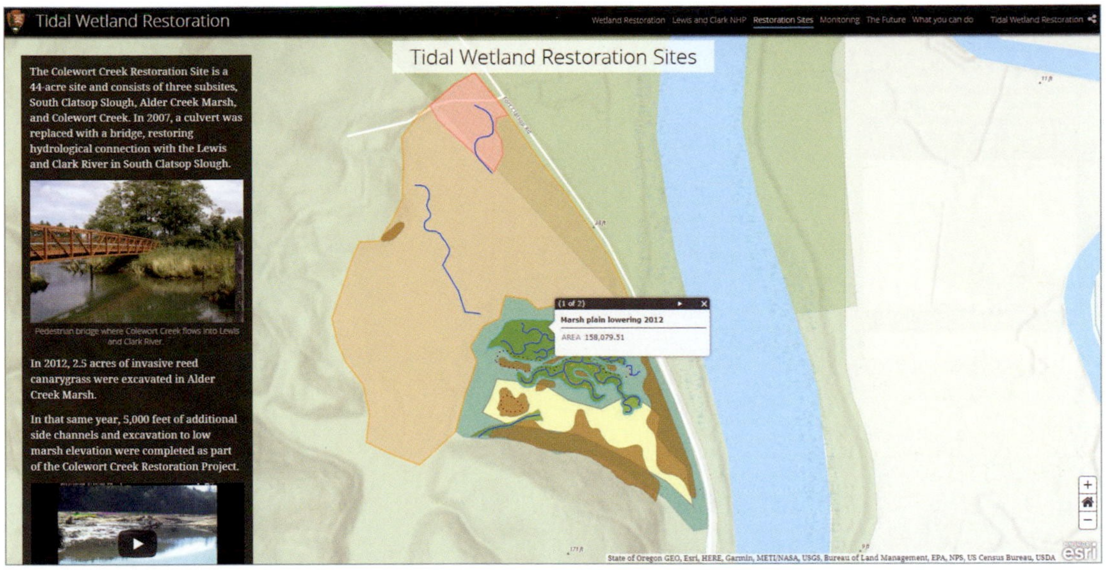
Tidal Wetland Restoration
Wetland Restoration Lewis and Clark NHP Restoration Sites Monitoring The Future What you can do Tidal Wetland Restoration
Tidal Wetland Restoration Sites
The Colewort Creek Restoration Site is a 44-acre site and consists of three subsites, South Clatsop Slough, Alder Creek Marsh, and Colewort Creek. In 2007, a culvert was replaced with a bridge, restoring hydrological connection with the Lewis and Clark River in South Clatsop Slough.
Pedestrian bridge where Colewort Creek flows into Lewis and Clark River.
In 2012, 2.5 acres of invasive reed canarygrass were excavated in Alder Creek Marsh.
In that same year, 5,000 feet of additional side channels and excavation to low marsh elevation were completed as part of the Colewort Creek Restoration Project.
(1 of 2)
Marsh plain lowering 2012
AREA 158,079.51
State of Oregon GEO, Esri, HERE, Garmin, METI/NASA, USGS, Bureau of Land Management, EPA, NPS, US Census Bureau, USDA
esri

Interactive maps with pictures and descriptions of each of the tidal wetland restoration sites are shown.

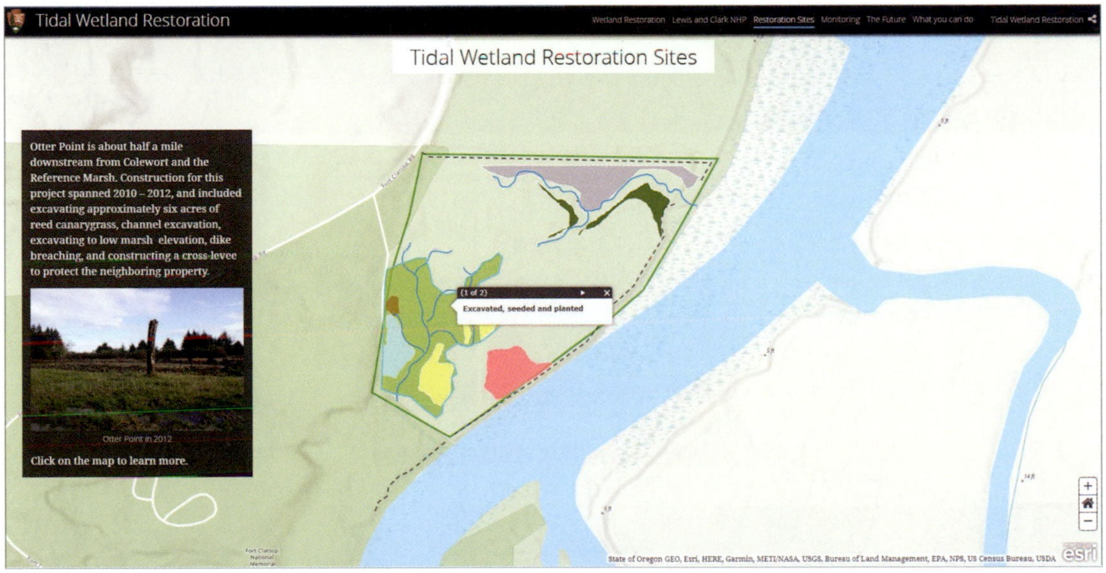
Tidal Wetland Restoration
Wetland Restoration Lewis and Clark NHP Restoration Sites Monitoring The Future What you can do Tidal Wetland Restoration
Tidal Wetland Restoration Sites
Otter Point is about half a mile downstream from Colewort and the Reference Marsh. Construction for this project spanned 2010 – 2012, and included excavating approximately six acres of reed canarygrass, channel excavation, excavating to low marsh elevation, dike breaching, and constructing a cross-levee to protect the neighboring property.
Otter Point in 2012
Click on the map to learn more.
(1 of 2)
Excavated, seeded and planted
State of Oregon GEO, Esri, HERE, Garmin, METI/NASA, USGS, Bureau of Land Management, EPA, NPS, US Census Bureau, USDA
esri

Interactive maps with pictures and descriptions of each of the tidal wetland restoration sites are shown.

Data sources for maps on pages 96-97: World Hillshade: National Park Service, Esri, Airbus DS, Garmin, OpenStreetMap contributors, USGS, NGA, NOAA, FAO, HERE, NASA, CGIAR, N Robinson, NCEAS, NLS, OS, NMA, Geodatastyrelsen, Rijkswaterstaat, GSA, Geoland, FEMA, Intermap, and the GIS user community. World Topographic Map: Esri, HERE, Garmin, FAO, NOAA, USGS, © OpenStreetMap contributors, and the GIS User Community.

Tidal Wetland Restoration
Wetland Restoration Lewis and Clark NHP Restoration Sites Monitoring The Future What you can do Tidal Wetland Restoration
Tidal Wetland Restoration Sites
After construction, native plantings and seeding techniques were implemented throughout the sites.
Biological Technician Jennifer Bell planting at Otter Point, 2011
Volunteers planting native shrubs at South Clatsop Slough, 2012

Volunteers plant native shrubs throughout the sites.

Tidal Wetland Restoration
Wetland Restoration
Lewis and Clark NHP
Restoration Sites
Monitoring
The Future
What you can do
Tidal Wetland Restoration
Herbaceous Vegetation Monitoring
Plants were identified down to species and cover was recorded in the 1 meter by 1 meter plots. Monitoring was conducted in the late summer months during peak vegetative growth.
Monitoring plot example
Mosaics intern identifying slough sedge (Carex lynbyei) at South Clatsop Slough, 2017

Vegetation monitoring is conducted every year to measure each projects' success.

Tidal Wetland Restoration
Wetland Restoration
Lewis and Clark NHP
Restoration Sites
Monitoring
The Future
What you can do
Tidal Wetland Restoration
Wetlands are an important habitat and food source for birds, small mammals, amphibians, fish and more. These projects provide a critical transition zone for juvenile salmon in the Columbia River estuary.
Great Blue Heron along Lewis and Clark River captured by volunteer: Dennis Adams

A great blue heron perches on a piling in the Lewis and Clark River.

AQUATIC INVASIVE SPECIES: MUSSEL MONITORING AT GLEN CANYON NATIONAL RECREATION AREA

VANESSA GLYNN-LINARIS, GLEN CANYON NATIONAL RECREATION AREA
ANETH WIGHT, NORTHERN COLORADO PLATEAU INVENTORY AND MONITORING NETWORK

Glen Canyon National Recreation Area (GLCA) encompasses over 1.25 million acres in Utah and Arizona, including Lake Powell. Lake Powell has been monitored for invasive mussels since 1999 by the NPS, Bureau of Reclamation, and Utah Division of Wildlife Resources. In 2012, mussel larvae, or veligers, were confirmed in Lake Powell after routine water monitoring tests. Since then, thousands of adult quagga mussels have been found in Lake Powell.

GLCA Wahweap Environmental Laboratory employees obtain monthly water samples from routine sites lake-wide, and then use microscopic analysis to visually analyze plankton samples for *dreissenid veligers* to track the population and extent of mussel colonization in the lake. Measurements are also taken from mussels found on vessels, structures, and canyon walls. This data makes up one of the first studies of an early mussel infestation in a water body the size of Lake Powell.

Before 2012, routinely sampled sites were based upon likely locations of dreissenid mussel introduction and propagation. Random sampling sites are generated to reduce bias and are based on current lake elevation using ArcMap. As the population spread throughout the lake, sampling protocols have also evolved to routinely include areas of the lake only previously selected at random. Thales MobileMapper GPS units were previously used to confirm sampling locations, but GLCA recently transitioned to utilizing Collector offline in the field with mobile devices.

Many NPS employees have assisted with these efforts, including Mark Anderson (previous GLCA aquatic ecologist) who worked on monitoring efforts as early as 1999 and Colleen Allen (aquatic invasive species coordinator) who has worked on prevention and containment efforts since 2008. See https://www.nps.gov/glca for updates.

Kristin Lewis, NPS, conducts field work to collect dressenid mussels at Glen Canyon National Recreation Area.

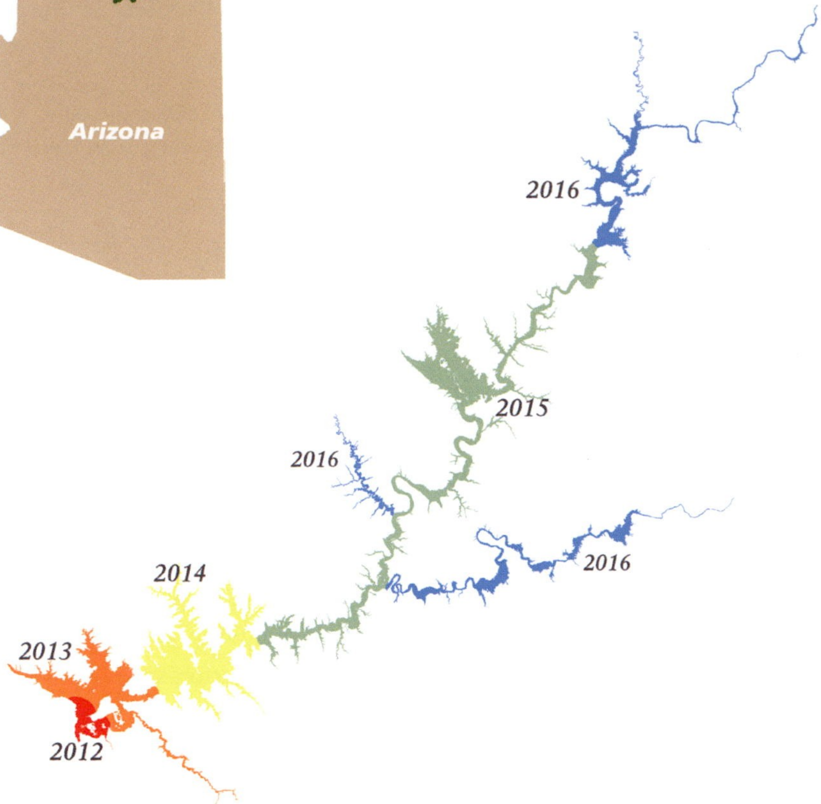

From left: Dropping water levels at Lake Powell expose invasive quagga mussels clinging to rocks and branches at Lone Rock. NPS photo/David Rankin. The geographic location of Glen Canyon National Recreation Area.The map shows the estimated rate of veliger spread based on samples from 2012 to 2016 for Lake Powell.

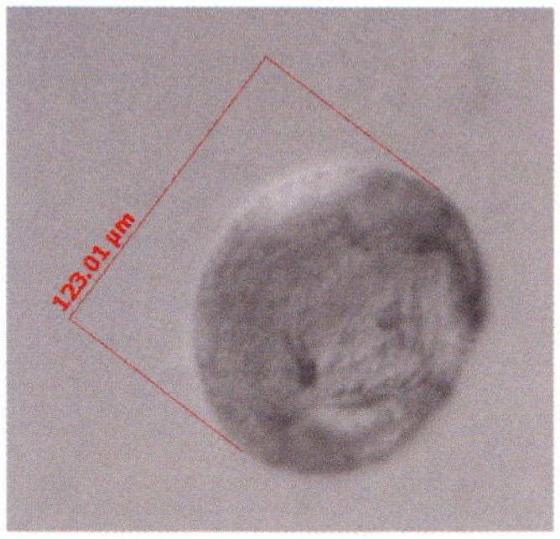

(*Above*) The photo shows veliger as seen magnified 10x with compound microscope (non-polarized). (*Right*) This model generates random sampling sites to reduce bias. Parameters are based on current lake elevation, visitor use zones, temperature, minimum water depth, and number of samples. Model originally created by Todd Lickfett and maintained by many NPS employees.

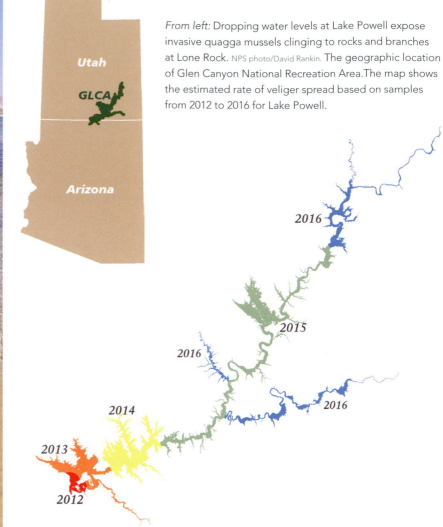

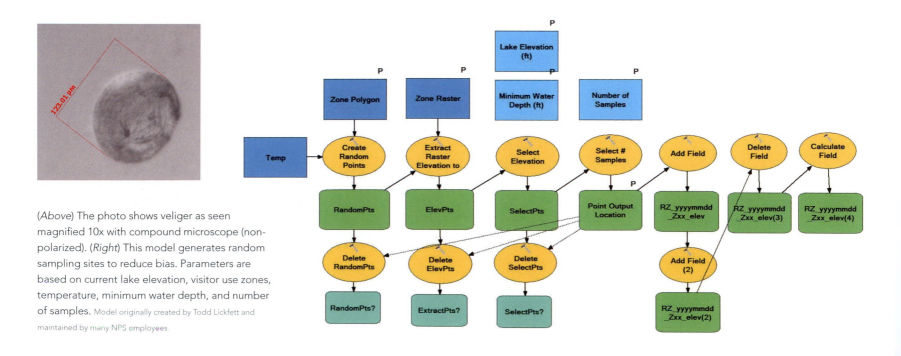

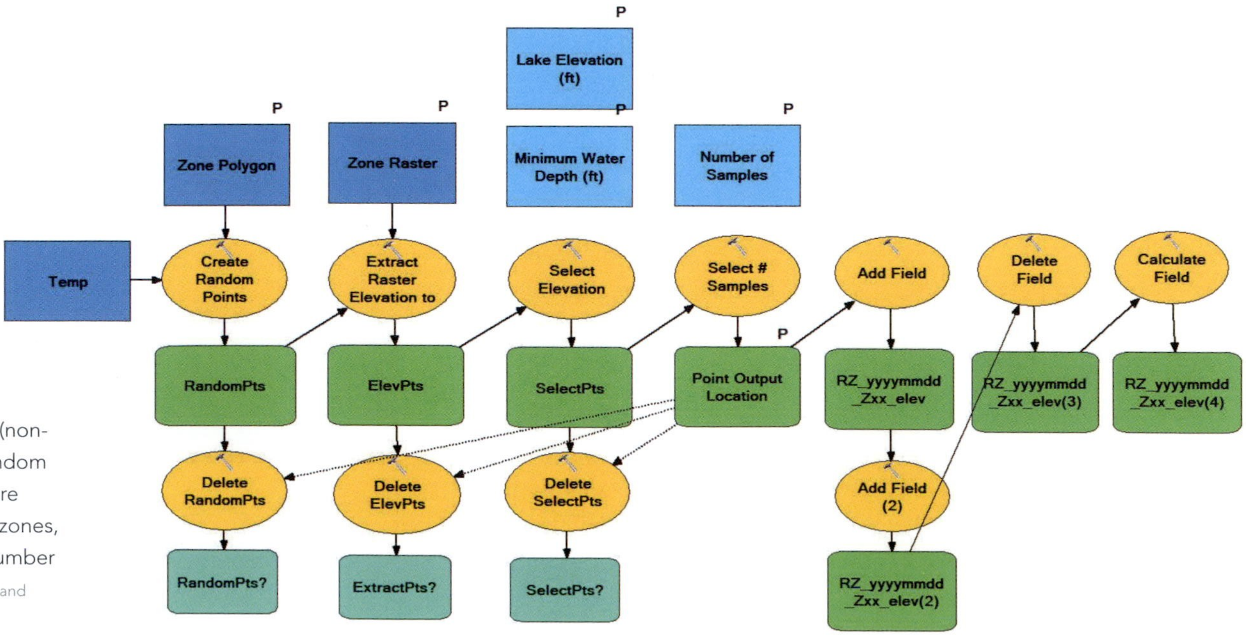

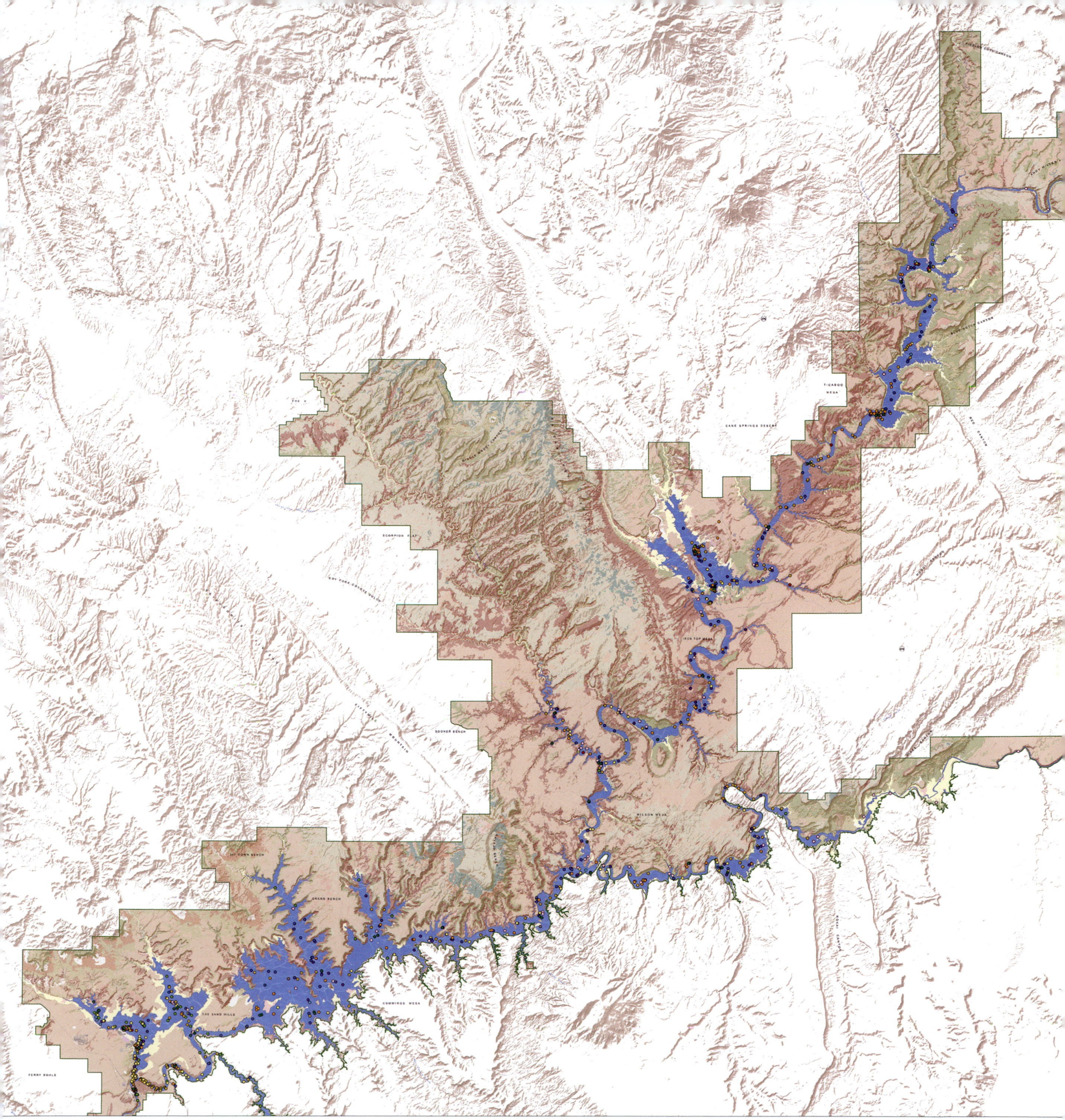

Monitoring locations are shown for mussels and other aquatic invasive species in Glen Canyon National Recreation Area.

Data sources: NPS, Esri Terrain Basemap, ESRI Terrain basemap, Glen Canyon National Recreation Area, Northern Colorado Plateau Inventory and Monitoring Networks (NPS).

BEYOND THE WATER'S EDGE: BENTHIC MAPPING IN THE GREAT LAKES NATIONAL PARKS

JAMIE HOOVER, NATURAL RESOURCE STEWARDSHIP AND SCIENCE, WATER RESOURCES DIVISION

Unlike terrestrial areas that have been explored and researched for decades, many parks lack detailed spatial information beyond the water's edge. Consequently, many underwater areas are often depicted as flat and blue. We all intuitively know that the benthic zone (ocean floor or lake bottom) is not nearly that simple. Likewise, managing a park without information covering the full spatial extent is extraordinarily challenging. Many park managers face this obstacle in 88 ocean and Great Lakes parks, which encompass more than 11,000 shoreline miles and 2.5 million water acres. These parks protect intertidal and submerged habitats such as salt marshes, seagrass, kelp forests, tidewater glaciers, coral reefs, beaches, and rocky intertidal areas, and they preserve valuable cultural resources such as shipwrecks, forts, and historic sites.

The NPS and partners in the Great Lakes region have collected bathymetric data using lidar, multibeam sonar, and other technologies. The data has revealed several unknown sunken ships, interesting geological features, and baseline habitat information. Benthic habitat maps will provide critically important information to quantify and monitor park resources.

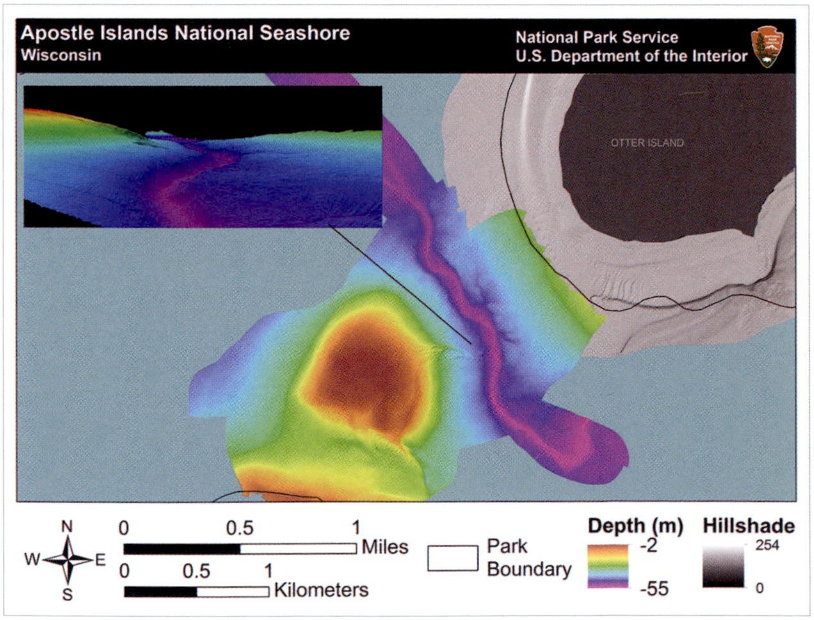

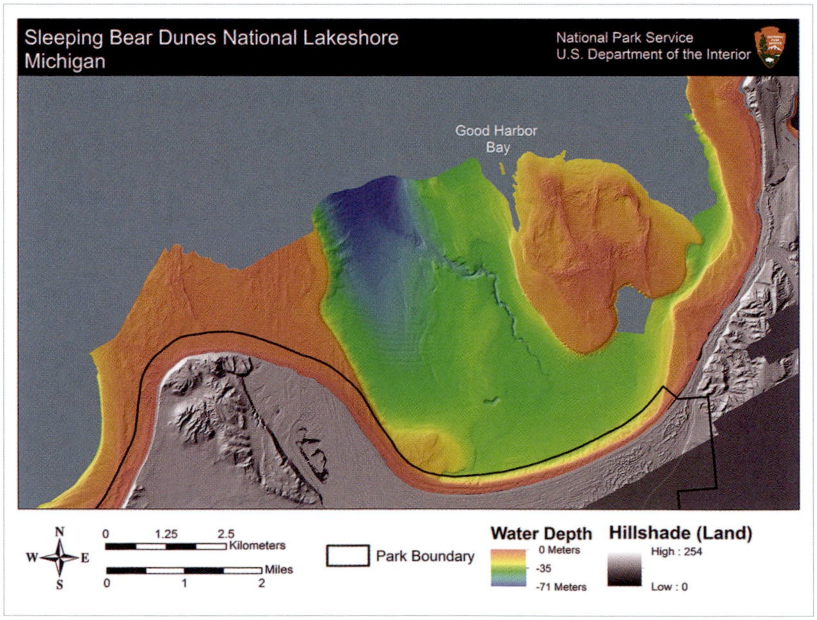

(*Left*) A previously undocumented submarine channel is shown at a depth of -40 meters at Apostle Island National Lakeshore. (*Right*) This map shows lake bottom depth and characteristics of Good Harbor Bay, Sleeping Bear National Lakeshore. Collecting bathymetric data within and beyond park boundaries is critical, as resources and resiliency of parks are affected by internal and external factors. Data sources: NPS, Esri Dark Grey Basemap, Hans VanSumeren at Northwestern Michigan College.

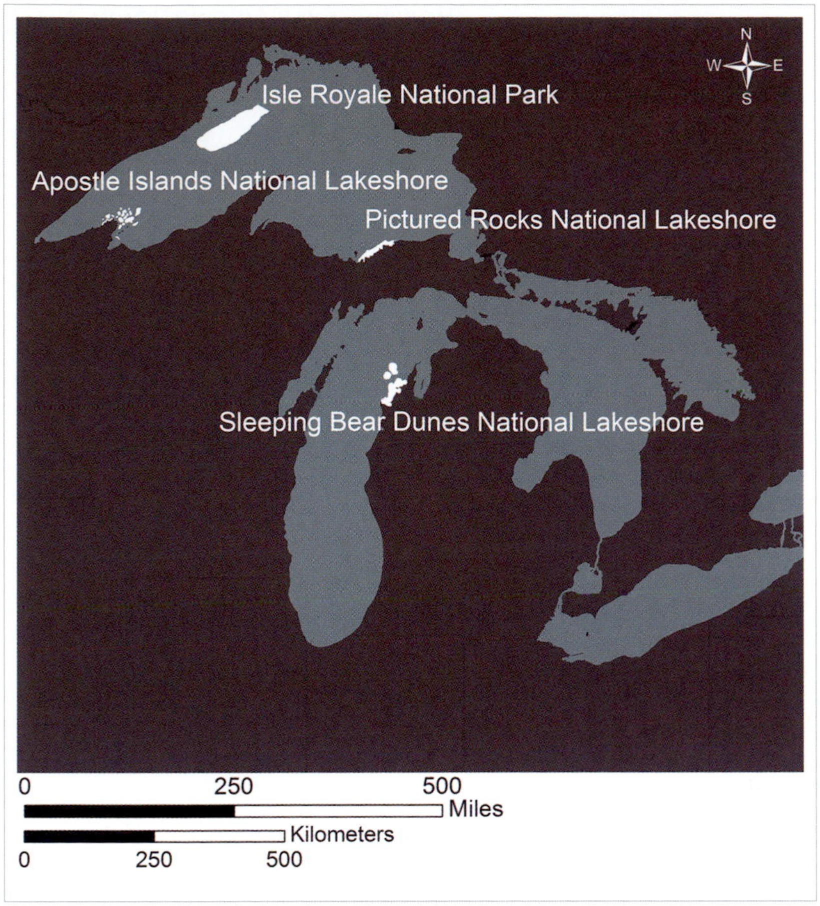

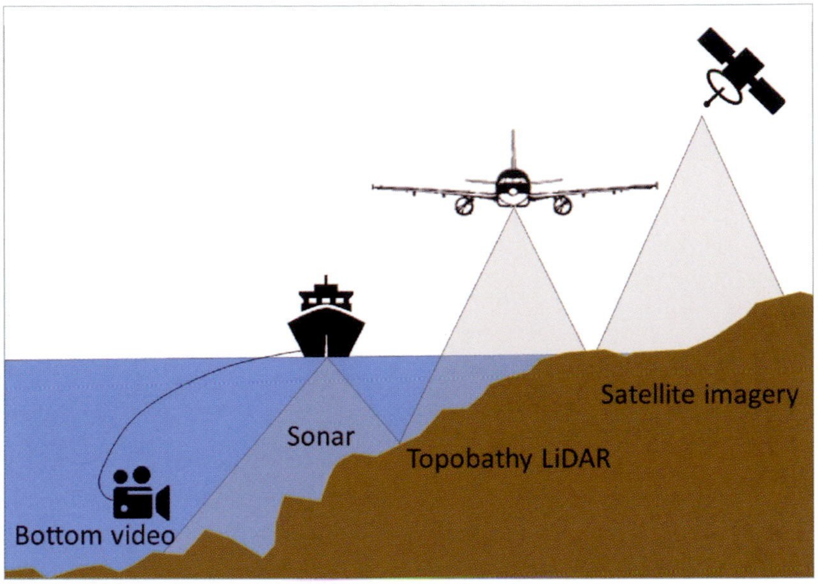

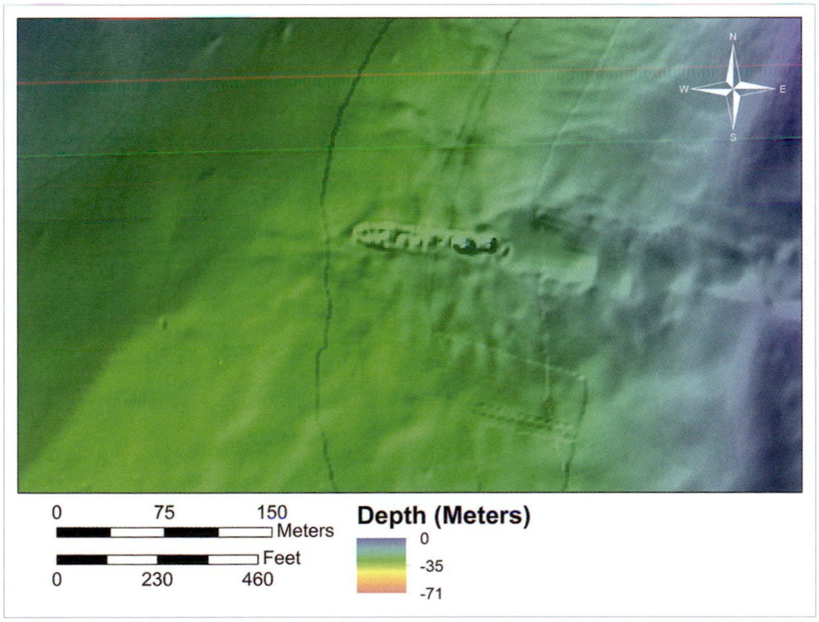

(*Top left*) This map shows Great Lakes Parks where bathymetric data were collected. (*Bottom left*) Historic shipwreck near Pictured Rocks National Lakeshore, Lake Superior.

(*Top right*) A combination of remote sensing datasets is needed to develop benthic habitat maps such as topobathymetric lidar (a specialized green laser for water penetration), sonar, backscatter, and aerial imagery. Water inhibits direct observation of submerged features; thus, physical sampling is also required and includes underwater video and sediment samples. (*Center right*) Bottom videos at Good Harbor Bay show invasive mussels and algae. Photo/Hans VanSumeren, Northwestern Michigan College. (*Bottom right*) Bottom videos at Good Harbor Bay show invasive mussels, algae, and a burbot. Photo/Hans VanSumeren, Northwestern Michigan College. Data sources: NPS, Esri Dark Grey Basemap, Hans VanSumeren at Northwestern Michigan College.

VEGETATION MAPPING INVENTORY PROJECT

JEFFREY W. MALLINSON, PACIFIC ISLAND INVENTORY AND MONITORING NETWORK

The Pacific Island Network (PACN) is one of 32 inventory and monitoring networks within the NPS. These vegetation maps provide a snapshot in time of the vegetative communities within the PACN parks. These maps integrate field data, vegetation classification, expert park knowledge, spatial analysis, and complex computer models. The US National Vegetation Classification System (USNVCS) was followed to ensure nationwide data compatibility and to allow for comparisons across NPS parks.[1] The maps also meet the National Mapping Accuracy Standards (NMAS) for positional accuracy to maintain a high-quality, accurate product across each park. These vegetation maps support resource management decisions, research, and conservation needs. Specifically, the maps have been used for fire mapping and monitoring, researching sampling designs, studying change in forest composition and structure, projecting costs for invasive species removal, assessing resource conditions, and prioritizing management objectives.

NOTE

1. United States National Vegetation Classification System—Community Element Global 8536.

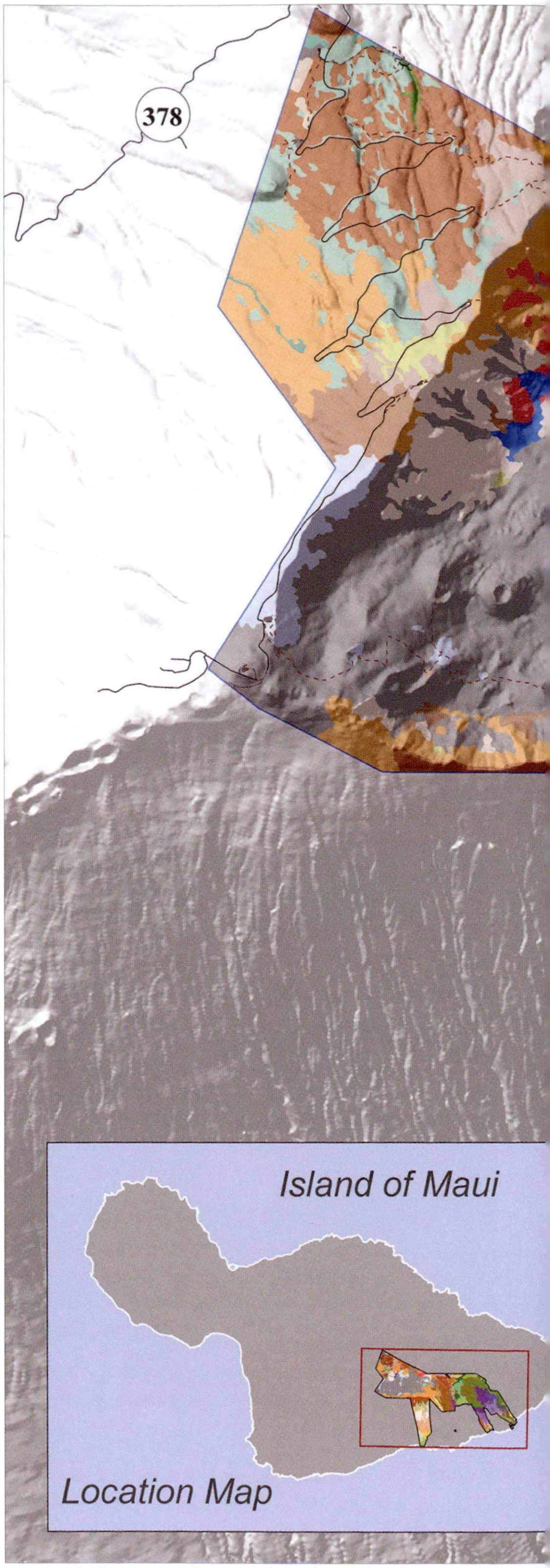

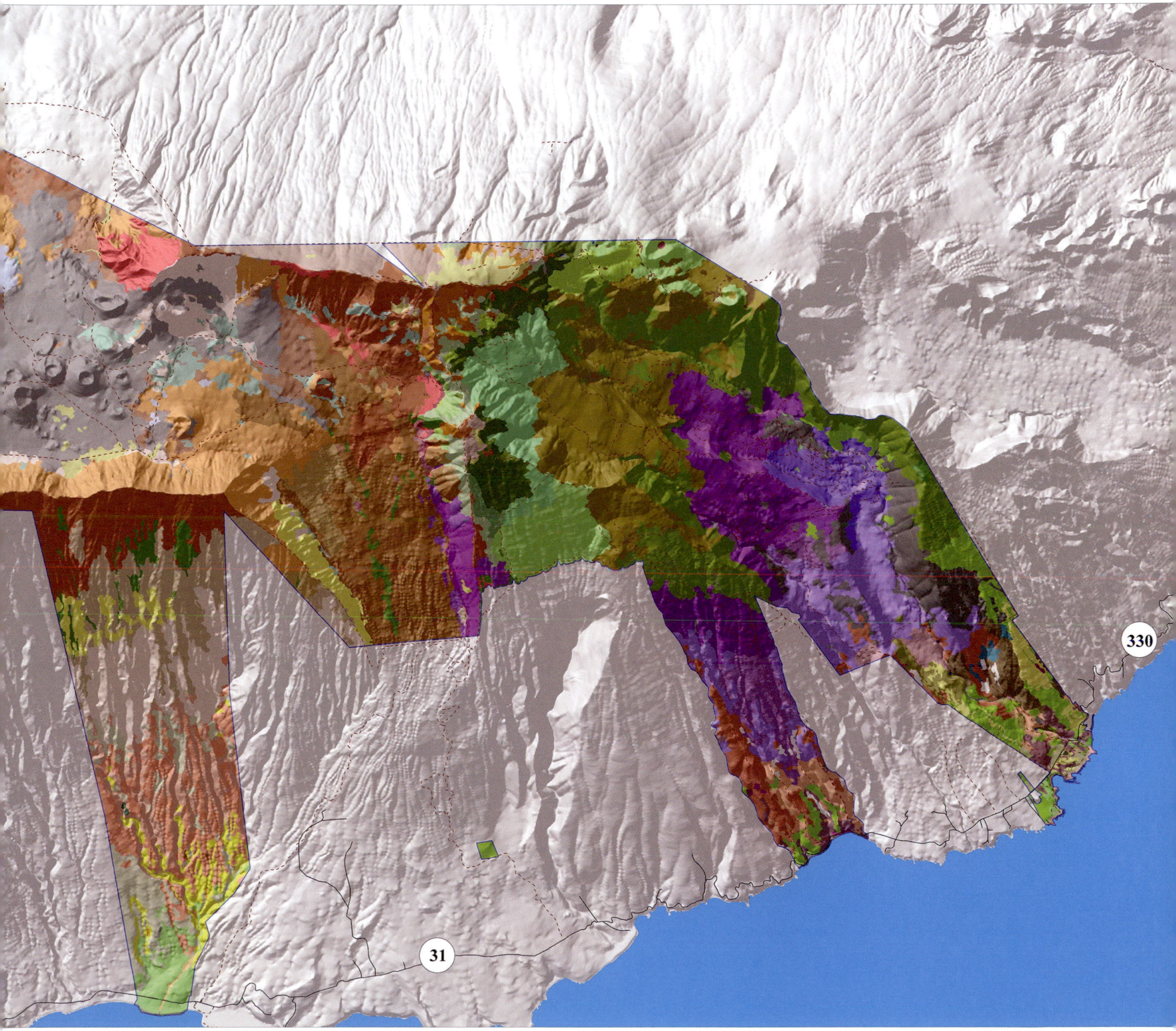

The Haleakalā National Park vegetation inventory. NPS, Kass Green & Associates, Chad Lopez.

(*Above*) Park Service ecologist records vegetation data in the subalpine shrubland at Haleakalā National Park. Data sources: NPS, USGS, Kass Green & Associates, Chad Lopez. (*Right*) The Kalaupapa National Historical Park vegetation inventory. NPS, Kass Green & Associates, Chad Lopez.

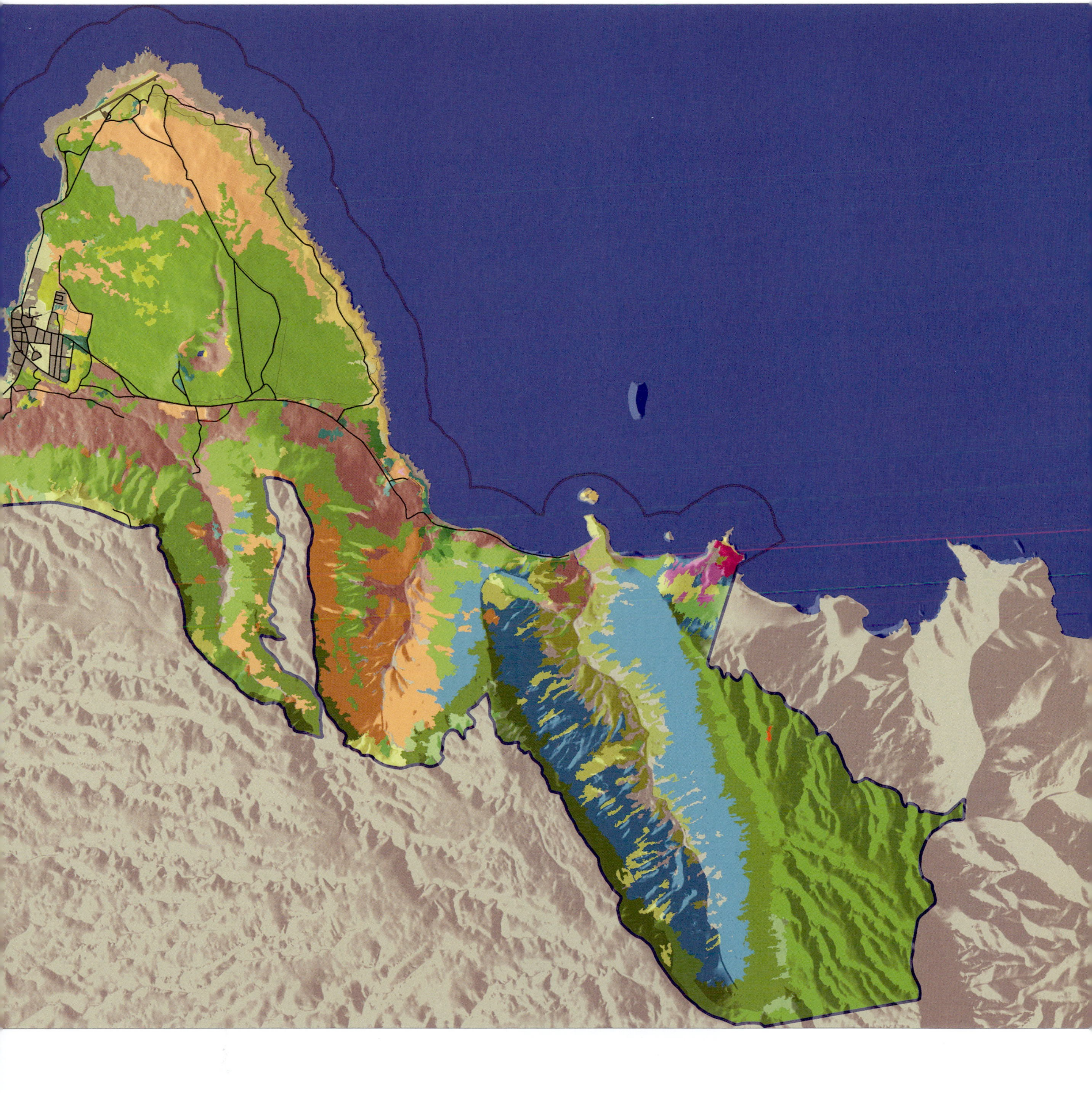

WILDLIFE CONNECTIVITY IN THE CROWN OF THE CONTINENT ECOSYSTEM

MORGAN VOSS, GLACIER NATIONAL PARK
RICHARD MENICKE, GLACIER NATIONAL PARK

The US Highway 2 corridor along Glacier National Park's south boundary is an important ecological linkage zone, connecting Glacier with the Bob Marshall Wilderness to the south. Together, this large mass of protected land provides core habitat for several large mammal species (including grizzly bears, lynx, mountain lion, wolves, wolverine, bighorn sheep, mountain goats, elk, and moose) within the Crown of the Continent ecosystem. The Highway 2 corridor is the primary east-west transportation route in northern Montana for both rail and vehicles and, as such, the highway and rail line present a formidable challenge for wildlife movement across the landscape.

The increased mortality of grizzly bears within the Highway 2 corridor in years past was a result of train derailments and grain spills that subsequently became a food attractant. This led to the creation of the Great Northern Environmental Stewardship Area, or GNESA, a diverse stakeholder group focused on wildlife conservation in the Highway 2 corridor. Mixed land ownership leading to habitat fragmentation is yet another challenge for wildlife migration within the corridor. This map, constructed initially in 2008 and updated early in 2018, assimilates wildlife observation data from multiple sources over a multiyear timeframe to identify important areas along the corridor where conservation actions such as easements and highway modification (for example, enlarged culverts) might be possible to facilitate animal movement between the Great Bear Wilderness and Glacier National Park.

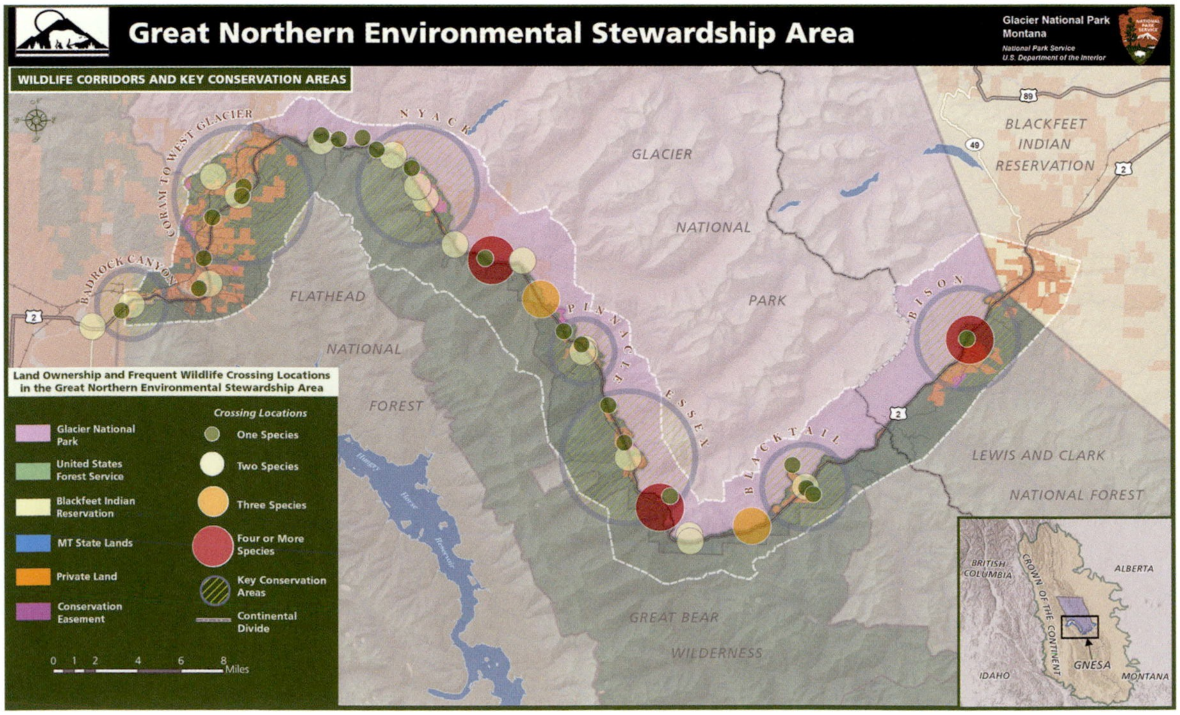

This map shows wildlife observations and related conservation targets to facilitate wildlife migration in the US Highway 2 Corridor adjacent to Glacier National Park. Data sources: NPS, Glacier National Park, Montana State Library (https://geoinfo.msl.mt.gov/data) , Open Street Map Contributors, Esri (Terrain: Multi-directional hillshade basemap).

BLACK CANYON HYDROGRAPHY 3D MAP

JOE MILBRATH, NPS HARPERS FERRY CENTER FOR MEDIA SERVICES

This 3D map of the Gunnison River Basin illustrates the hydrographic headwaters of the Black Canyon of the Gunnison National Park in Colorado. The map was developed as an informational graphic for the park's brochure in 2016. Geology and the power of water are paramount to park interpretation and this graphic weaves the two stories together. Snowmelt from four nearby mountain ranges flows downstream until it's channeled into the Black Canyon's narrow walls. Rendered in Natural Scene Designer, the graphic mixes terrain texture shading, multidirectional light sources, and hypsometric tints to create a clean terrain.

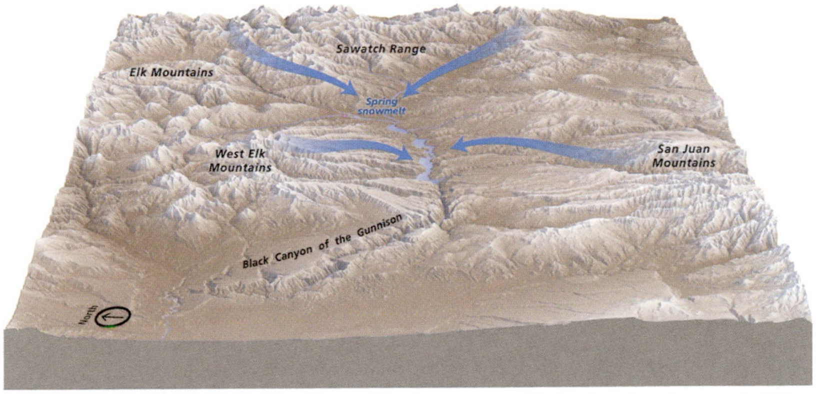

This 3D map of the Gunnison River Basin illustrates the hydrographic source of the Black Canyon of the Gunnison. Data sources: USGS and NPS.

SEQUOIA AND KINGS CANYON ECOSYSTEM MAP

TOM PATTERSON (RETIRED) AND JIM EYNARD,
NPS HARPERS FERRY CENTER FOR MEDIA SERVICES

This 3D oblique map depicts the varied ecosystem environments within Sequoia and Kings Canyon National Parks and was created for a new park brochure in 2018. The Sierra Nevada ecosystems include alpine, subalpine, coniferous forests, and the canyon and foothills ecosystem. The ecosystems are defined by elevation and the map uses hypsometric tints and natural colors to define these regions, showing snowy, white areas in the alpine ecosystem and a deep green in the forested areas. The elevation range of the map is nearly 13,000 feet ranging from over 14,000 feet on Mount Whitney down to 1,000 feet in the canyons and foothills. This ecosystem map of Sequoia and Kings National Parks supplements a more detailed inset map found on the other side of the park's brochure.

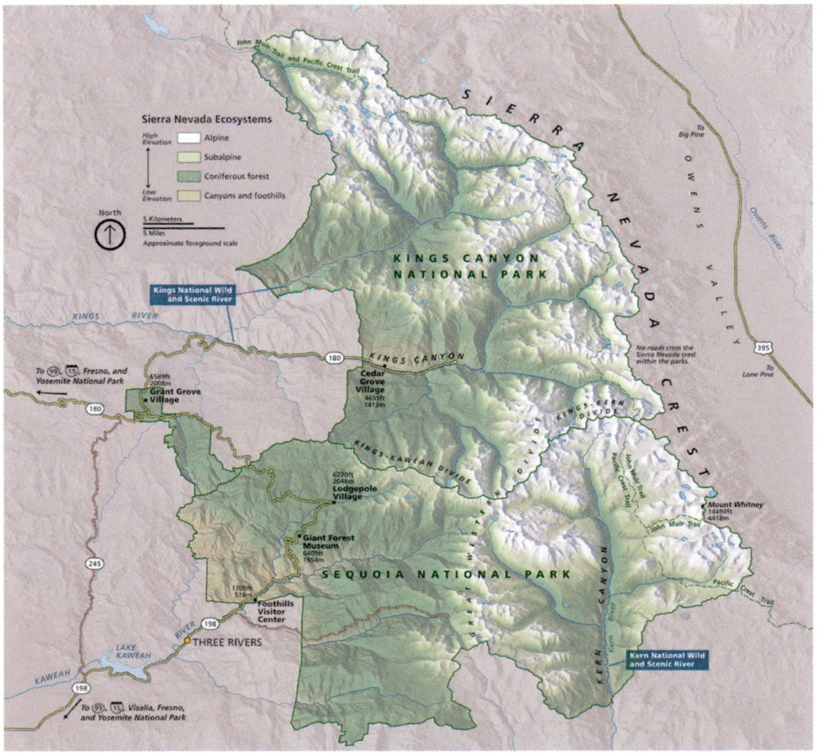

This 3D oblique map depicts the varied ecosystem environments within Sequoia and Kings Canyon National Parks. Data sources: USGS and NPS.

CROWN OF THORNS SEASTAR POPULATION CONTROL SUCCESSFUL AT THE NATIONAL PARK OF AMERICAN SAMOA

IAN MOFFITT, NATIONAL PARK OF AMERICAN SAMOA
KERSTEN SCHNURLE, NATIONAL PARK OF AMERICAN SAMOA
SARAH LUMMIS, NATIONAL PARK OF AMERICAN SAMOA

A crown of thorns seastar outbreak was observed in National Park of American Samoa (NPSA) jurisdictional waters in 2013. Carnivorous and poisonous, crown of thorns seastars can decimate reef ecosystems, requiring decades to recover. NPSA mounted a rapid response control program utilizing help from American Conservation Experience, the Coral Reef Advisory group, and other local partners. These maps show how seastar density changed before and after park control efforts. More than 26,000 of these seastars were killed by park divers and partners. The images depict park divers Bert Fuiava and Ian Moffitt controlling seastar populations using scuba and closed-circuit rebreather (CCR) technologies.

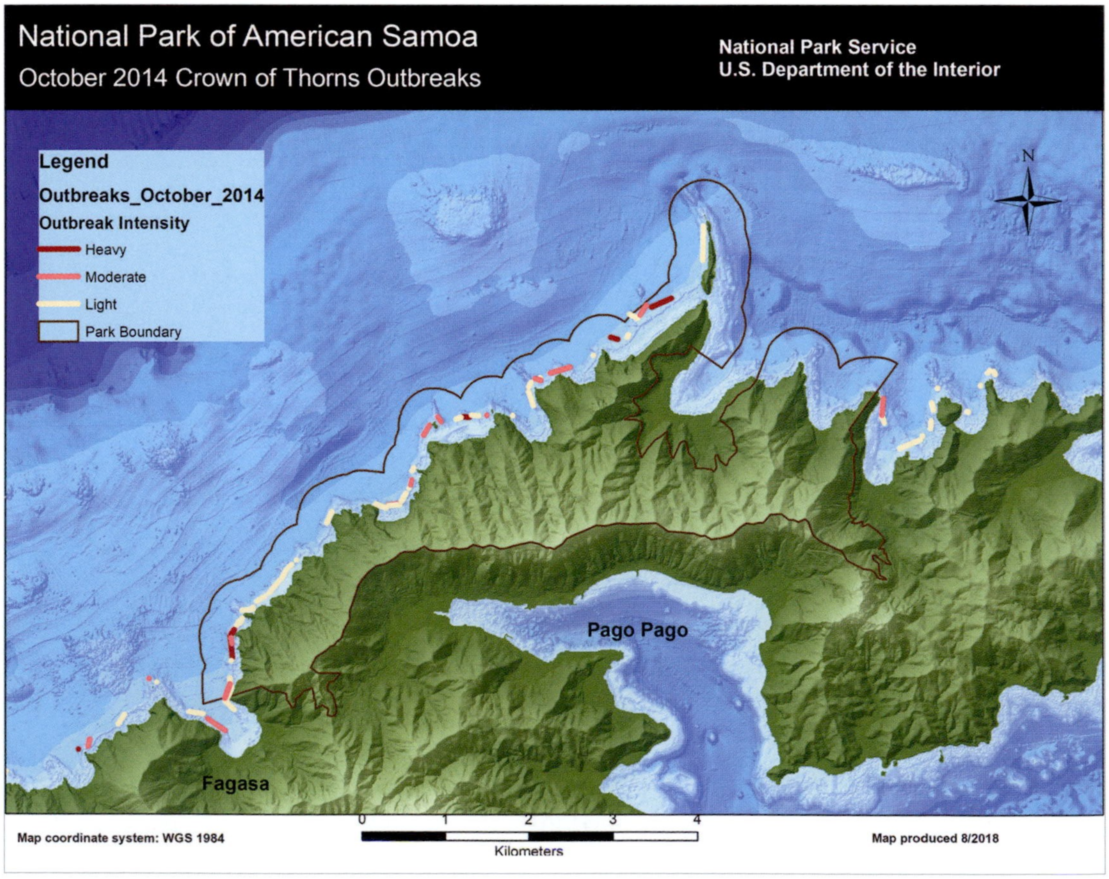

This map shows the October 2014 crown of thorns seastar outbreak intensity. Data sources: National Park of American Samoa.

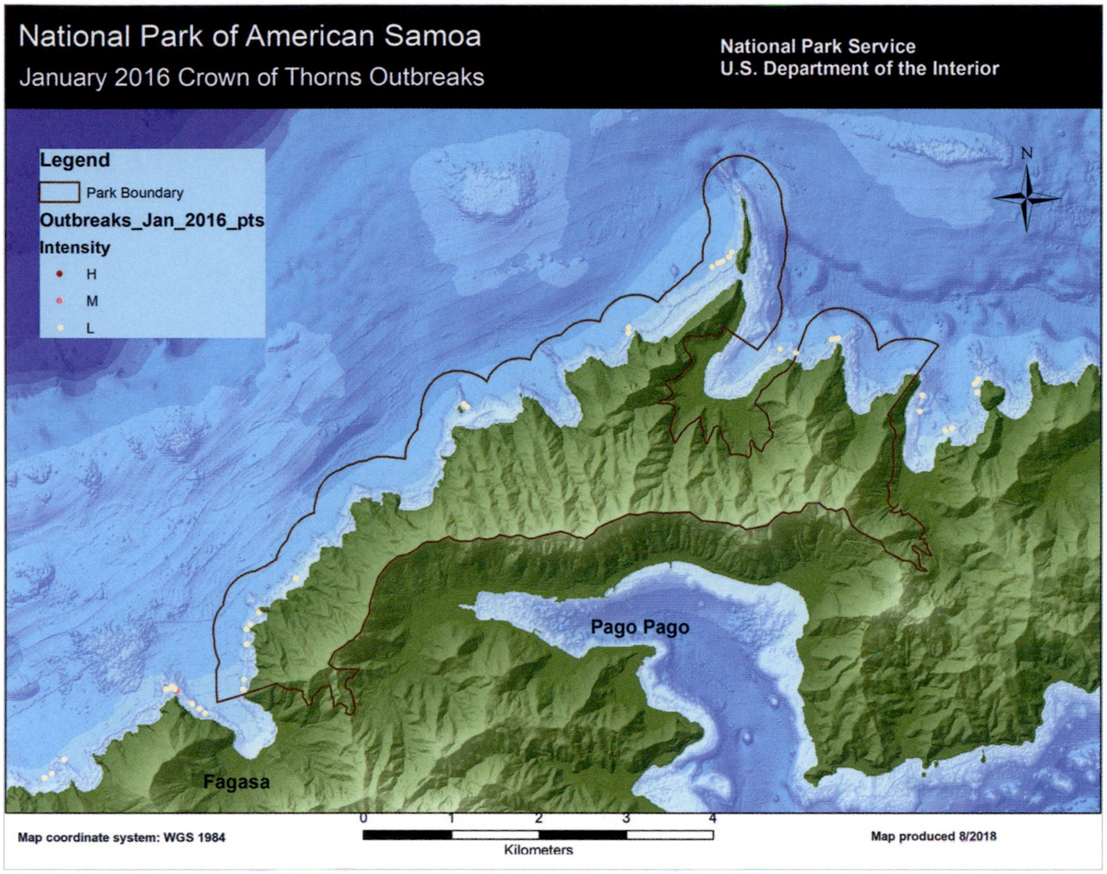

This map shows the January 2016 crown of thorns seastar outbreak intensity after control treatment.

Data sources: National Park of American Samoa.

Ian Moffit, NPS diver, controls the crown of thorns seastar populations. NPS photo.

Bert Fuiava, NPS diver, controls crown of thorns seastar populations. NPS photo.

MAPPING GROUNDWATER VULNERABILITY TO PROTECT GRAND CANYON NATIONAL PARK'S GROUNDWATER AND DRINKING WATER RESOURCES

MARK NEBEL, JERED HANSEN, NATALIE JONES, BEN TOBIN, AND CYNTHIA VALLE,
GRAND CANYON NATIONAL PARK

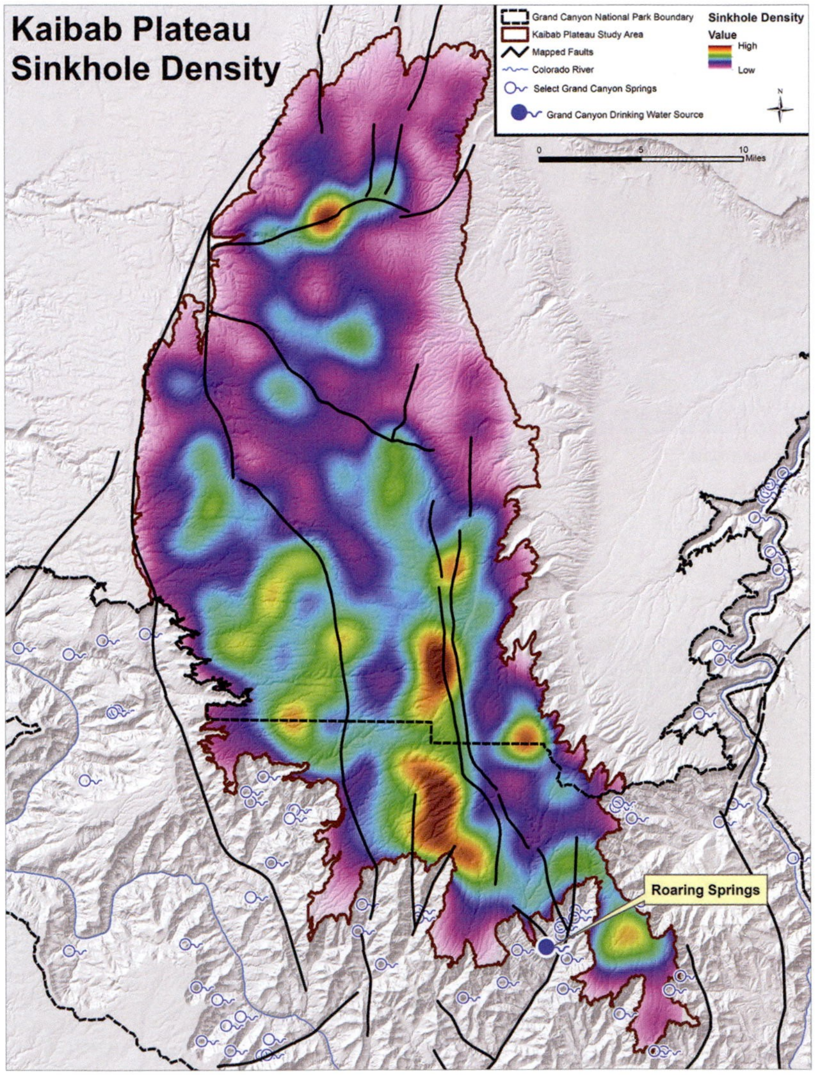

This map shows sinkhole density in the Kaibab Plateau study area of Grand Canyon National Park. Data sources: USGS, NPS, Grand Canyon Monitoring and Research Center.

This map represents the density of karst sinkhole features on the Kaibab Plateau on the north rim of the Grand Canyon in Arizona, as a component of groundwater vulnerability to contamination. Snowmelt is a major contributor to groundwater recharge on the Kaibab Plateau. It enters aquifers through sinkholes, flows along faults, and discharges as springs within the Grand Canyon, including the park's current source of drinking water, Roaring Springs. Sinkhole density on the Kaibab Plateau is highly correlated with the presence of faults and fractures and may provide insights into the location of unmapped faults.

Topographic depressions on the plateau were delineated from a lidar-derived digital elevation model. Thirteen characteristic variables, based on shape, depth, orientation, and concavity, were derived for a subset of training features that could be clearly identified as sinkhole or non-sinkhole depressions. These variables were used in a Random Forests regression model to characterize sinkhole versus non-sinkhole depressions across the plateau. Of the total 257,519 mapped depressions, 6,973 features (2.7 percent) were classified as true sinkholes. Field validation showed that 87.5 percent of all depressions were correctly classified by the regression modeling. Sinkhole density was calculated using a Kernel Density function.

Grand Canyon National Park uses sinkhole density, mapped faults, and other data to delineate areas where groundwater is especially susceptible to contamination from surface factors, including development, transportation corridors, and animal impacts (for example, non-native bison herds). These results contribute to the information needed to help protect groundwater resources and springs that provide drinking water for park residents and millions of visitors annually.

TWENTY-FIVE YEARS OF LANDSCAPE CHANGE MONITORING IN THE PACIFIC NORTHWEST NATIONAL PARKS

NATALYA ANTONOVA, NORTH COAST AND CASCADES INVENTORY AND MONITORING NETWORK
CATHARINE COPASS, OLYMPIC NATIONAL PARK

Every year, natural events such as landslides, floods, and fires alter the landscapes in the large remote wilderness parks of the Pacific Northwest in Washington State: Mount Rainier, Olympic, and the North Cascades NPS Complex. Gusts of strong wind or avalanches topple patches of trees. Glaciers can melt rapidly, flooding rivers, and erode the adjacent forested terraces, leaving behind rocks and water where tall forests stood. When these events occur in the remote wilderness, how do we know when or how often they occur? Climate change in the parks is predicted to alter the intensity and frequency of fires and floods, reduce forest health, and change our landscapes in ways that are currently difficult to foresee but will likely affect park facilities, natural resources, and visitor recreational opportunities.

The three parks cover a combined area of more than 2,800 square miles, an area larger than the state of Delaware. Landscape change in these parks cannot be monitored by foot—or even by airplane—but using Landsat satellite data makes tracking changes in park ecosystems efficient and economical. The NPS Landscape Change monitoring program detects changes in the satellite images that occur due to changes in the land cover. For example, a patch of forest that appears dark green in the Landsat image one year and bright grey in a subsequent year indicates that a disturbance event has removed or significantly damaged the forest. The NPS used this approach to detect and map changes due to avalanches, fires, landslides, windthrow, insects and disease, and riparian flooding that occurred between 1985 and 2009.

These maps will allow researchers to identify links between Pacific Northwest weather patterns and the frequency and timing of landscape change, potentially leading to more accurate predictions of when these events may occur. Over decades, the program will give researchers evidence to assess the effects of climate change on natural disturbance events. With this study, the parks serve as frontiers for some of the most expansive and detailed natural resource monitoring in the Pacific Northwest.

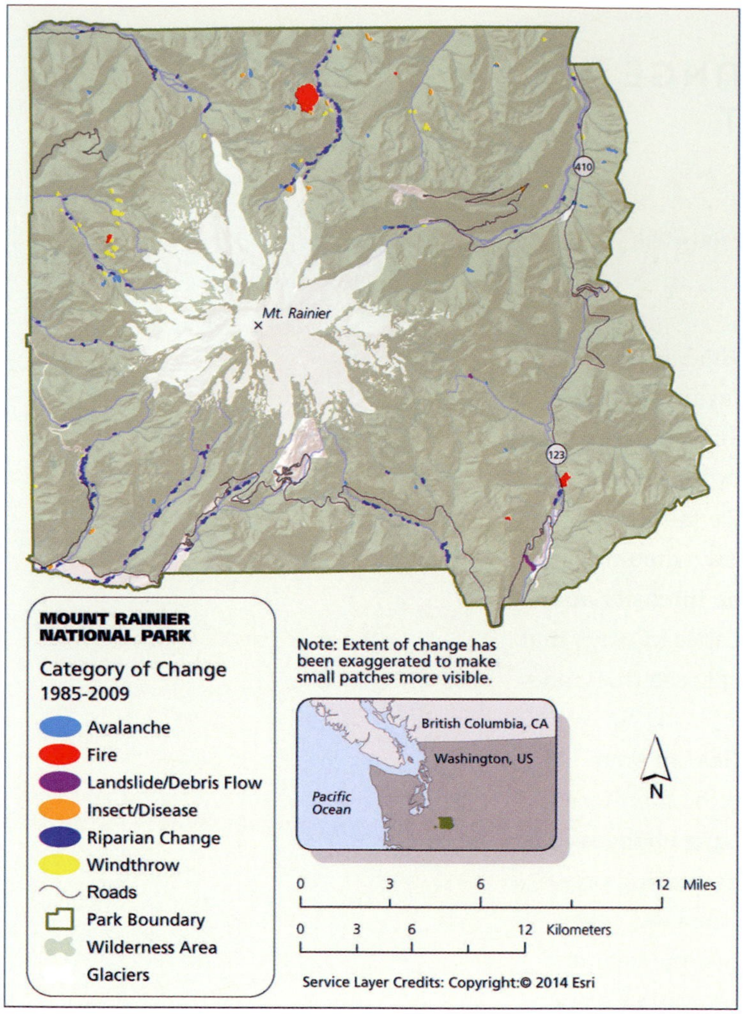

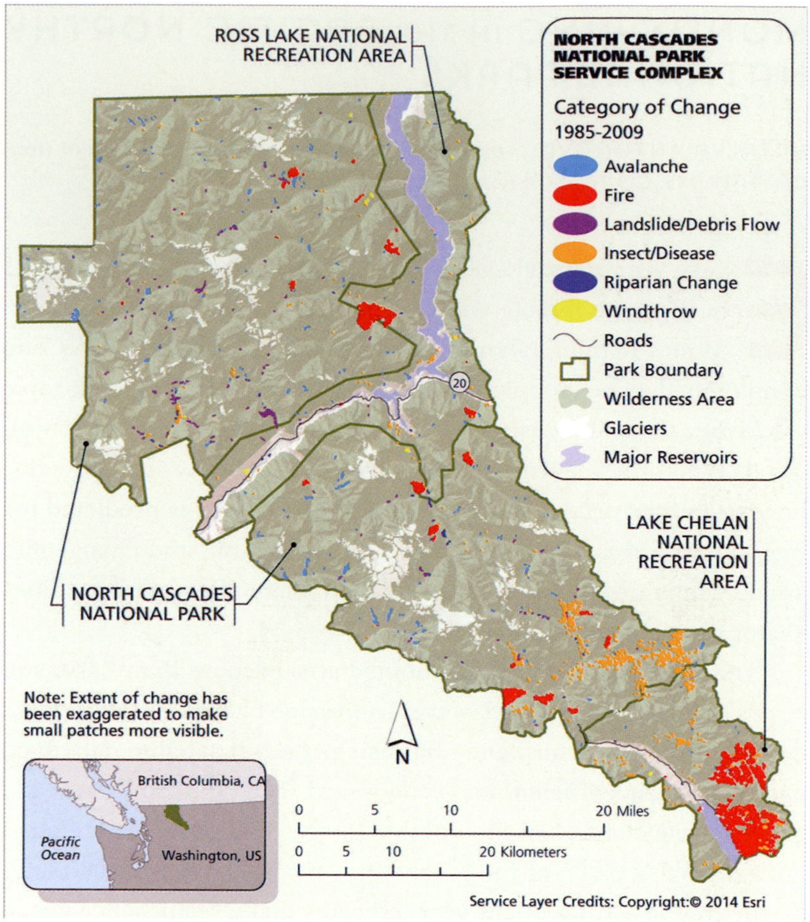

Clockwise from above:

This map shows the category of change for Mount Rainier National Park.

This map shows the category of change for North Cascades National Park.

This map shows the category of change for Olympic National Park. Data sources: NPS.

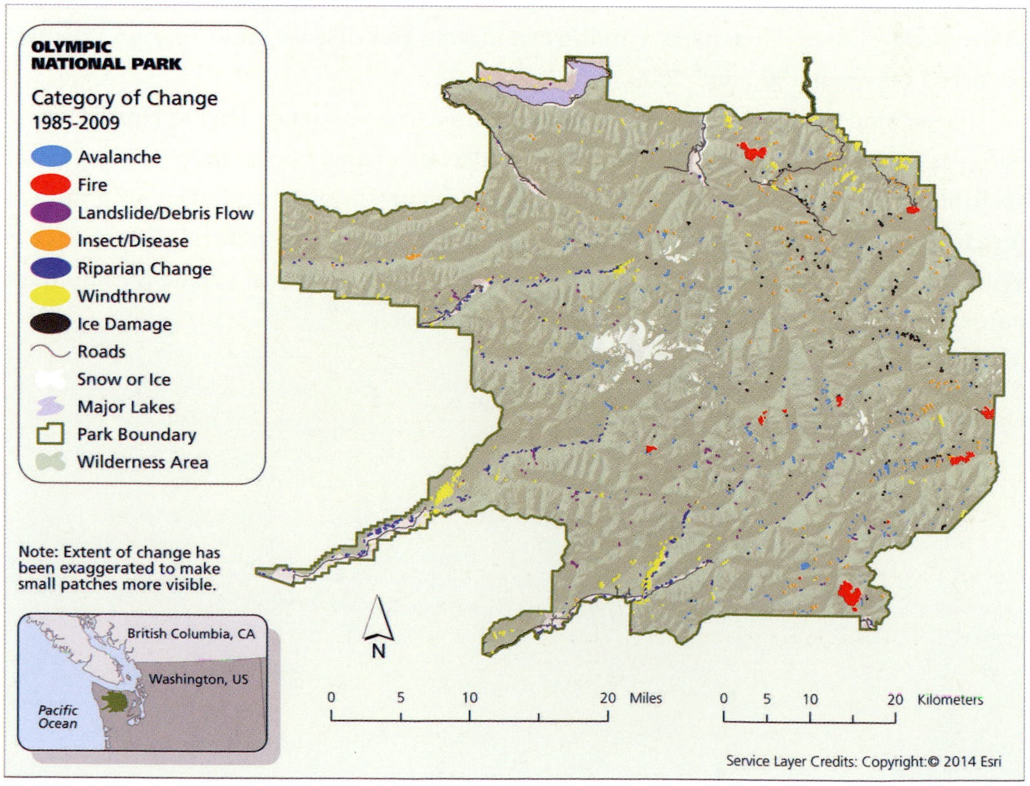

(*Above*) The Nisqually River carves away at the bank next to the Emergency Operations Center in Longmire, Mount Rainier National Park. (*Right*) Remnants of a 2004 wildfire in the wilderness of Mount Rainier National Park are shown.

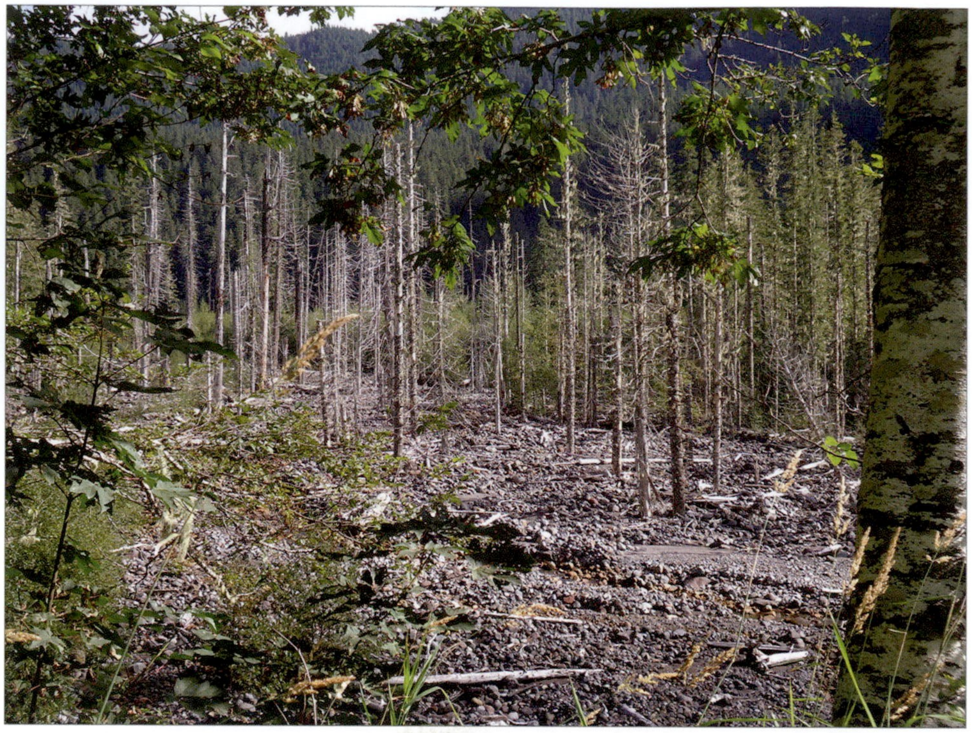

(*Right*) Only ghosts remain where forest once stood: tree mortality several years after, and rocks and debris deposited by the Tahoma Creek at Mount Rainier National Park during the 2006 floods. (*Below*) Road to nowhere: the sign to the Sunshine Point Campground at Mount Rainier National Park stands as a reminder of what the Nisqually River removed.

(*Above*) Toppled and broken trees were a common sight at Mount Rainier and Olympic National Parks following the Great Coastal Gale of December 1–3, 2007. (*Left*) Debris from a winter 2008–2009 avalanche blocks a trail near Dagger Lake in North Cascades NPS Complex.

(*Right*) These whitebark pine trees were killed by blister rust, a fungal disease, in the wilderness of the North Cascades NPS Complex. (*Below*) Smoke billows above the community of Stehekin, North Cascades National Park Complex, in the summer of 2006.

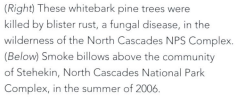

(*Left*) Destruction of a private cabin following 2003 Stehekin River floods in North Cascades NPS Complex. (*Below*) This road was damaged by riparian flooding following the December 2007 storm along Graves Creek in Olympic National Park.

Wind-thrown trees block this road after the December 2007 storm in Olympic National Park.

6
CULTURAL RESOURCES

DEIDRE MCCARTHY, NPS CULTURAL RESOURCE GIS PROGRAM

CARING FOR CULTURAL RESOURCES

The National Historic Preservation Act became law in 1966, creating the basis for cultural resource management practices used today. In the legislation, Congress articulated that "the historical and cultural foundations of the Nation should be preserved as a living part of our community life and development in order to give a sense of orientation to the American people." As part of its mission, the National Park Service (NPS) builds on this theme, working to inventory, manage, and interpret archeological sites, historic buildings, landscapes, and other cultural features for the benefit of the public, as well as the various communities they inhabit.

Over 35 years ago, Dr. Lawrence Aten, chief of the Interagency Resources Division of the NPS, recognized that locational information is a key factor in understanding cultural resources and how to properly curate them within appropriate contexts. Aten pioneered the application of GIS to cultural resources in the NPS, establishing the Cultural Resource GIS Facility. Today, the NPS continues to emphasize these concepts, exploring how geographic clues provide data about the human and environmental influences on cultural resources—helping explain why they exist, how they interrelate, and what historical significance each might possess individually or collectively. Park and resource managers persist in expanding on Aten's legacy, increasingly relying on GIS to achieve their mission, as well as protect sensitive resources for the future enrichment of all the communities they represent.

As these technologies expand their capabilities, parks find innovative ways to apply them to cultural resource management and stewardship. At Glen Canyon National Recreation Area in Utah and Arizona, statistical and modeling functionality help predict areas within the vast landscape that may contain cultural resources, never surveyed. Using photogrammetry also helps the park meet its requirements to monitor sites for change or damage, assisting in developing preservation strategies. Great Smoky Mountains National Park in Tennessee and North Carolina uses GIS to assess site conditions and identify documentation priorities following natural disasters, such as wildfires. Stories created with ArcGIS and other GIS-based tools help the park provide public interpretation of hard-to-reach regions, or in the case of Hot Springs National Park in Arkansas, create virtual tours of sites considered sensitive or unavailable to the public, increasing their exposure to new communities.

The Cultural Resource GIS Facility tells stories using ArcGIS and related software to highlight the importance of national programs, such as the Heritage Documentation Programs, which create baseline documentation forming a permanent record housed at the Library of Congress in Washington, DC. In the event those resources are destroyed due to large-scale disasters, such as hurricanes, climate change, or sea level rise, they will continue to contribute to our collective understanding of architecture, engineering, and landscapes. Further, the creation of an enterprise cultural resource GIS dataset by the Cultural Resource GIS Facility provides an authoritative data source contributing to better resource management across the NPS in addition to integrating all the park cultural resource databases through spatial representations.

As Aten intrinsically understood in the 1980s, GIS and spatial tools endure in allowing the NPS to meet park and program mandates, resulting in better planning, disaster preparedness, and protection for all the cultural resources under its care. The use of these tools in turn opens the door to enhanced public access and better understanding of these unique sites within all our communities.

OCMULGEE NATIONAL MONUMENT: CELEBRATING ARCHEOLOGY

JILL HALCHIN, NPS SOUTHEAST ARCHEOLOGICAL CENTER

The federal work programs of the Great Depression era enabled American archeologists to make major progress in learning about prehistoric cultures as well as to sharpen their techniques for excavation and analysis. From late 1933 until October 1941, archeologists and large crews of laborers excavated miles of trenches in the area that is now Ocmulgee Mounds National Historical Park in Macon, Georgia. Millions of artifacts and thousands of pages of field notes, drawings, maps, and photographs from those excavations are now preserved at the NPS Southeast Archeological Center in Tallahassee, Florida.

In 2015, the center spent more than a month poring over the field records and digitizing the excavations, trench by trench. By converting the information to a digital format, researchers no longer must wade through the archives for location information.

Additionally, the GIS data became the basis of a story prepared using ArcGIS for the NPS Centennial in 2016. The application illustrates how the archeologists recorded the features and artifacts that they found at Ocmulgee, and how they used that information to discover changes in American Indian culture through time. The story links the GIS data with original photographs, field drawings, and other documents that can be expanded for closer examination. The results of this project have made this important archeological data and information available to a broad audience.

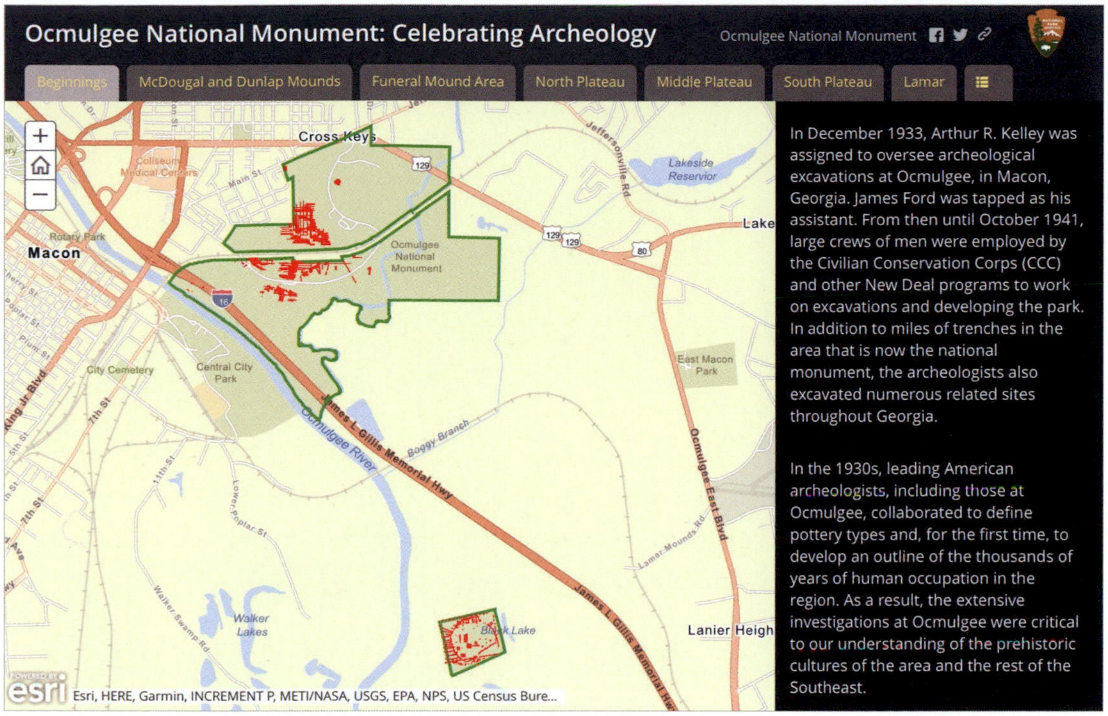
Ocmulgee National Monument: Celebrating Archeology
Ocmulgee National Monument
Beginnings
McDougal and Dunlap Mounds
Funeral Mound Area
North Plateau
Middle Plateau
South Plateau
Lamar
Cross Keys
Macon
Lakeside Reservoir
Ocmulgee National Monument
City Cemetery
Central City Park
East Macon Park
Ocmulgee River
James L Gillis Memorial Hwy
Lanier Heigh
Walker Lakes
Esri, HERE, Garmin, INCREMENT P, METI/NASA, USGS, EPA, NPS, US Census Bure...
In December 1933, Arthur R. Kelley was assigned to oversee archeological excavations at Ocmulgee, in Macon, Georgia. James Ford was tapped as his assistant. From then until October 1941, large crews of men were employed by the Civilian Conservation Corps (CCC) and other New Deal programs to work on excavations and developing the park. In addition to miles of trenches in the area that is now the national monument, the archeologists also excavated numerous related sites throughout Georgia.
In the 1930s, leading American archeologists, including those at Ocmulgee, collaborated to define pottery types and, for the first time, to develop an outline of the thousands of years of human occupation in the region. As a result, the extensive investigations at Ocmulgee were critical to our understanding of the prehistoric cultures of the area and the rest of the Southeast.

The story about the park takes the viewer to each area where excavations were undertaken. The map and sidebar provide the spatial and historical context. This view of the application and the others in this article are screenshots of the story open in the Safari browser on an iPad Pro screen.

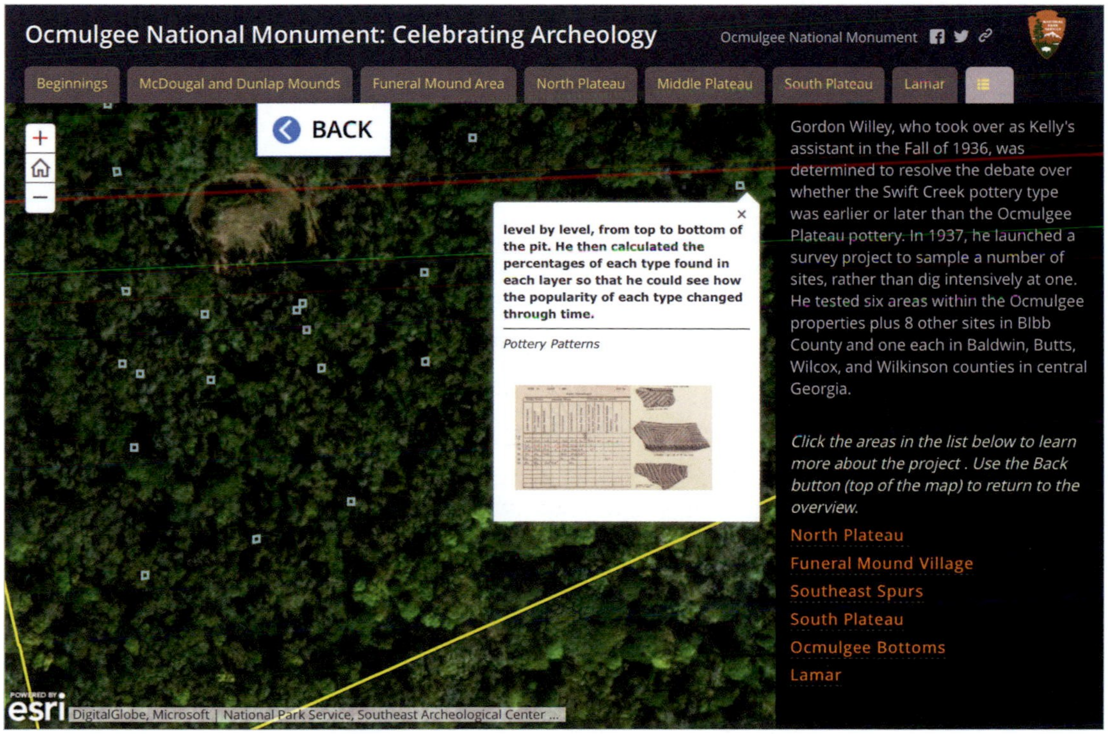
Ocmulgee National Monument: Celebrating Archeology
Ocmulgee National Monument
Beginnings
McDougal and Dunlap Mounds
Funeral Mound Area
North Plateau
Middle Plateau
South Plateau
Lamar
BACK
level by level, from top to bottom of the pit. He then calculated the percentages of each type found in each layer so that he could see how the popularity of each type changed through time.
Pottery Patterns
DigitalGlobe, Microsoft | National Park Service, Southeast Archeological Center ...
Gordon Willey, who took over as Kelly's assistant in the Fall of 1936, was determined to resolve the debate over whether the Swift Creek pottery type was earlier or later than the Ocmulgee Plateau pottery. In 1937, he launched a survey project to sample a number of sites, rather than dig intensively at one. He tested six areas within the Ocmulgee properties plus 8 other sites in Bibb County and one each in Baldwin, Butts, Wilcox, and Wilkinson counties in central Georgia.
Click the areas in the list below to learn more about the project . Use the Back button (top of the map) to return to the overview.
North Plateau
Funeral Mound Village
Southeast Spurs
South Plateau
Ocmulgee Bottoms
Lamar

One tab in the story is about a special project, the Stratigraphic Survey that sampled several parts of the park. To show all those places with one tab, Story Actions (highlighted in the side panel) zoom and pan the map to the specific locations. Like the other tabs, the excavation polygons have pop-ups that showcase more examples from the massive archives of photographs, field records, and in this case, a chart used to analyze changes in pottery decorations.

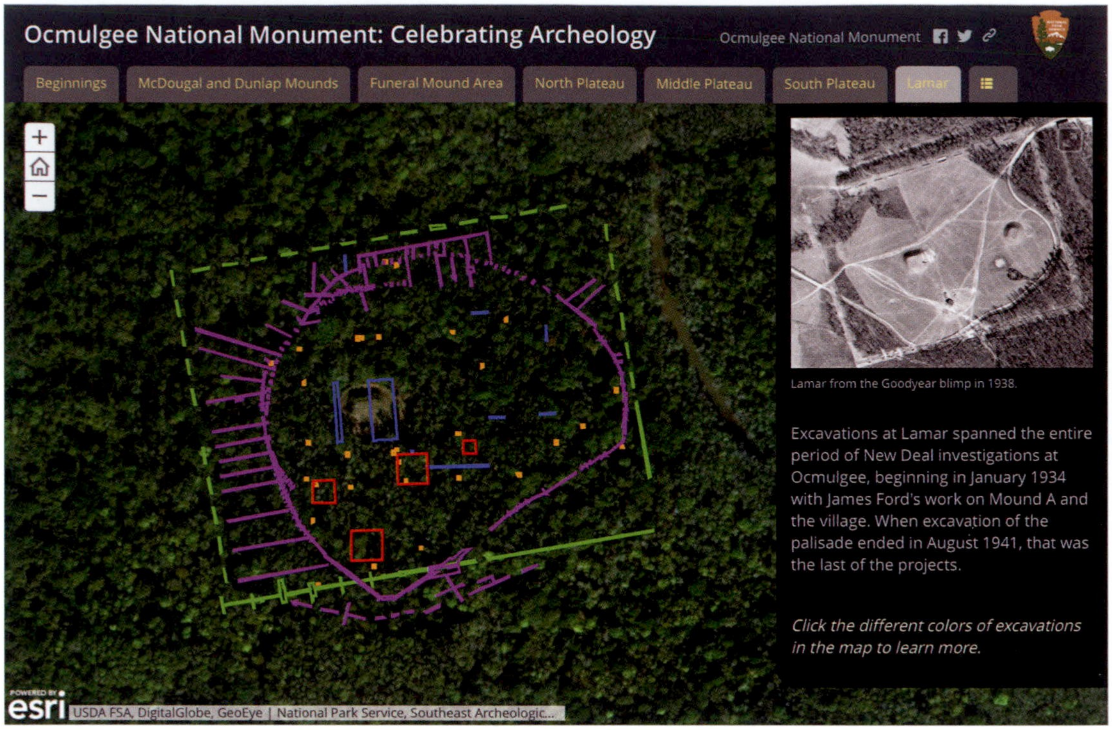
Ocmulgee National Monument: Celebrating Archeology
Ocmulgee National Monument
Beginnings
McDougal and Dunlap Mounds
Funeral Mound Area
North Plateau
Middle Plateau
South Plateau
Lamar
Lamar from the Goodyear blimp in 1938.
Excavations at Lamar spanned the entire period of New Deal investigations at Ocmulgee, beginning in January 1934 with James Ford's work on Mound A and the village. When excavation of the palisade ended in August 1941, that was the last of the projects.
Click the different colors of excavations in the map to learn more.
esri
USDA FSA, DigitalGlobe, GeoEye | National Park Service, Southeast Archeologic...

The work of local photographer Joseph Coke, who photographed the Ocmulgee excavations, is featured in the story. In 1938, he had the chance to fly over the Lamar site when the Goodyear Blimp visited Macon, Georgia. His striking aerial (in the sidebar) images displayed beside the map of excavations helps us to visualize the site as the archeologists experienced it, rather than the forested landscape today. Taken in winter, the photographs show the excavations into the side of Mound B, open trenches around the perimeter, and the scars of back-filled excavations inside the site.

Among the field records displayed in the story are hand-drawn field maps. Artist James Jackson added slopes and other details to this field map, giving a little flair to what would normally be a basic drawing. The excavations were funded through government agencies that were dedicated to employing people who had lost their jobs due to the Great Depression in the 1930s, including archeologists, professional surveyors, artists, photographers, and clerical staff.

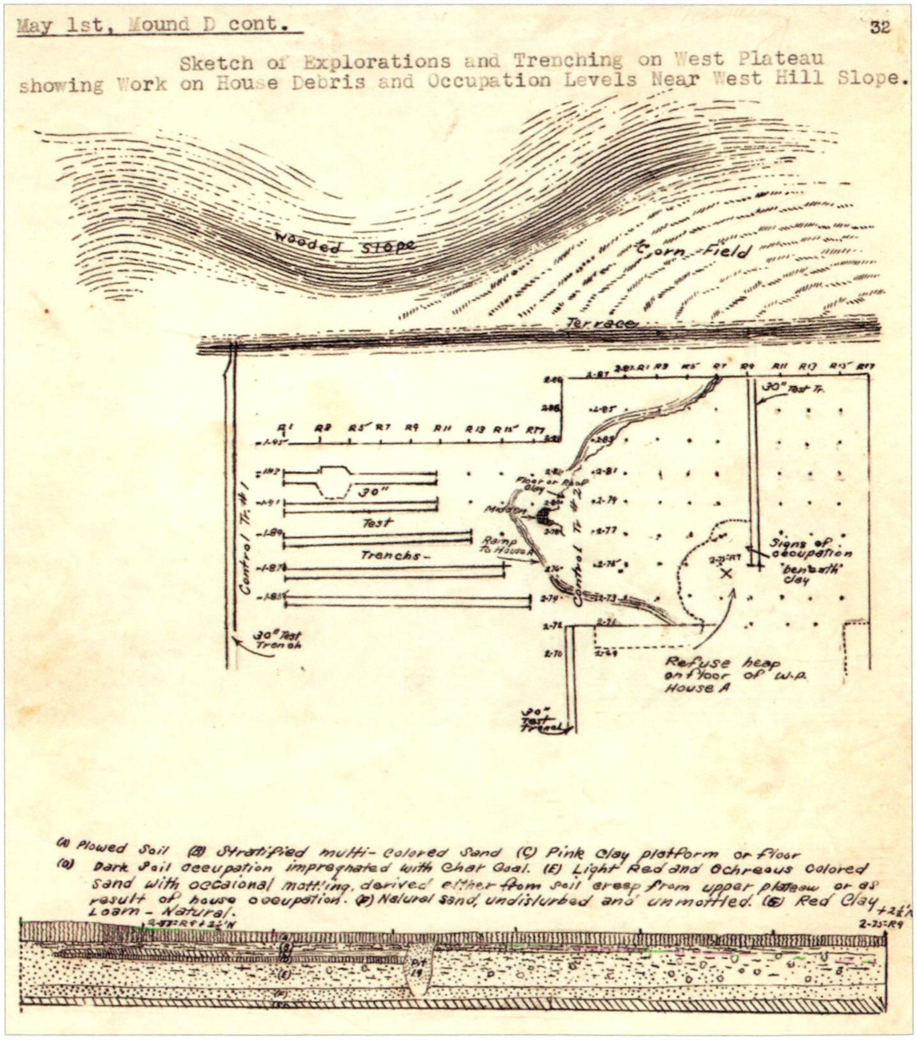
May 1st, Mound D cont.
32
Sketch of Explorations and Trenching on West Plateau showing Work on House Debris and Occupation Levels Near West Hill Slope.
Wooded Slope
Corn Field
Terrace
30" Test Tr.
30" Test Trenchs
Control Tr. #1
Control Tr. #2
30" Test Trench
Ramp to House A
Signs of occupation beneath clay
Refuse heap on floor of W.P. House A
(A) Plowed Soil (B) Stratified multi-colored Sand (C) Pink Clay platform or floor
(D) Dark Soil occupation impregnated with Char Coal. (E) Light Red and Ochreous colored sand with occaional mottling, derived either from soil creep from upper plateau or as result of house occupation. (F) Natural sand, undisturbed and unmottled. (G) Red Clay Loam - Natural.
Pit 19

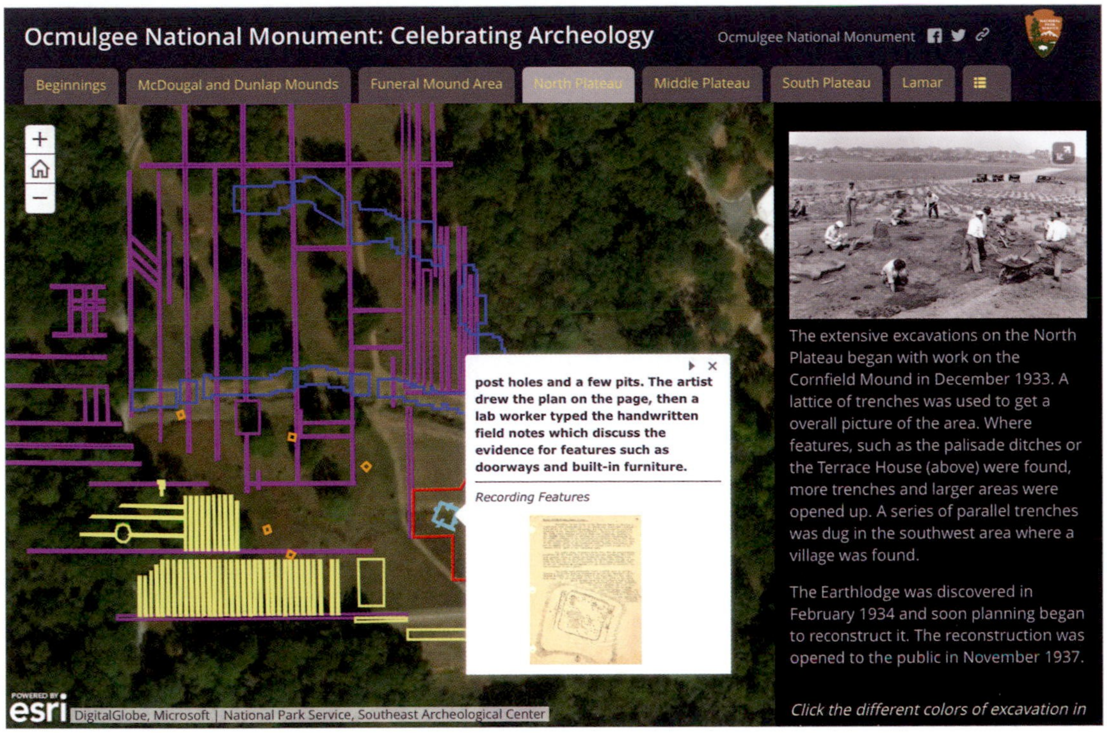
Ocmulgee National Monument: Celebrating Archeology
Ocmulgee National Monument
Beginnings
McDougal and Dunlap Mounds
Funeral Mound Area
North Plateau
Middle Plateau
South Plateau
Lamar
post holes and a few pits. The artist drew the plan on the page, then a lab worker typed the handwritten field notes which discuss the evidence for features such as doorways and built-in furniture.
Recording Features
The extensive excavations on the North Plateau began with work on the Cornfield Mound in December 1933. A lattice of trenches was used to get a overall picture of the area. Where features, such as the palisade ditches or the Terrace House (above) were found, more trenches and larger areas were opened up. A series of parallel trenches was dug in the southwest area where a village was found.
The Earthlodge was discovered in February 1934 and soon planning began to reconstruct it. The reconstruction was opened to the public in November 1937.
Click the different colors of excavation in
esri
DigitalGlobe, Microsoft | National Park Service, Southeast Archeological Center

Each tab zooms to one part of the park, showing the excavated areas and giving a brief description of the work there, illustrated with a photograph from the project archives. Clicking a polygon on the map opens a pop-up about techniques and tools used during the excavations. The image in the pop-up is linked to a larger view that is stored on the NPS website NPGallery (https://npgallery.nps.gov), where it is available to the public.

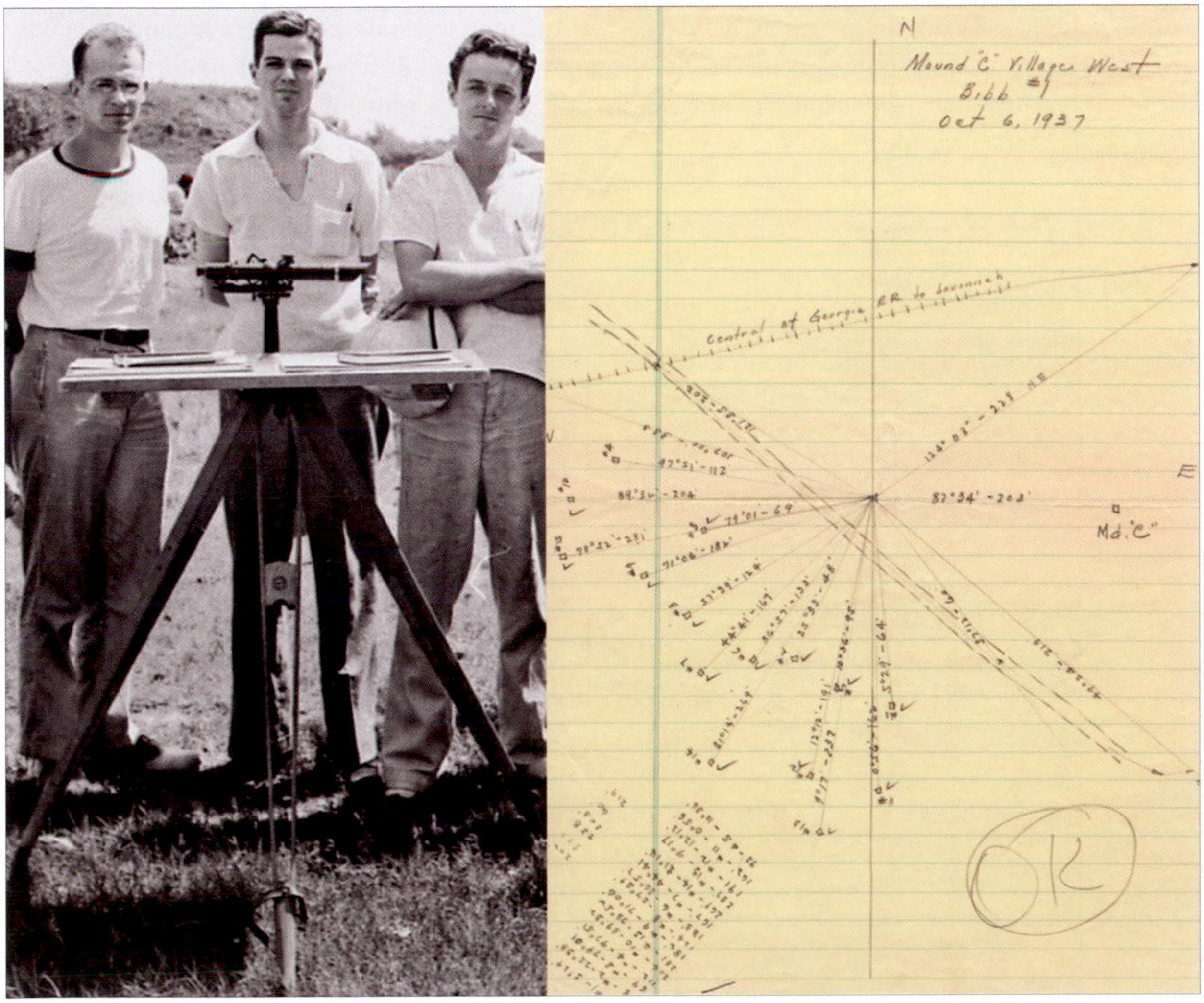
N
Mound "C" Village West
Bibb #1
Oct 6, 1937
Md. "C"
E
OK

For decades, before total stations and high-accuracy GPS, archeologists used the plane table and alidade for mapping sites. The Ocmulgee project archives include examples of plane table diagrams that record angles and distances. There is also a photograph of three graduate students who spent the summer of 1933 at Ocmulgee. Left to right: Joseph Birdsell, Charles Wagley, and Gordon Willey, who later directed much of the work at Ocmulgee and went on to a distinguished career in archeology. Birdsell and Wagley became noted cultural anthropologists. Data sources: NPS, Esri.

MONITORING EROSION OF ARCHEOLOGICAL SITES USING PHOTOGRAMMETRY

ALAHNA MOORE AND VANESSA GLYNN-LINARIS, GLEN CANYON NATIONAL RECREATION AREA

The landscape of Glen Canyon National Recreation Area was drastically altered by the construction of Glen Canyon Dam in 1956. The dam restricts the flow of water and sediment from Lake Powell into the Colorado River, causing unforeseen erosion south of the dam and threatening the longevity of prehistoric archeological sites.

In response, the cultural resources team at Glen Canyon National Recreation Area has employed aerial photogrammetry as a means of monitoring terrain change. Photogrammetry, or the science of generating 3D spatial data from overlapping 2D photographs, is a cost-effective method of data collection that produces high-resolution orthophotos and elevation models. For large sites, archeologists at Glen Canyon utilized a 5.5-foot helium balloon mapping kit supplied by Public Lab to collect data. A camera was attached to the balloon and pointed toward the ground, and then the area was transected, with the camera continuously shooting. For a two-acre site, approximately 1,000 photographs were captured over the course of one hour. This method produced imagery with a ground sampling distance of 5 centimeters per pixel. Such high-resolution data has greatly enhanced the resource manager's ability to monitor the volume and location of sediment gained or lost at archeological sites.

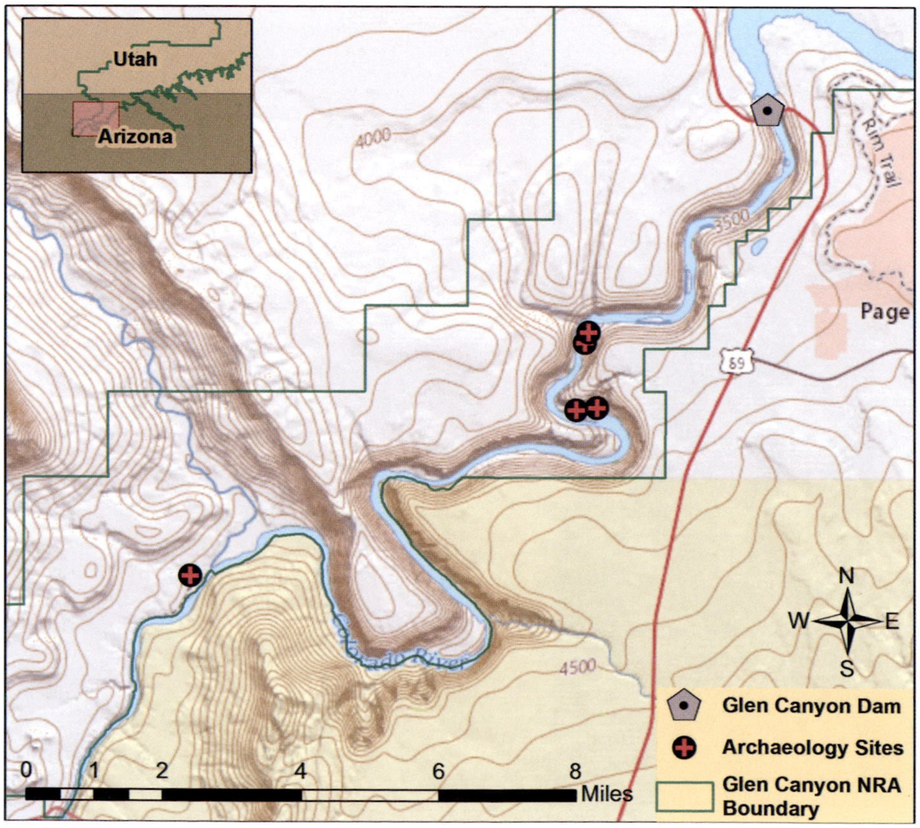

This map shows archeological study site locations.

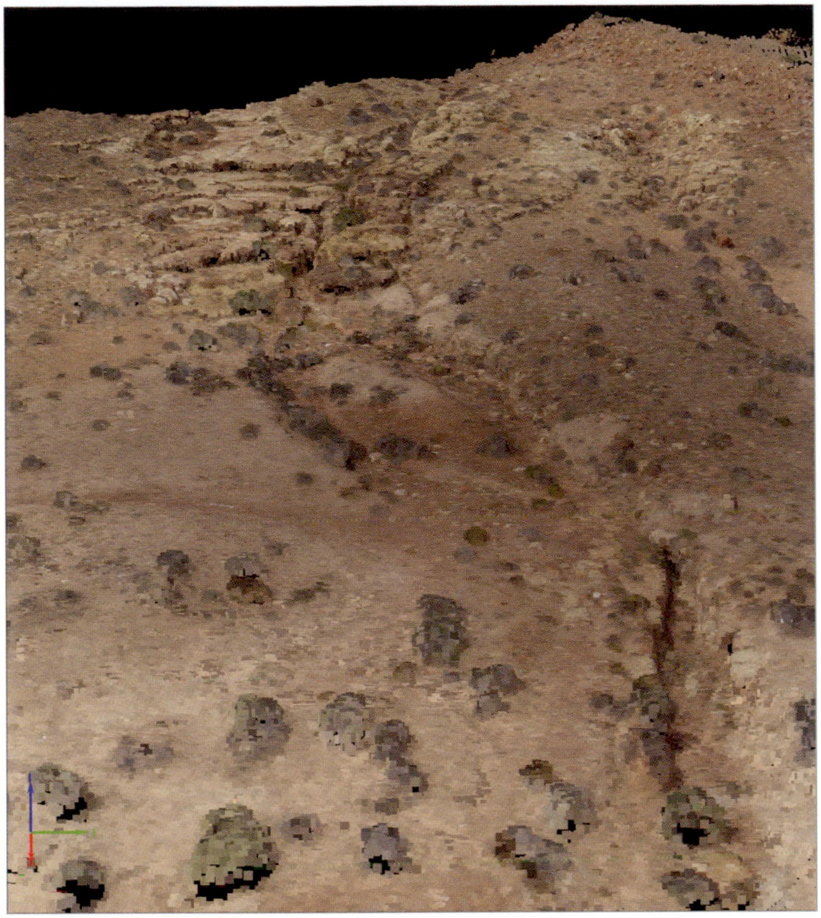

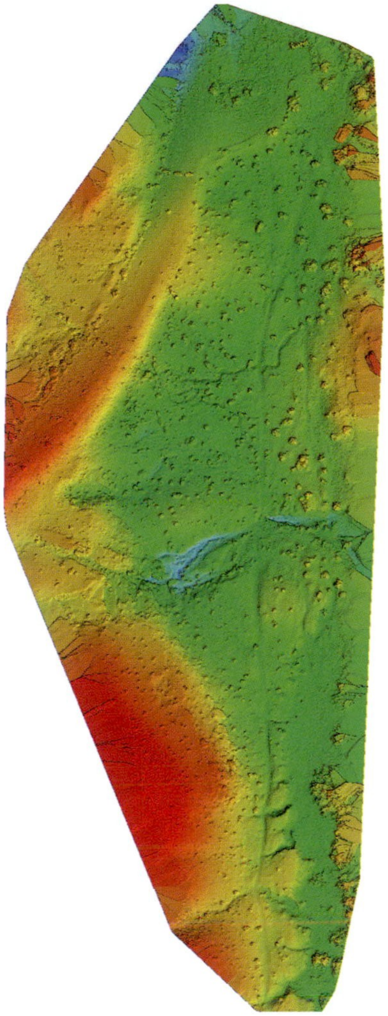

(*Above*) An example of a digital surface model. (*Right*) An example of orthoimagery.

(*Top left*) A 3D rendering of an arroyo. (*Left*) NPS staff prepare a helium balloon to collect data. NPS photo/Vanessa Glynn-Linaris. Data sources: NPS, USGS.

PREDICTING CULTURAL RESOURCE PROBABILITY AT GLEN CANYON NATIONAL RECREATION AREA

JERED HANSEN AND MARK NEBEL, GRAND CANYON NATIONAL PARK

These maps represent a series of predictive spatial models of cultural resource probability at Glen Canyon National Recreation Area (GLCA). GLCA comprises an area of approximately 1.25 million acres in Arizona and Utah containing evidence of multiple and diverse prehistoric cultures spanning at least 10,000 years. With a vast landscape and areas submerged below Lake Powell, large portions of GLCA have never been surveyed for cultural resources. Known and undiscovered cultural resources are potentially subject to irreparable damage along roads as well as receding Lake Powell shorelines that are accessible to off-road vehicle use.

ArcGIS and the statistical regression modeling method Random Forests were used to develop spatially explicit predictive spatial models from cultural resource site (dependent variable) data and a series of independent variables. Random Forests leverages classification and regression tree (CART) modeling to discover correlations between dependent and independent variables to, in this case, predict areas with the greatest and least potential for containing cultural resources.

Modeling challenges included a diverse cultural history, cultural resource site data inconsistencies, highly variable physiographic environments, and limited available model variable data. This work demonstrates how classifying sites based on their inferred usage as habitation, rock art, or scatter and dividing the GLCA landscape into more homogenous uplands, midlands, and lowlands regions significantly improves model performance. These models and maps are used by GLCA resource managers to inform their Off-Road Vehicle Management Plan and accompanying Cultural Resources Stewardship Plan and provide a basis for identifying and prioritizing the inventory and protection of cultural resource locations.

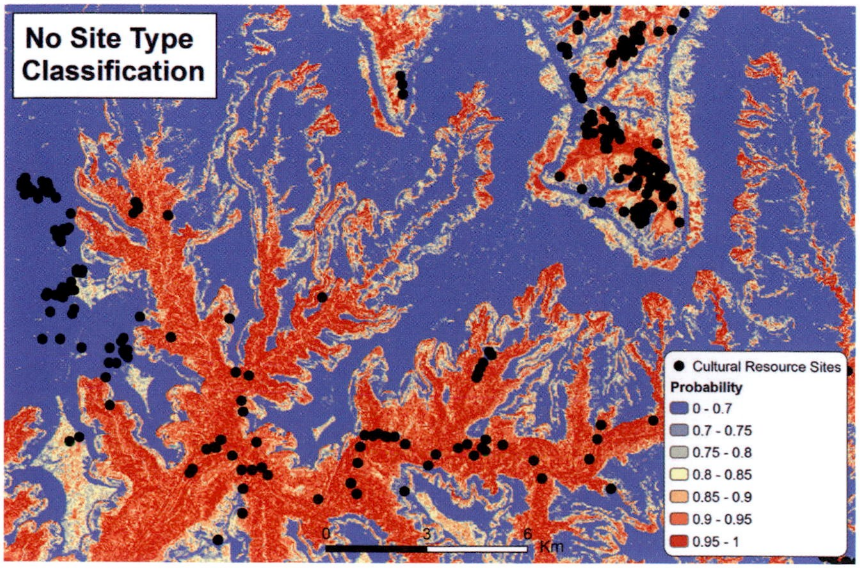

This map shows cultural resource probability for any site type.

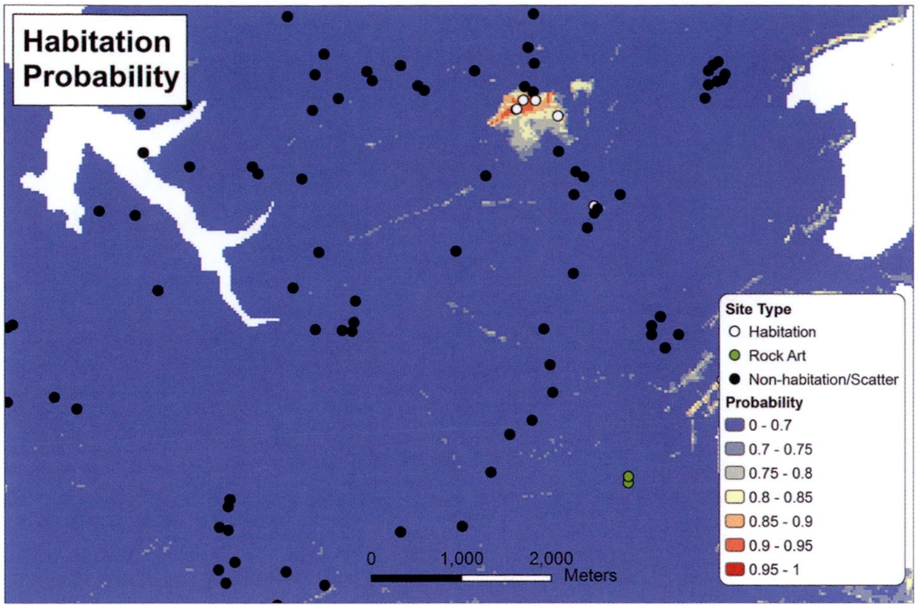

This map shows cultural resource probability for habitation sites.

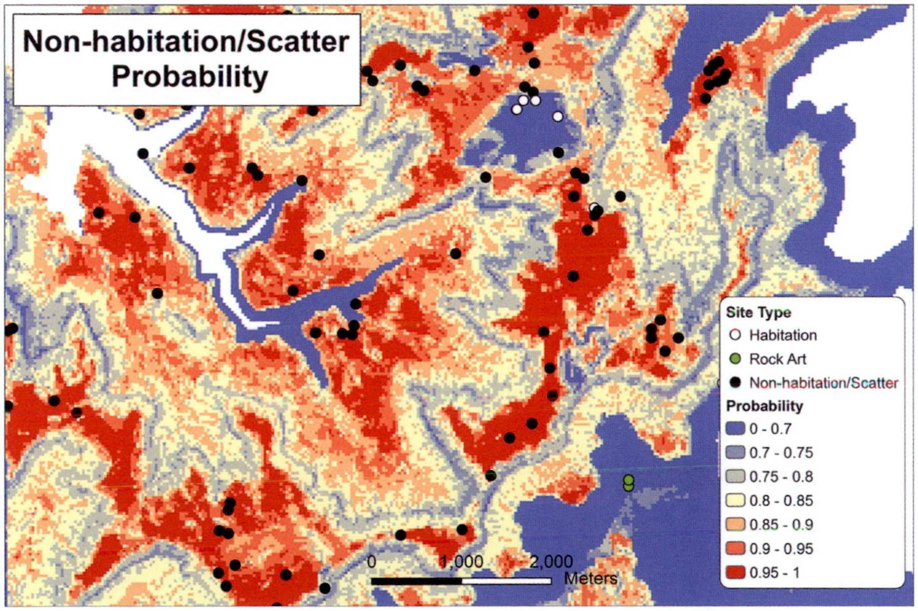

This map shows cultural resource probability for non-habitation sites.

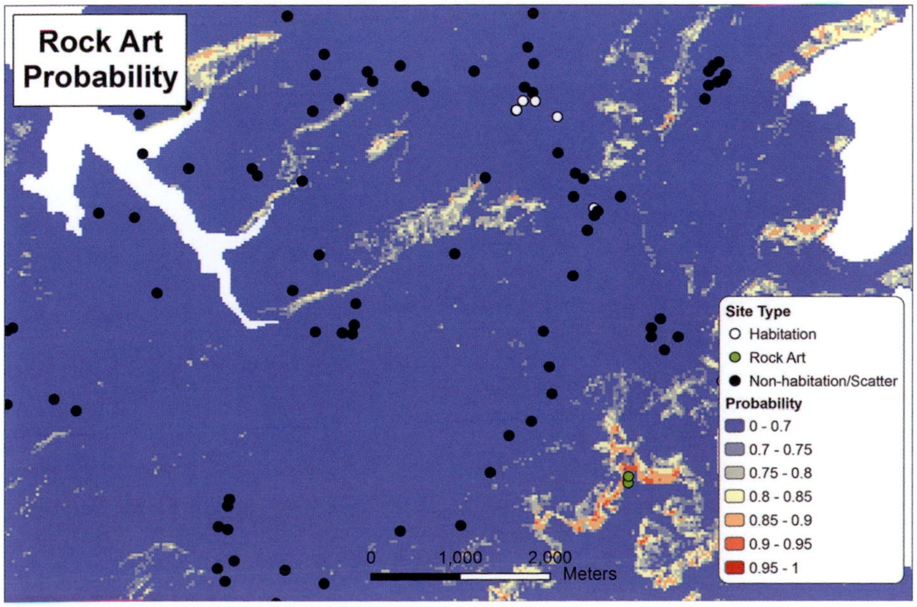

This map shows cultural resource probability for rock art sites. Data sources: NPS Cultural Resource Probability Models Dataset, NPS Archeological Sites Management Information System (ASMIS) Dataset.

SHARING THE HISTORY AND EVOLUTION OF THE ELKMONT HISTORIC DISTRICT IN GREAT SMOKY MOUNTAINS NATIONAL PARK

KATELYN HILLMEYER AND RAVI VENKATARAMAN, GREAT SMOKY MOUNTAINS NATIONAL PARK

Great Smoky Mountains National Park in the Appalachian Mountains of North Carolina and Tennessee is well known for hiking, natural resources, and abundance of historic structures and cultural landscapes. In the heart of the park lies Elkmont, a once-bustling timber camp turned tourist destination. After the creation of the park in 1934, life in Elkmont changed. The town of Elkmont, Tennessee, became the site of a Civilian Conservation Corps (CCC) camp and later a campground. Interest in the resort cabins found in the Appalachian Club and Wonderland areas prompted a preservation movement, and Elkmont was listed as a National Historic District in 1994. In recent years, the park began to restore several of the resort-era buildings and to provide interpretive opportunities to explore the once-lively community. Through the dynamic use of ArcGIS, the park created a visual guide to the everchanging story of Elkmont. It uses maps created with the park's Cultural Resource GIS dataset, and multimedia from the park's archives to tell about the area's rich history and share information about on-going restoration work. By combining storytelling and geography, the park placed historical items from the museum collections in time and space. The visualizations added a clearer caption of the past, present, and future of the Elkmont area.

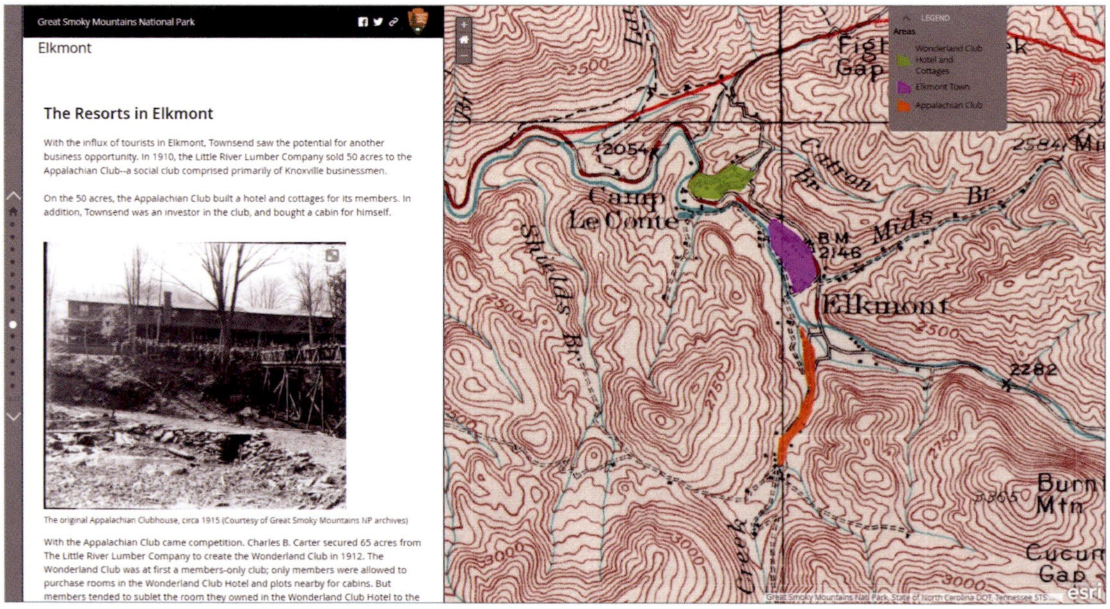
Great Smoky Mountains National Park
Elkmont
The Resorts in Elkmont
With the influx of tourists in Elkmont, Townsend saw the potential for another business opportunity. In 1910, the Little River Lumber Company sold 50 acres to the Appalachian Club--a social club comprised primarily of Knoxville businessmen.
On the 50 acres, the Appalachian Club built a hotel and cottages for its members. In addition, Townsend was an investor in the club, and bought a cabin for himself.
The original Appalachian Clubhouse, circa 1915 (Courtesy of Great Smoky Mountains NP archives)
With the Appalachian Club came competition. Charles B. Carter secured 65 acres from The Little River Lumber Company to create the Wonderland Club in 1912. The Wonderland Club was at first a members-only club; only members were allowed to purchase rooms in the Wonderland Club Hotel and plots nearby for cabins. But members tended to sublet the room they owned in the Wonderland Club Hotel to the
LEGEND
Areas
Wonderland Club Hotel and Cottages
Elkmont Town
Appalachian Club
Camp Le Conte
Elkmont
2054
2146
2282
2584
Burnt Mtn
esri

The map shows various resorts in the Elkmont Historic District.

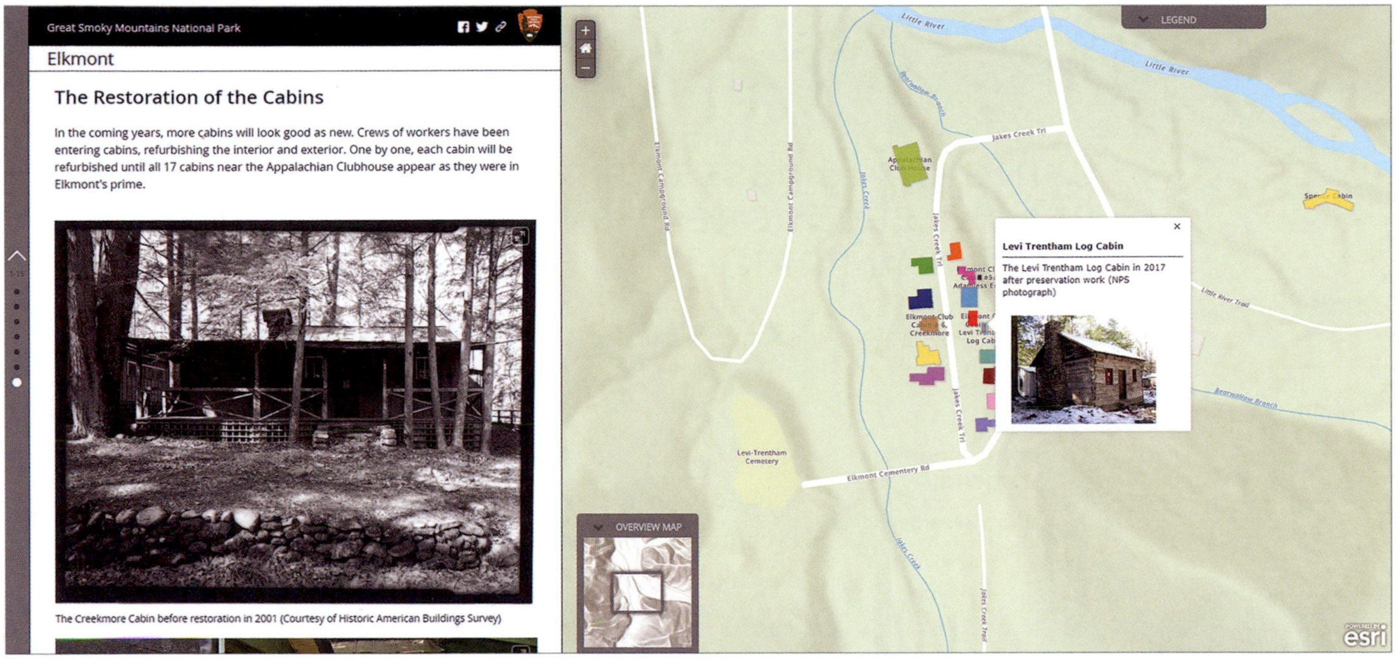
Great Smoky Mountains National Park
Elkmont
The Restoration of the Cabins
In the coming years, more cabins will look good as new. Crews of workers have been entering cabins, refurbishing the interior and exterior. One by one, each cabin will be refurbished until all 17 cabins near the Appalachian Clubhouse appear as they were in Elkmont's prime.
The Creekmore Cabin before restoration in 2001 (Courtesy of Historic American Buildings Survey)
LEGEND
Little River
Jakes Creek Trl
Elkmont Campground Rd
Levi Trentham Log Cabin
The Levi Trentham Log Cabin in 2017 after preservation work (NPS photograph)
Little River Trail
Levi-Trentham Cemetery
Elkmont Cemetery Rd
OVERVIEW MAP
esri

This map used ArcGIS to display the efforts to restore cabins.

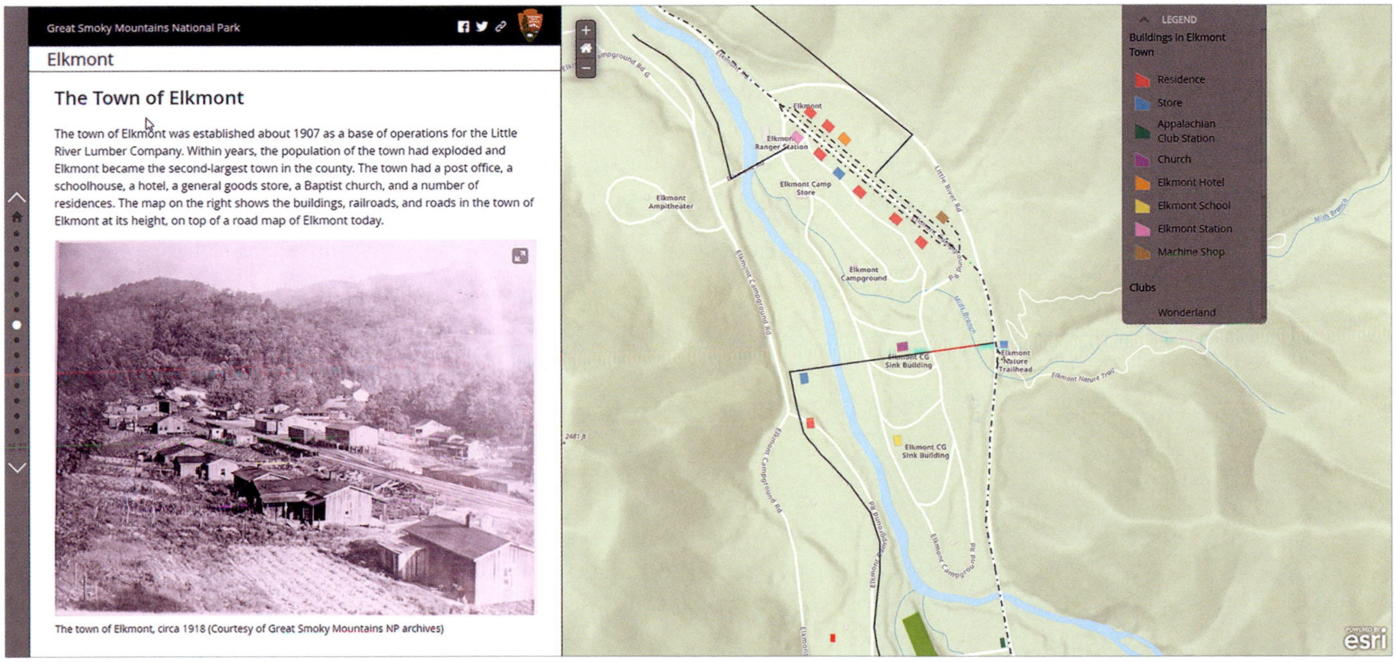
Great Smoky Mountains National Park
Elkmont
The Town of Elkmont
The town of Elkmont was established about 1907 as a base of operations for the Little River Lumber Company. Within years, the population of the town had exploded and Elkmont became the second-largest town in the county. The town had a post office, a schoolhouse, a hotel, a general goods store, a Baptist church, and a number of residences. The map on the right shows the buildings, railroads, and roads in the town of Elkmont at its height, on top of a road map of Elkmont today.
The town of Elkmont, circa 1918 (Courtesy of Great Smoky Mountains NP archives)
LEGEND
Buildings in Elkmont Town
Residence
Store
Appalachian Club Station
Church
Elkmont Hotel
Elkmont School
Elkmont Station
Machine Shop
Clubs
Wonderland
esri

The map shows the town of Elkmont, Tennessee.

Great Smoky Mountains National Park
Elkmont
Tourism in Elkmont
CONDENSED TIME TABLE.

(*Left*) The Little River Railroad Company provided early tourism in Elkmont. (*Right*) A historic preservation crew member paints a structure.

Data sources: NPS Elkmont Club Areas NPS map notes—cabins, buildings in Elkmont Town, clubs, railroad (before 1925), roads, Parton_Store, Esri World Topographic Map Basemap.

USING GIS TO ASSESS FIRE IMPACTS ON CULTURAL RESOURCES AT GREAT SMOKY MOUNTAINS NATIONAL PARK

KATELYN HILLMEYER AND HEATH BAILEY, GREAT SMOKY MOUNTAINS NATIONAL PARK

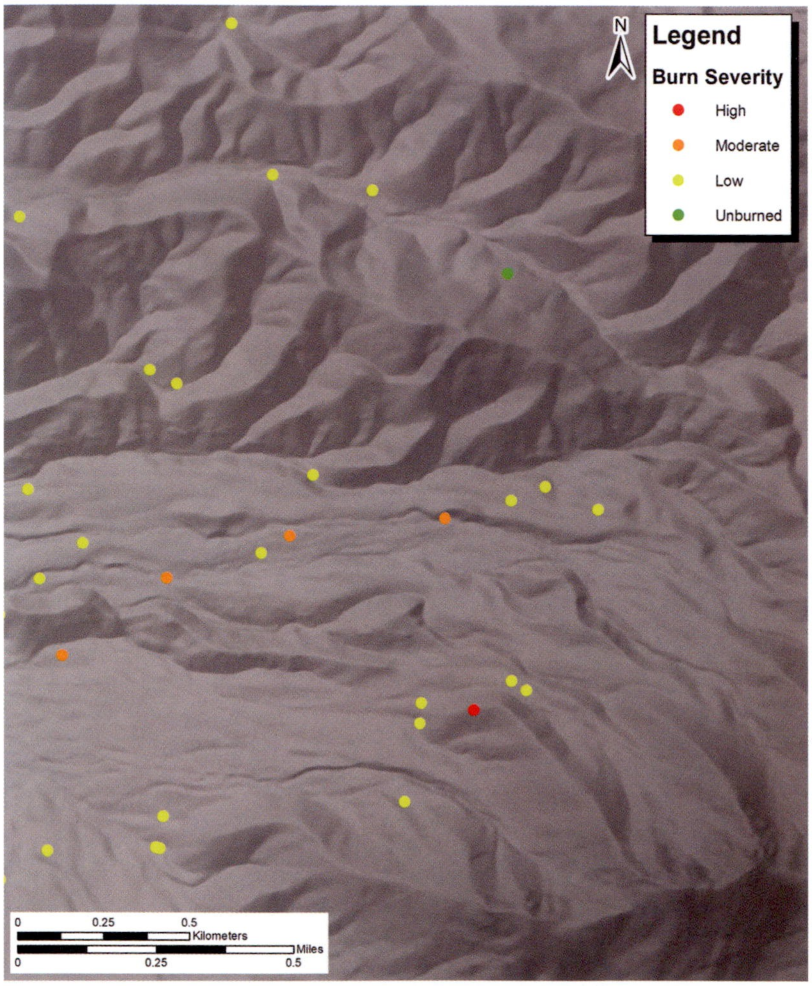

This map shows cultural resource burn severity.

In Great Smoky Mountains National Park, the effects of climate change, fire, and normal erosion have a major impact on cultural resources. In the aftermath of a recent wildland fire, the Cultural Resources division at the park utilized GIS in work to assess the condition of cultural sites within the burn perimeter, complete condition assessment documentation, restore or rehab affected sites, and record new site information to monitor further impacts. Just days after fire containment, burn severity maps were produced to create a predictive model of the condition of known cultural resources within the burn area.

These predictive models help focus work and allow staff to concentrate first on sites most likely to have been severely impacted. In areas usually heavily vegetated, fire also removes the vegetation and helps uncover site features that may have previously been undocumented. Using GPS and GIS, these features were located, documented, and mapped. Adding these newly documented resources to the park's larger Cultural GIS database allows for continued monitoring of their post-fire condition and study of how they interact with other features in the cultural landscape. By utilizing GIS, the work in the field can proceed more efficiently and accurately. These tools have allowed us to better document and preserve park history.

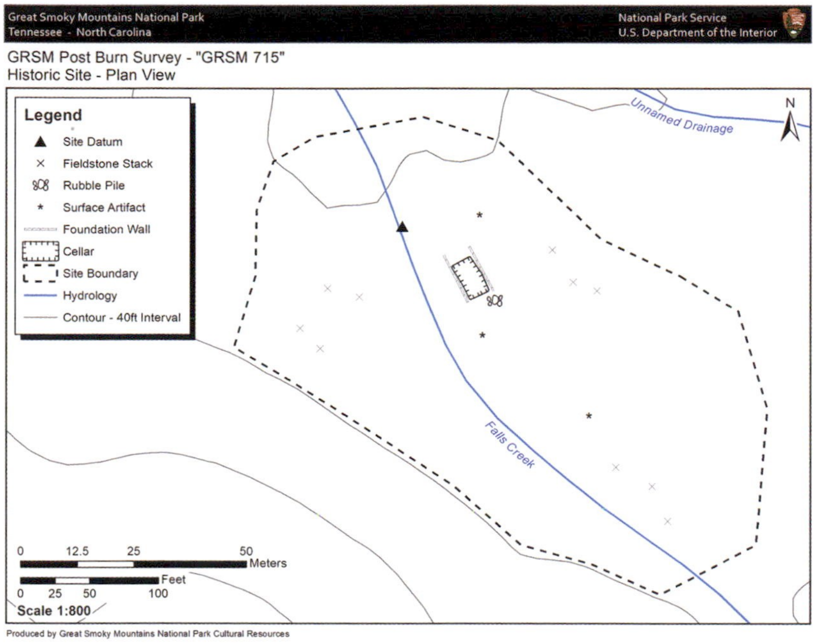

Clockwise from top left:

The plan view map shows the site recorded during the post-burn survey.

A burned stone house. NPS photo/Heath Bailey. Data sources: NPS Cultural Resource Site and Feature Dataset, NPS Burn Severity Dataset, NPS GRSM Park Hill shade, USGS GSRM park hydrology.

A burned cemetery. NPS photo/Heath Bailey.

ARCHEOLOGICAL SITE CONDITION ASSESSMENTS AT OZARK NATIONAL SCENIC RIVERWAYS

AMANDA RENNER, NPS MIDWEST ARCHEOLOGICAL CENTER
ALLISON YOUNG, OZARK NATIONAL SCENIC RIVERWAYS

In May 2017, Ozark National Scenic Riverways in southeastern Missouri experienced extensive flooding. Following the initial disaster response, archeologists needed to rapidly assess the current condition of known archeological sites within the park boundary. Since NPS staff were already using ArcGIS Collector in their archeological site monitoring program, they determined they could apply the same workflow to assess site conditions in the aftermath of the flood.

Due to the sensitive nature of the archeological site location information, the feature services and web map were hosted on the internal NPS portal in ArcGIS Online. A new geodatabase for data collection was created with fields that mirrored the standard archeological site condition assessment form used in the region. Staff used the maps in Collector to navigate to a known site location, update the associated condition information, and take photos. The information synced automatically and was then used to update the condition for each site in the NPS Archeological Sites Management Information System database.

This data collection workflow helped NPS staff quickly and efficiently record important information during disaster response. Archeologists at Ozark National Scenic Riverways continue to use the Collector application for their archeological site monitoring program and for conducting site condition assessments as they work to protect these special resources. Collector will continue to be a useful go-to tool in future disaster response situations that require rapid response at the park because of its ability to provide key information for park managers.

Facing page:
(*Top*) This map shows an archeological site condition assessment for Ozark National Scenic Riverways hosted on the NPS portal in ArcGIS Online. (*Bottom*) This map shows an archeological site condition assessment for Ozark National Scenic Riverways hosted on the NPS portal in ArcGIS Online, including Culpepper Cemetery. Data sources: Esri, Digital Globe, Earthstar Geographies; CNES/Airbus DS, GeoEye, USDA FSA, USGS, Getmapping, Aerogrid, IGN, IGP, swisstopo, and the GIS user community, NPS Lands Resources Division, NPS-Midwest Archeological Center, World Imagery (for export) map service. This layer presents satellite imagery for the world and high-resolution imagery for many locations worldwide. This layer is designed to support export of base map tiles for offline use.

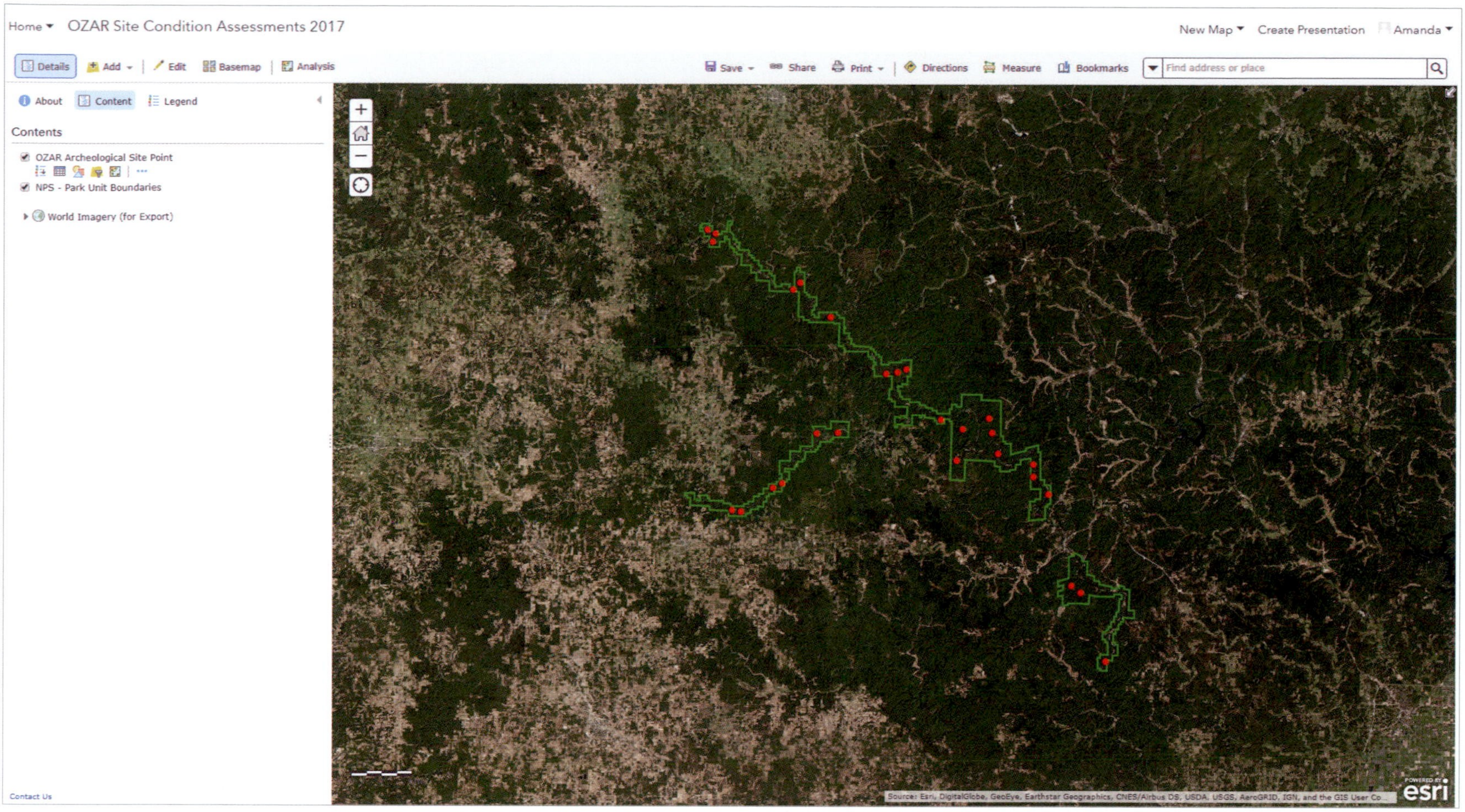
Home ▾ OZAR Site Condition Assessments 2017
New Map ▾ Create Presentation Amanda ▾
Details Add ▾ Edit Basemap Analysis
Save ▾ Share Print ▾ Directions Measure Bookmarks
Find address or place
About Content Legend
Contents
OZAR Archeological Site Point
NPS - Park Unit Boundaries
World Imagery (for Export)
Contact Us
Source: Esri, DigitalGlobe, GeoEye, Earthstar Geographics, CNES/Airbus DS, USDA, USGS, AeroGRID, IGN, and the GIS User Co...
esri

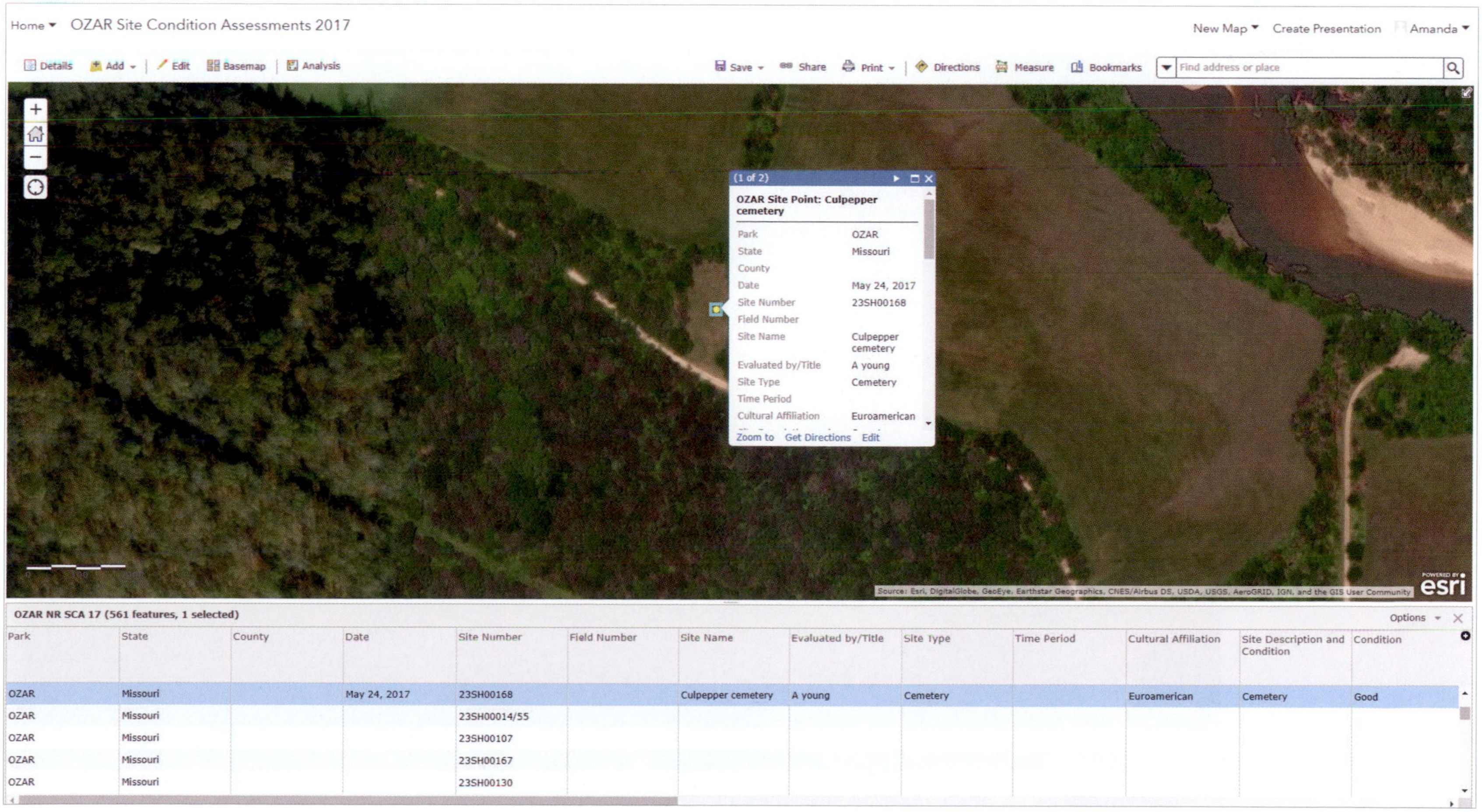
Home ▾ OZAR Site Condition Assessments 2017
New Map ▾ Create Presentation Amanda ▾
Details Add ▾ Edit Basemap Analysis
Save ▾ Share Print ▾ Directions Measure Bookmarks
Find address or place
(1 of 2)
OZAR Site Point: Culpepper cemetery
Park OZAR
State Missouri
County
Date May 24, 2017
Site Number 23SH00168
Field Number
Site Name Culpepper cemetery
Evaluated by/Title A young
Site Type Cemetery
Time Period
Cultural Affiliation Euroamerican
Zoom to Get Directions Edit
Source: Esri, DigitalGlobe, GeoEye, Earthstar Geographics, CNES/Airbus DS, USDA, USGS, AeroGRID, IGN, and the GIS User Community
esri
OZAR NR SCA 17 (561 features, 1 selected)
Options ▾
Park State County Date Site Number Field Number Site Name Evaluated by/Title Site Type Time Period Cultural Affiliation Site Description and Condition Condition
OZAR Missouri May 24, 2017 23SH00168 Culpepper cemetery A young Cemetery Euroamerican Cemetery Good
OZAR Missouri 23SH00014/55
OZAR Missouri 23SH00107
OZAR Missouri 23SH00167
OZAR Missouri 23SH00130

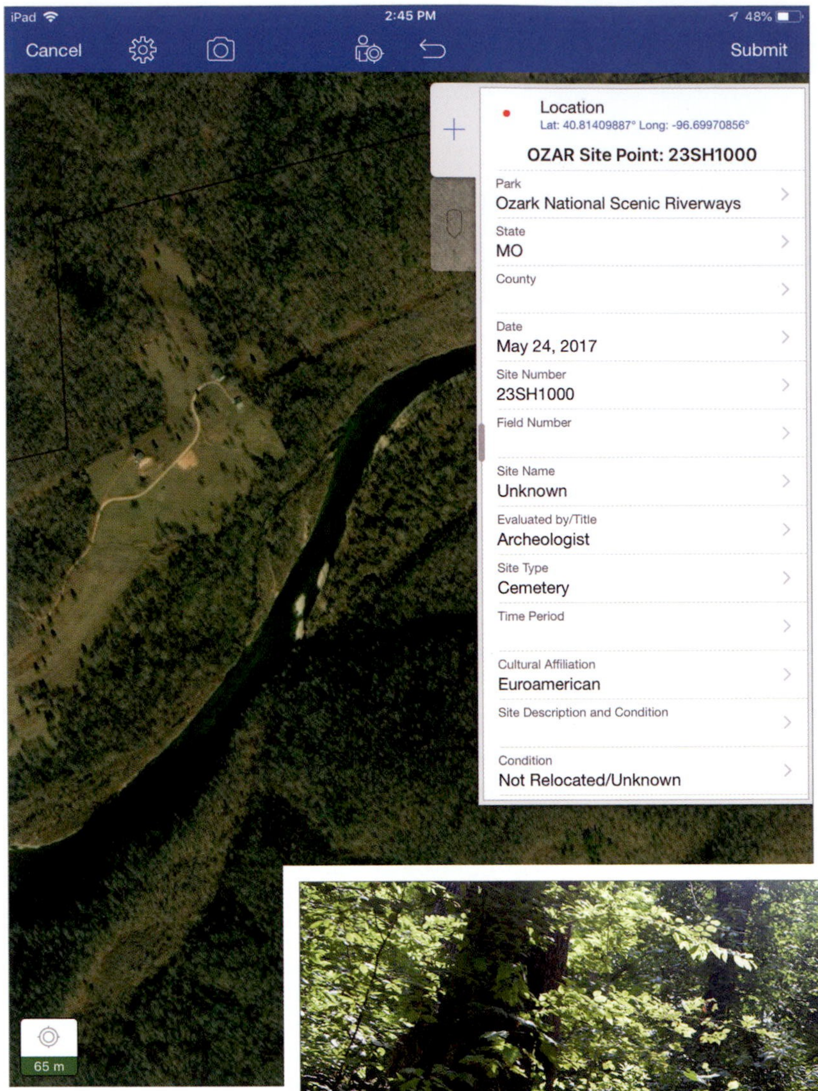

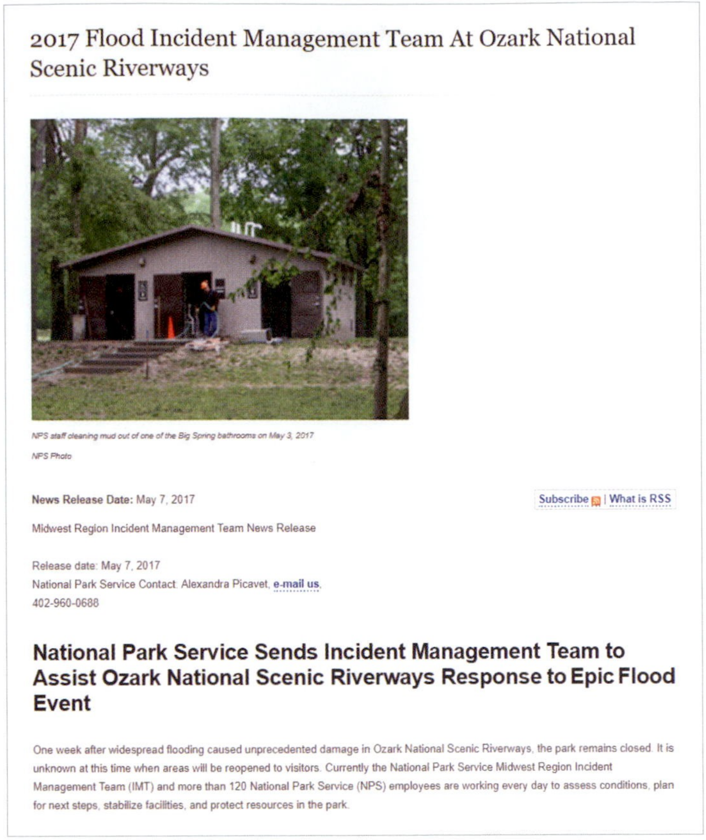

2017 Flood Incident Management Team At Ozark National Scenic Riverways

NPS staff cleaning mud out of one of the Big Spring bathrooms on May 3, 2017

NPS Photo

News Release Date: May 7, 2017

Subscribe | What is RSS

Midwest Region Incident Management Team News Release

Release date: May 7, 2017
National Park Service Contact: Alexandra Picavet, e-mail us,
402-960-0688

National Park Service Sends Incident Management Team to Assist Ozark National Scenic Riverways Response to Epic Flood Event

One week after widespread flooding caused unprecedented damage in Ozark National Scenic Riverways, the park remains closed. It is unknown at this time when areas will be reopened to visitors. Currently the National Park Service Midwest Region Incident Management Team (IMT) and more than 120 National Park Service (NPS) employees are working every day to assess conditions, plan for next steps, stabilize facilities, and protect resources in the park.

An NPS archeologist uses Collector to record the condition of a historic cemetery marker impacted by the 2017 flood event at Ozark National Scenic Riverways. NPS photo/Midwest Archeological Center.

Facing page, clockwise from top left:
An archeological site condition assessment data entry form in the Collector mobile application.

A news release details the NPS response to epic flooding in the Ozark National Scenic Riverways.

An NPS archeologist uses Collector to record the condition of an archeological site impacted by erosion caused during a 2017 flood event at Ozark National Scenic Riverways. NPS photo/Midwest Archeological Center.

HISTORIC BATHHOUSE TOUR AT HOT SPRINGS NATIONAL PARK

AMANDA RENNER, ASHLEY BARNETT, AND LAURA BENDER,
NPS MIDWEST ARCHEOLOGICAL CENTER

Bathhouse Row, the largest remaining collection of 20th-century bathhouses in the United States, was designated as a National Historic Landmark on May 28, 1987. Located in downtown Hot Springs, Arkansas, the bathhouses were originally included in the Hot Springs Reservation, established by Congress in 1832 to protect 47 natural hot springs. The Hot Springs Reservation became Hot Springs National Park on March 4, 1921, and now encompasses more than 5,000 acres.

The story of the Historic Bathhouse Tour was created by the Midwest Archeological Center in 2015 to share 360-degree panoramic photos of historic bathhouse interiors, some of which are not open to the public. The panoramic photos were taken in 2013 by Midwest Archeological Center staff as part of an effort, partially funded by the NPS Southeast Archeological Center, to provide web-based access to historical and archeological properties under NPS stewardship to the public through an online web map. In the first year the map was available to the public (2015–2016), it received 2,961 views. Subsequent years show continued interest with 1,805 (2016–2017) and 1,927 (2017–2018) views.

Each tour point shows the location of the featured bathhouse and either a photo from the NPGallery (https://npgallery.nps.gov) or a 360-degree panoramic photo. The 360-degree photos rotate automatically in an image display panel and can be manually rotated by the viewer. Historical bathhouse details are provided in the caption as well as a link to the Hot Springs National Park web page with additional information about each bathhouse.

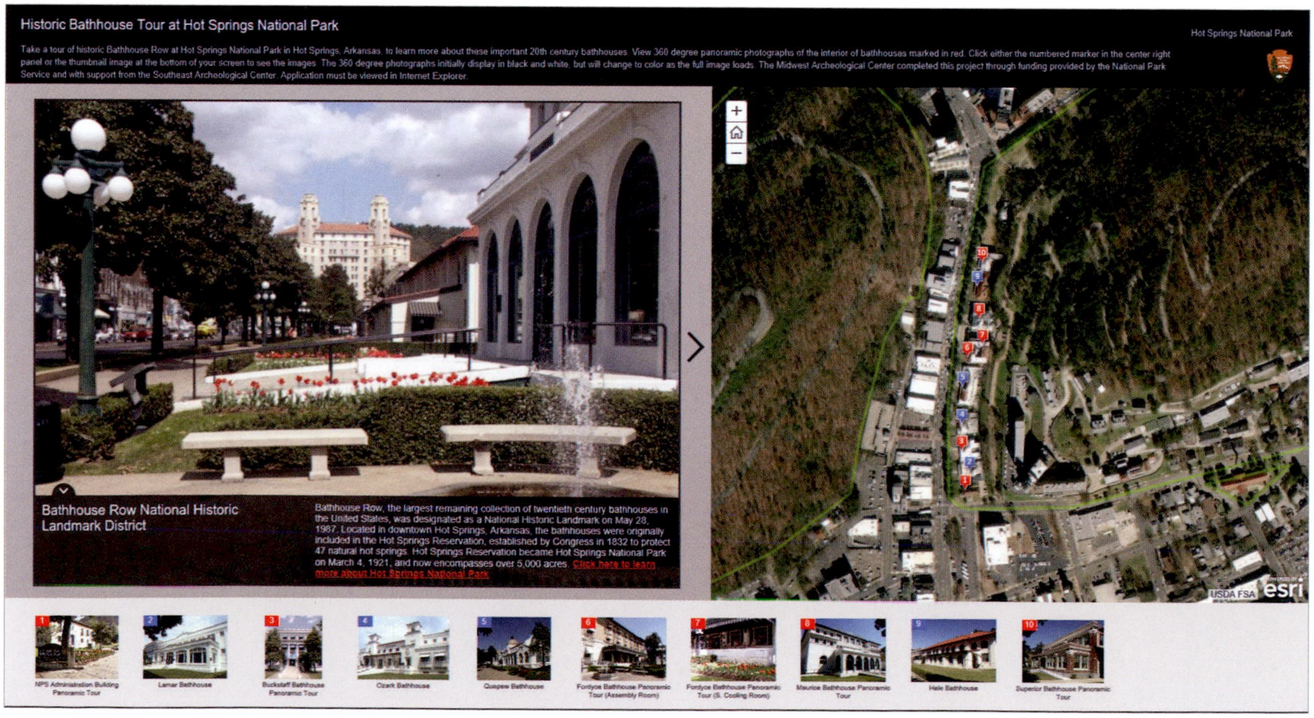
Historic Bathhouse Tour at Hot Springs National Park
Hot Springs National Park
Take a tour of historic Bathhouse Row at Hot Springs National Park in Hot Springs, Arkansas, to learn more about these important 20th century bathhouses. View 360 degree panoramic photographs of the interior of bathhouses marked in red. Click either the numbered marker in the center right panel or the thumbnail image at the bottom of your screen to see the images. The 360 degree photographs initially display in black and white, but will change to color as the full image loads. The Midwest Archeological Center completed this project through funding provided by the National Park Service and with support from the Southeast Archeological Center. Application must be viewed in Internet Explorer.
Bathhouse Row National Historic Landmark District
Bathhouse Row, the largest remaining collection of twentieth century bathhouses in the United States, was designated as a National Historic Landmark on May 28, 1987. Located in downtown Hot Springs, Arkansas, the bathhouses were originally included in the Hot Springs Reservation, established by Congress in 1832 to protect 47 natural hot springs. Hot Springs Reservation became Hot Springs National Park on March 4, 1921, and now encompasses over 5,000 acres. Click here to learn more about Hot Springs National Park
USDA FSA
esri
1 NPS Administration Building Panoramic Tour
2 Lamar Bathhouse
3 Buckstaff Bathhouse Panoramic Tour
4 Ozark Bathhouse
5 Quapaw Bathhouse
6 Fordyce Bathhouse Panoramic Tour (Assembly Room)
7 Fordyce Bathhouse Panoramic Tour (S. Cooling Room)
8 Maurice Bathhouse Panoramic Tour
9 Hale Bathhouse
10 Superior Bathhouse Panoramic Tour

An aerial map shows Hot Springs National Park. Tulips brighten Bathhouse Row as seen from the park historic formal entrance fountain and north past the Maurice Bathhouse and Hale Bath.

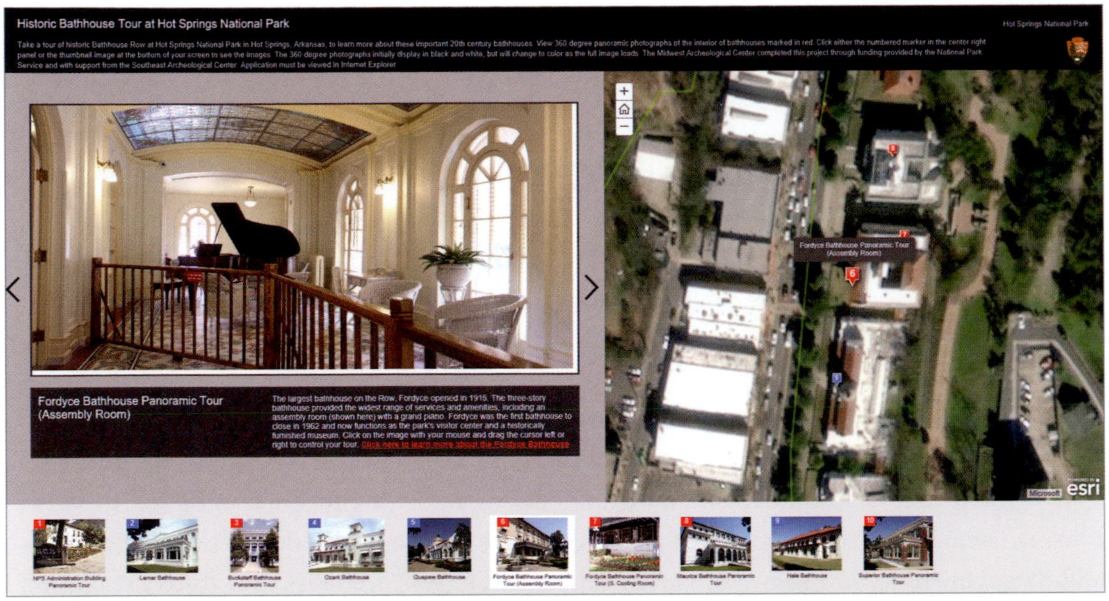
Historic Bathhouse Tour at Hot Springs National Park
Hot Springs National Park
Fordyce Bathhouse Panoramic Tour (Assembly Room)
The largest bathhouse on the Row, Fordyce opened in 1915. The three-story bathhouse provided the widest range of services and amenities, including an assembly room (shown here) with a grand piano. Fordyce was the first bathhouse to close in 1962 and now functions as the park's visitor center and a historically furnished museum. Click on the image with your mouse and drag the cursor left or right to control your tour. Click here to learn more about the Fordyce Bathhouse
esri

An aerial map of Hot Springs National Park and a panoramic photo show the largest bathhouse in the park, Fordyce Bathhouse, which opened in 1915 and was the first to close in 1962. It is now a visitor center and a museum.

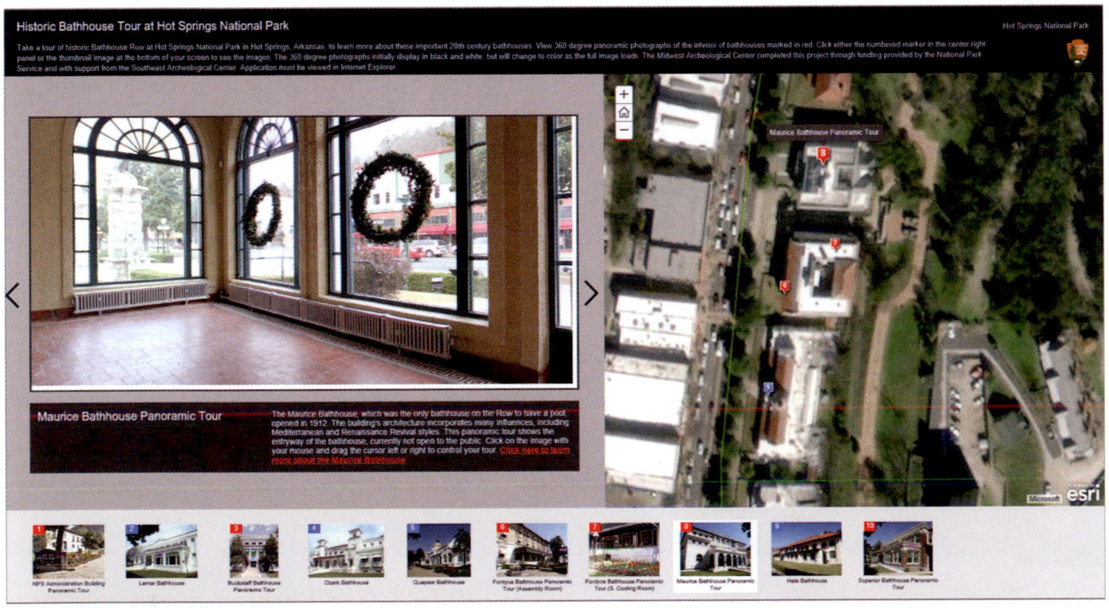
Historic Bathhouse Tour at Hot Springs National Park
Hot Springs National Park
Maurice Bathhouse Panoramic Tour
The Maurice Bathhouse, which was the only bathhouse on the Row to have a pool, opened in 1912. The building's architecture incorporates many influences, including Mediterranean and Renaissance Revival styles. This panoramic tour shows the entryway of the bathhouse, currently not open to the public. Click on the image with your mouse and drag the cursor left or right to control your tour. Click here to learn more about the Maurice Bathhouse
esri

An aerial map of Hot Springs National Park alongside a panoramic photo shows Maurice Bathhouse, which opened in 1912.

HOT SPRINGS
NATIONAL PARK
United States Department of the Interior
National Park Service

(*Left*) An NPS archeologist documents a historic bathhouse at Hot Springs National Park with 360-degree panoramic photography. NPS photo/Midwest Archeological Center. (*Right*) Red tulips in front of the Hot Springs National Park administration building grow next to the park's Bathhouse Row entrance sign. NPS photo/Midwest Archeological Center. Data sources: NPS, Esri.

DOCUMENTED RESOURCES IMPACTED OR THREATENED BY HURRICANES IRMA AND MARIA

JAMES STEIN, NPS CULTURAL RESOURCES GIS PROGRAM

The 2017 Atlantic hurricane season saw a total of 17 named storms, 7 tropical storms, and 10 hurricanes. Three of those hurricanes—Harvey, Irma, and Maria—made landfall on the mainland United States or US territories. In addition to causing tremendous destruction, the storms highlighted the likelihood that similar storms and damage will occur again. Recovery continues, and preparedness for future storms and sea level rise will take many forms. The Heritage Documentation Program focuses on documenting the physical places, especially those related to architecture, engineering, and cultural landscapes that illustrate American history. Because of their geographic location, resources in Florida, Puerto Rico, and the US Virgin Islands are especially vulnerable to damage from wind and storm surge. This map includes a selection of damaged or threatened cultural resources in those three places.

An ArcGIS application allowed the use of interactive maps to illustrate the tracks of the hurricanes and the locations of the documented resources in relation to the storms. In addition, the resource is fully illustrated with the use of text and photographs to document the importance, threats, and damage as well as linking to additional sources of information. The resources highlighted in this story were documented by the Historic American Buildings Survey, Historic American Engineering Record, and Historic American Landscapes Survey.

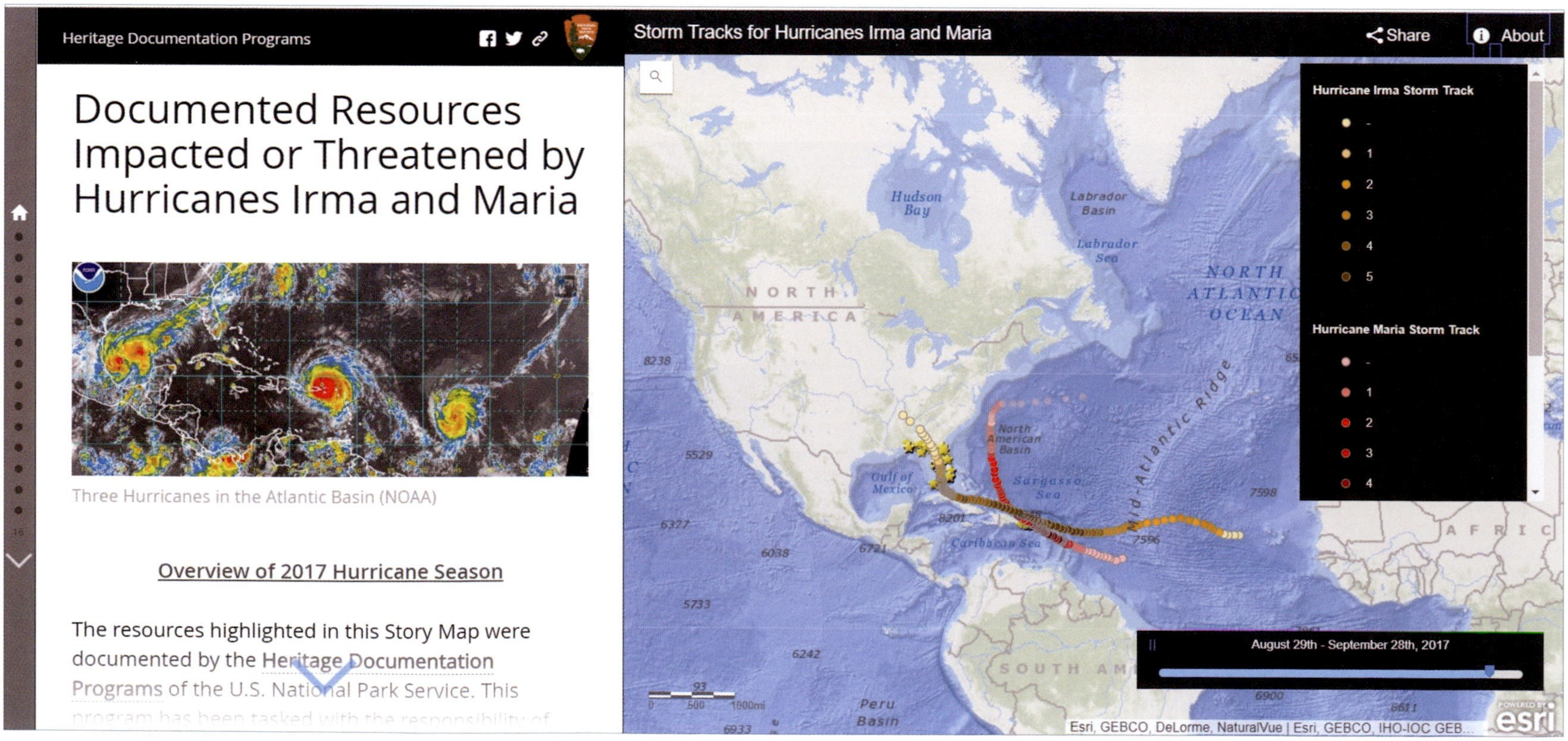

This map shows documented resources impacted or threatened by Hurricanes Irma and Maria.

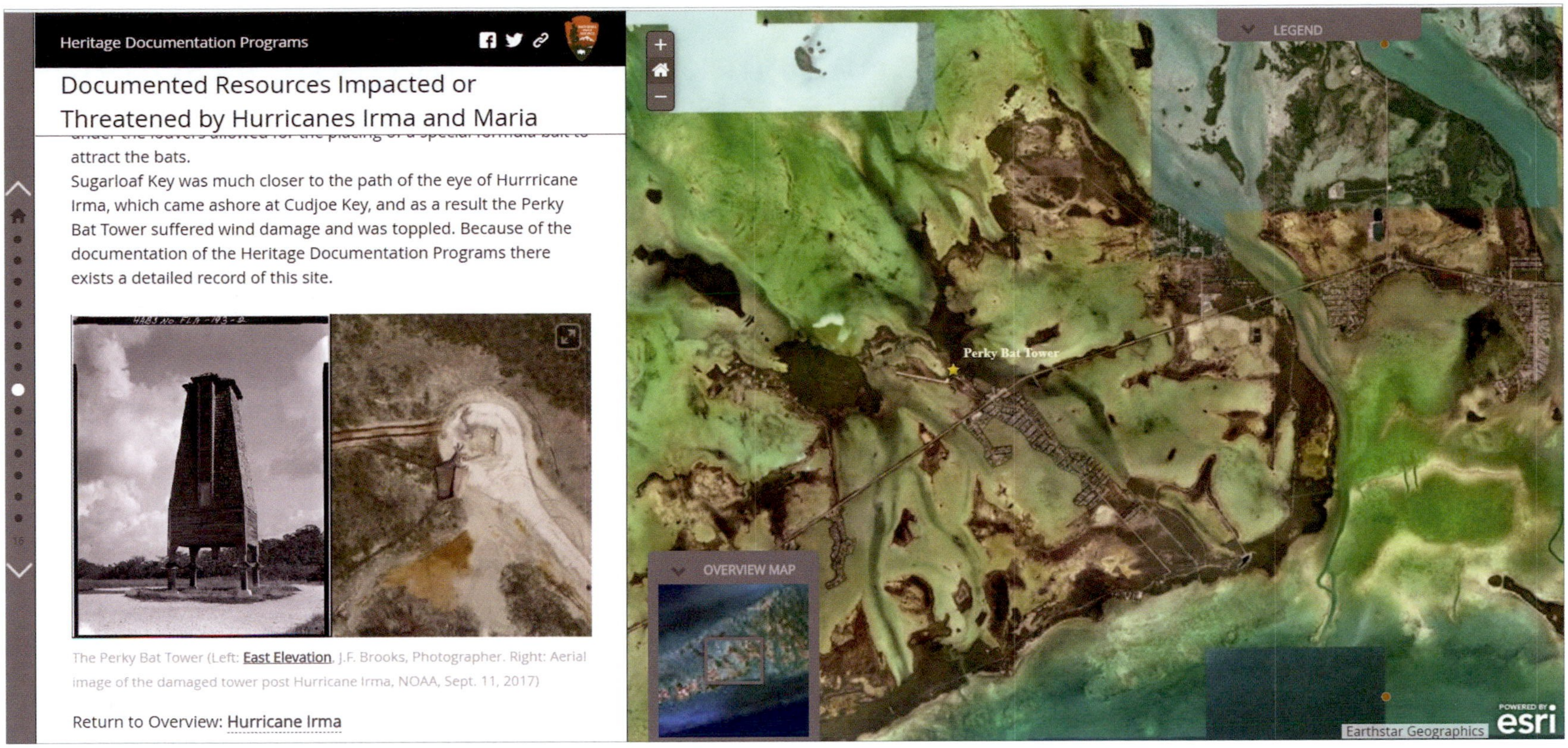

This image shows the Perky Bat Tower as an example of the resources impacted or threatened by Hurricanes Irma and Maria.

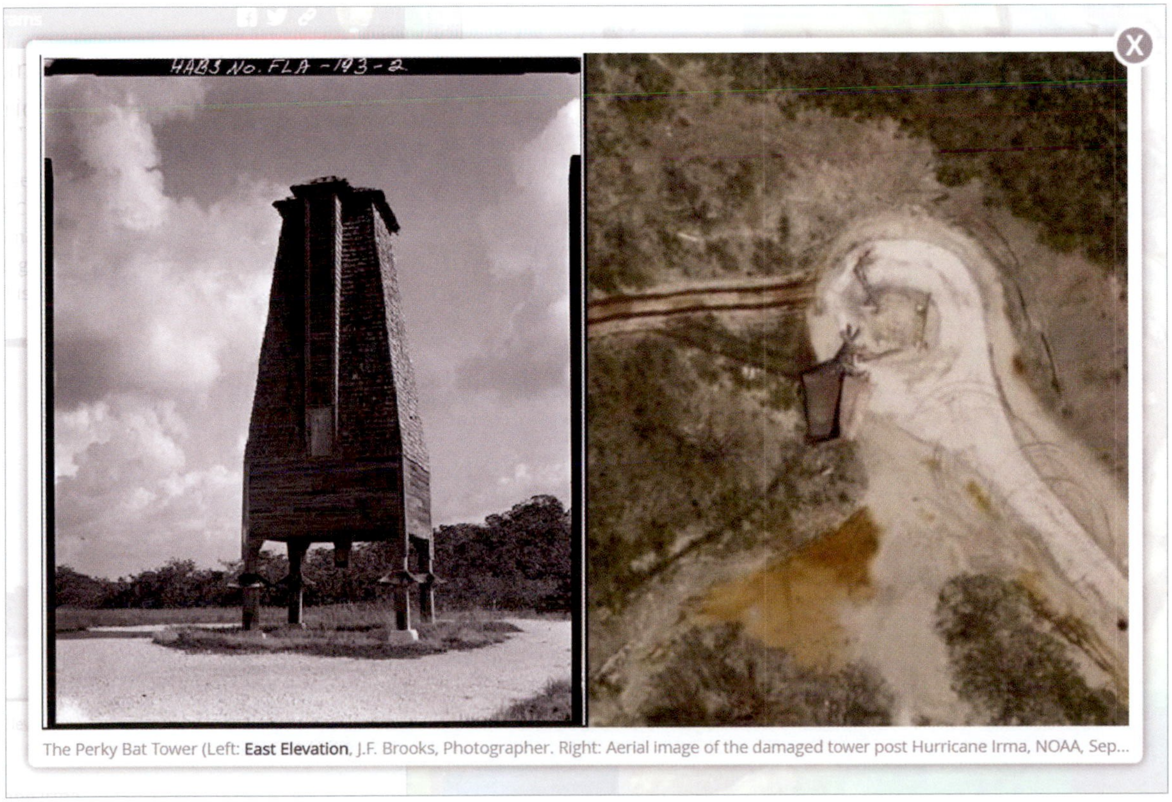

HABS No. FLA-193-2
The Perky Bat Tower (Left: East Elevation, J.F. Brooks, Photographer. Right: Aerial image of the damaged tower post Hurricane Irma, NOAA, Sep...

A pop-up window within the story shows the Perky Bat Tower, a historic site in Florida.

Data sources, pages 140–141: National Oceanic and Atmospheric Administration (NOAA) (Hurricane Irma/Maria Tracks), NPS (Heritage Documentation Program Sites), Esri, Digital Globe, GeoEye, Earthstar Geographics, CNES/Airbus DS, USDA, USGS, AeroGRID, IGN, and the GIS user community (World Imagery Basemap); Historic American Buildings Survey/Historic American Engineering Record/Historic American Landscapes photo, NOAA aerial image, Esri, Garmin, GEBCO, NOAA, NGDC, and other contributors (World Ocean Basemap); Esri, GEBCO, NOAA, National Geographic, Garmin, HERE, Geonames.org, and other contributors (World Ocean Reference); Historic American Buildings Survey/Historic American Engineering Record/Historic American Landscapes photo.

NATIONAL PARK SERVICE CULTURAL RESOURCES ENTERPRISE GIS

MATT STUTTS, NPS CULTURAL RESOURCES GIS PROGRAM

The NPS Cultural Resource Geographic Information Systems Facility (CRGIS)—in collaboration with other NPS programs, regional cultural resource and GIS staff, and the Resource Information Services Division (RISD)—has been developing an enterprise cultural resource GIS dataset. The purpose of the project is to provide an authoritative geographic dataset delineating features included in the national NPS cultural resource inventory databases, ensuring they comply with the NPS cultural resource spatial data transfer standards. The objective is to deliver better access to cultural resource information for inventory, planning, compliance, disaster preparedness, and resource stewardship through the integration of the existing cultural resource databases. While protecting sensitive information, the data will be made available through internet applications in conjunction with other park and regional GIS data and tools.

Coordinating with a network of regional data editors and subject matter experts who create the spatial data, CRGIS leads the project, provides technical support, develops tools to access and analyze the data, and works with RISD to maintain IT infrastructure. The cultural resource enterprise GIS dataset currently includes the National Register of Historic Places, the Cultural Landscapes Inventory (CLI), the List of Classified Structures (LCS), and Heritage Documentation Programs (HABS/HAER/HALS) as well as historic preservation tax incentives and grants. At this time, the GIS data is available only to NPS employees, and sensitive data has been restricted, available only by requests made to the responsible parks, programs, and regions that create the spatial data.

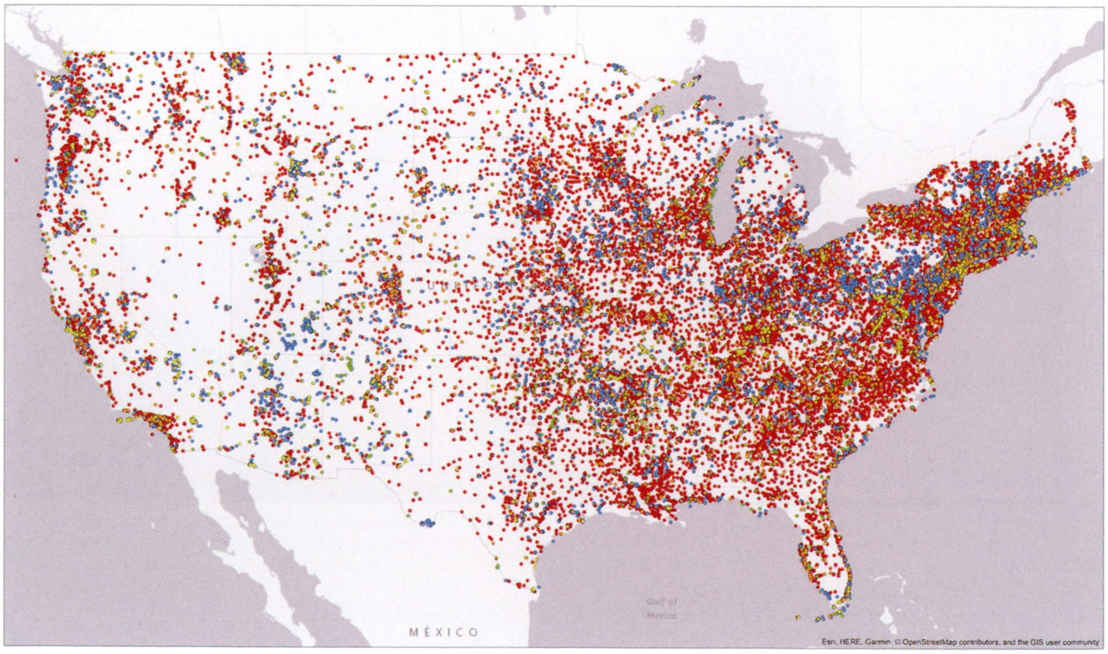

The NPS Cultural Resources (CR) National Dataset depicts locations and distribution of federal, state, local, and private cultural resource holdings inventoried in one or more databases maintained by NPS CR databases within the continental United States.

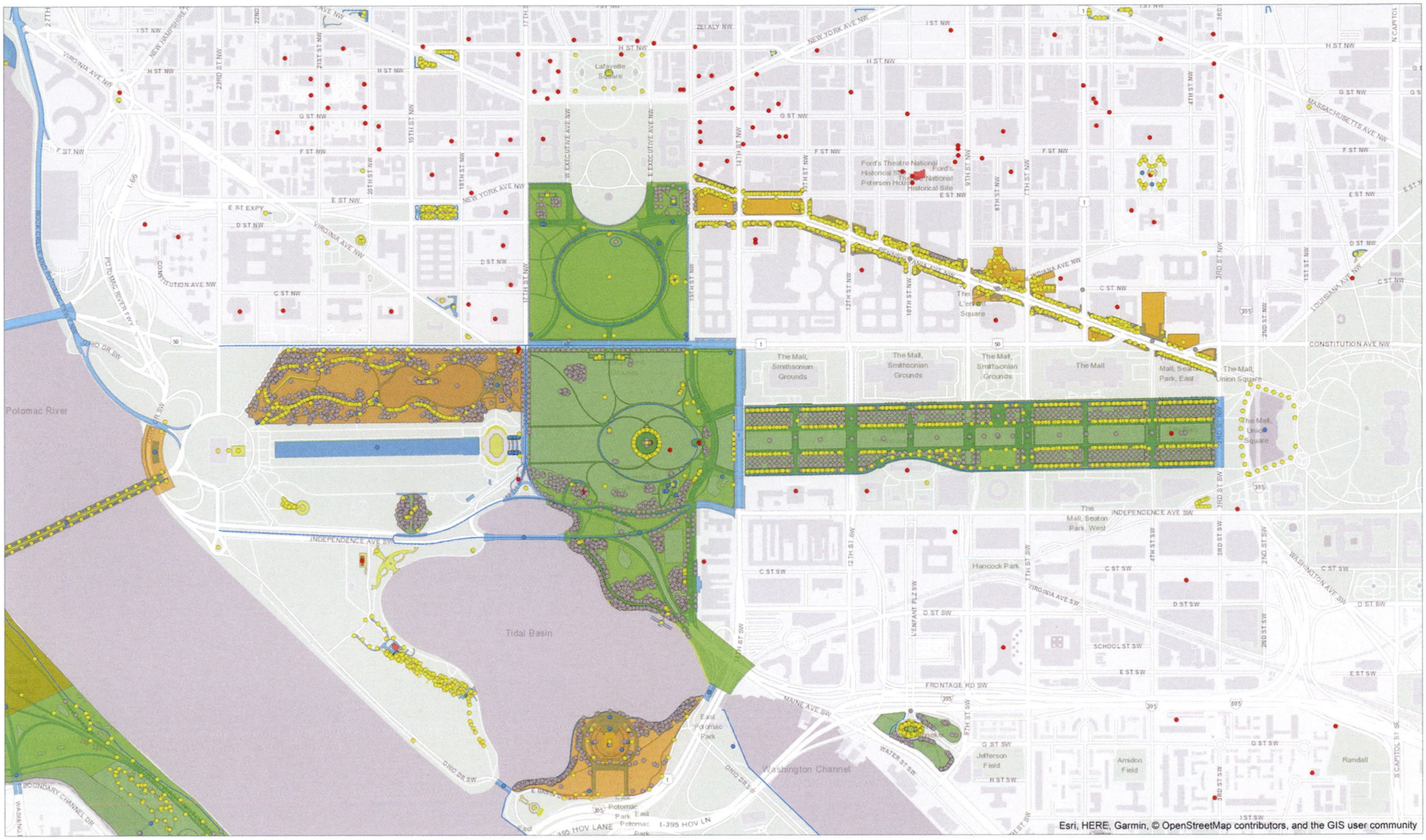

A view of the NPS Cultural Resources National Dataset in proximity to the National Mall in Washington, DC. The dataset contains the inventories of NPS cultural resources programmatic databases such as the National Register of Historic Places, Heritage Documentation Programs, and others. The dataset categorizes resources into five main feature types: buildings, districts, objects, sites, and structures. Data sources, pages 142–143: NPS Cultural Resources Enterprise Dataset, Esri, HERE, Garmin, © OpenStreetMap contributors, and the GIS user community (World Light Gray Canvas Base).

7 FACILITIES, INFRASTRUCTURE, AND TRANSPORTATION

DAWN FOY, NPS FEDERAL LANDS TRANSPORTATION PROGRAM

ESTABLISHING FACILITIES, INFRASTRUCTURE, AND TRANSPORTATION

GIS data is used by a wide range of National Park Service (NPS) employees and partners (including Federal Land Highways, other federal land management agencies, counties, cities, and states) in making critical decisions related to maintenance and development of facilities in the NPS.

Visitors and employees depend on a wide variety of facilities to enjoy and manage the 418 parks of the NPS. These facilities are required to meet the mission of the NPS and vary greatly: from a road that leads a visitor to a trail head in Olympic to a visitor center explaining the complex actions of the British and French troops at Fort Necessity to a water treatment plant that provides all potable water at the Grand Canyon.

The NPS Facility Management Mission is to "provide visitor and administrative facilities that are necessary, appropriate, and consistent with the conservation of park resources and values. Facilities will be harmonious with park resources, compatible with natural processes, esthetically pleasing, functional, energy- and water-efficient, cost-effective, universally designed, and as welcoming as possible to all segments of the population." To meet this mission, complex planning and management decisions are required to build and maintain facilities with minimum impact to the parks.

Facility, park, and program managers require multifaceted and varied data to make informed decisions that assure protection of resources, optimize long-term maintenance options, address safety concerns, and provide minimum disruption to the park visitor during the planning of a new facility or repair to existing facilities. Data takes on many forms such as cultural and natural resources, climate change, facility condition, traffic, and elevation, which are often disparate datasets that can be represented by individual layers.

GIS is a management tool available to managers that consolidates and visually displays numerous datasets, as represented in different data layers, and provides for analysis of "what-if" scenarios. Data layers can be customized to meet the primary mission of each park unit as laid out in its enabling legislation as well as investigating specific concerns within a park. For example, superimposing data layers of traffic crashes, the average number of visitors, critical resource areas, and potential flood areas creates an overarching visual display of critical elements when determining the optimum realignment of a road or placement of a parking lot.

SOLARWINDS NETWORK MONITORING MAPPING PROJECT

MATTHEW COLWIN, NPS MIDWEST REGION GEOSPATIAL RESOURCES

In February 2015, Midwest Region (MWR) IT began using SolarWinds to monitor network performance, analyze network traffic and bandwidth, and provide intelligent alerts for network issues. In 2017, SolarWinds monitoring capabilities were increased to include all regions (except Alaska), and MWR IT approached MWR GIS with what at first seemed a simple request: a park map with network switch locations. The work then evolved into a comprehensive visual overhaul of the SolarWinds maps with the goal of making the mapping aspects of SolarWinds easy to understand, navigate, and update. The first steps involved creating a nationwide map and a series of region maps, which were inserted into SolarWinds to provide a visual understanding of where network outages were occurring. From there, MWR IT obtained addresses of all network switches in each region, and MWR GIS geocoded these addresses and developed point layers. These layers were then published to the National Geospatial Portal, where they were added to customized editing applications. These applications were shared with region and park IT staff, allowing staff with expertise to change labels, move points, and export maps with useful names and accurate switch locations. These maps were then uploaded to SolarWinds, so a user could have a visual understanding of regional network traffic and zoom in further to visualize the network traffic at sites within the park itself. This work has helped enhance the capabilities and usefulness of SolarWinds to NPS IT.

Facing page:
The national map showing all the NPS regions is displayed on the home page of SolarWinds, with a single point for each region. If these points are green, all the network switches are running properly in that region. If the points are yellow or red, there is an issue with one or more switches. SolarWinds users can click these points to view a regional map that displays each park in the region, along with a green, yellow, or red point. From there, users can click a park's point to either get information about the switch, or, if there are multiple switches at the park, be taken to a park map with points displaying specific locations of switches along with their status.

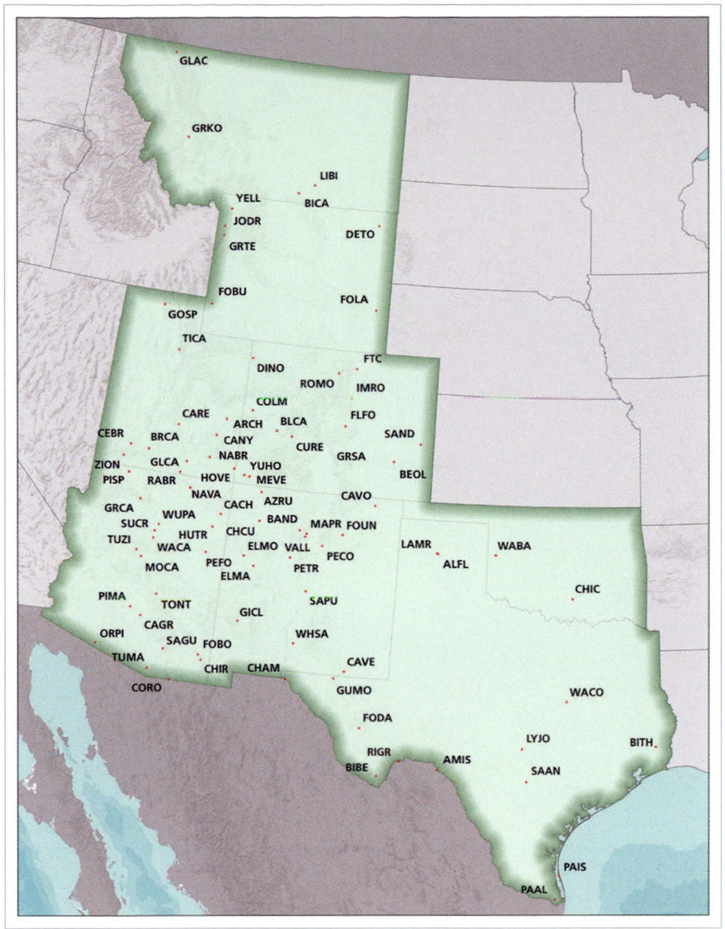
GLAC
GRKO
LIBI
YELL
BICA
JODR
DETO
GRTE
FOBU
FOLA
GOSP
TICA
FTC
DINO
ROMO
IMRO
COLM
CARE
FLFO
ARCH
BLCA
CEBR
BRCA
SAND
CANY
CURE
NABR
GRSA
ZION
GLCA
YUHO
PISP
RABR
HOVE
MEVE
BEOL
NAVA
AZRU
CAVO
GRCA
CACH
WUPA
BAND
SUCR
MAPR
FOUN
HUTR
CHCU
TUZI
WACA
ELMO
VALL
LAMR
WABA
PECO
PEFO
MOCA
ALFL
ELMA
PETR
CHIC
PIMA
TONT
SAPU
GICL
CAGR
ORPI
SAGU
FOBO
WHSA
TUMA
CHIR
CHAM
CAVE
CORO
GUMO
WACO
FODA
LYJO
BITH
RIGR
AMIS
BIBE
SAAN
PAIS
PAAL

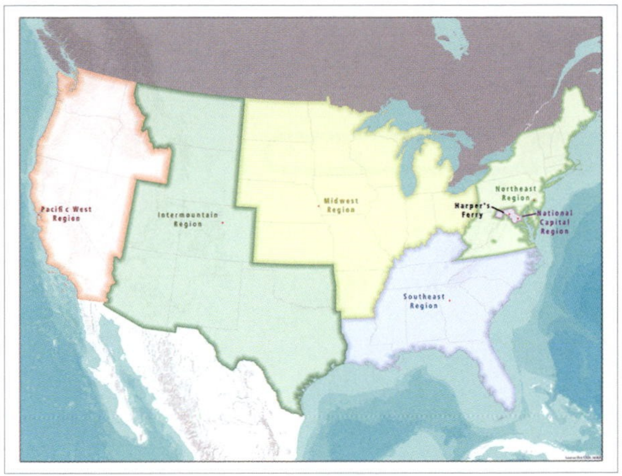
Pacific West Region
Intermountain Region
Midwest Region
Northeast Region
Harper's Ferry
National Capital Region
Southeast Region

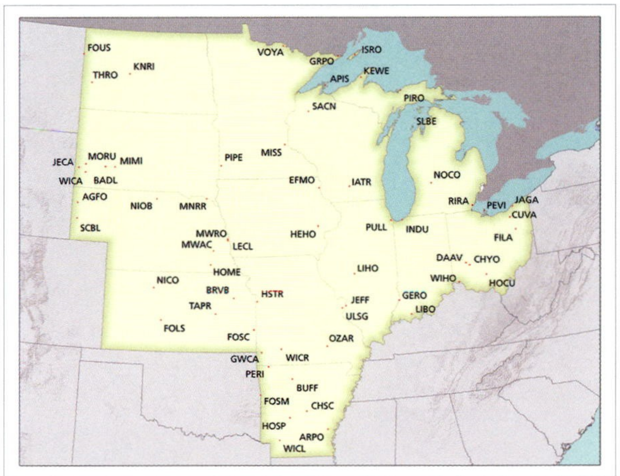
FOUS
VOYA
ISRO
KNRI
GRPO
THRO
KEWE
APIS
SACN
PIRO
SLBE
JECA
MORU
MIMI
PIPE
MISS
NOCO
WICA
BADL
EFMO
IATR
AGFO
NIOB
MNRR
RIRA
PEVI
JAGA
CUVA
SCBL
PULL
INDU
MWRO
MWAC
LECL
HEHO
FILA
DAAV
CHYO
HOME
LIHO
WIHO
HOCU
NICO
BRVB
HSTR
JEFF
GERO
TAPR
ULSG
LIBO
FOLS
FOSC
OZAR
GWCA
WICR
PERI
BUFF
FOSM
CHSC
HOSP
ARPO
WICL

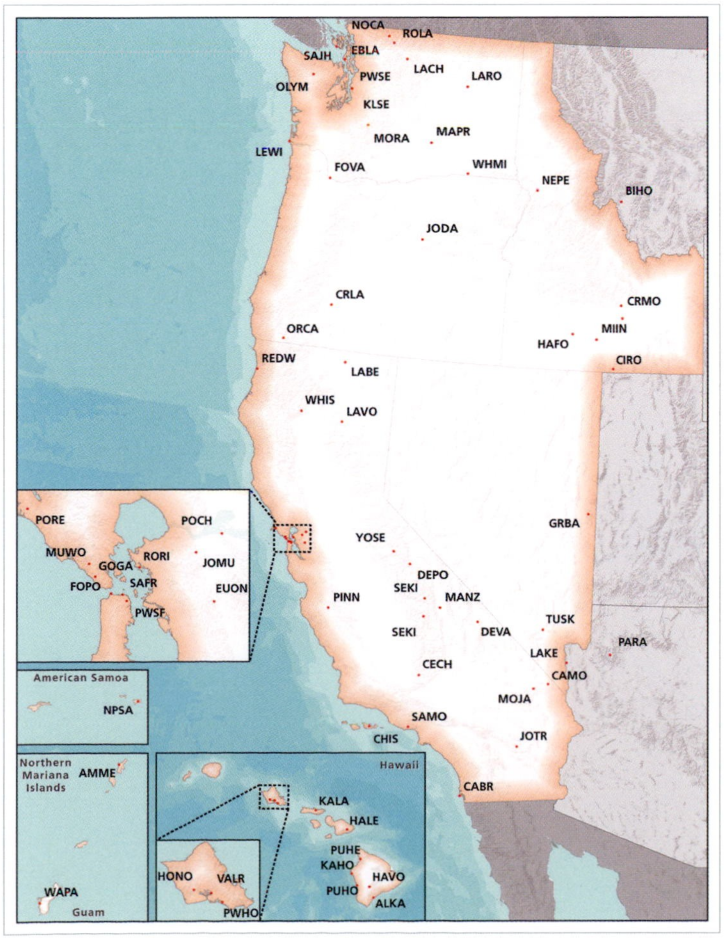
NOCA
ROLA
EBLA
SAJH
PWSE
LACH
LARO
OLYM
KLSE
MAPR
MORA
LEWI
WHMI
FOVA
NEPE
BIHO
JODA
CRLA
CRMO
ORCA
MIIN
HAFO
REDW
CIRO
LABE
WHIS
LAVO
PORE
POCH
GRBA
MUWO
RORI
JOMU
YOSE
GOGA
FOPO
SAFR
EUON
DEPO
PWSF
SEKI
PINN
MANZ
TUSK
SEKI
DEVA
PARA
LAKE
CECH
CAMO
American Samoa
NPSA
MOJA
SAMO
CHIS
JOTR
Northern Mariana Islands
AMME
Hawaii
KALA
CABR
HALE
PUHE
KAHO
HONO
VALR
HAVO
PUHO
WAPA
ALKA
Guam
PWHO

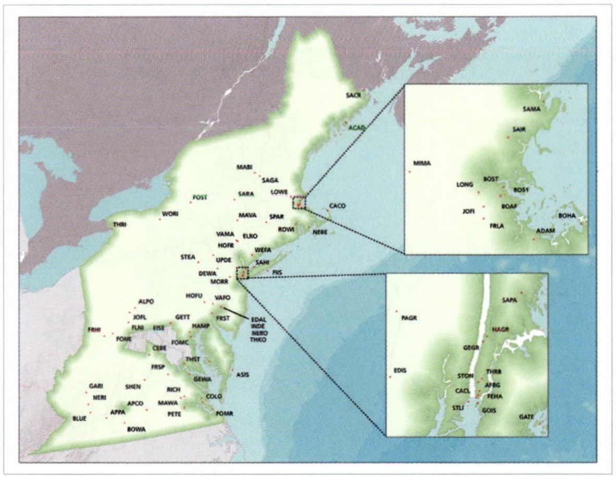
SACR
ACAD
SAMA
SAIR
MIMA
BOST
LONG
BOS1
BOAF
JOFI
BOHA
FRLA
ADAM
MABI
SAGA
FOST
SARA
LOWE
CACO
WORI
MAVA
SPAR
THRI
ROWI
NEBE
VAMA
ELRO
HOFR
WEFA
STEA
UPDE
SAHI
DEWA
FIIS
MORR
HOFU
VAFO
ALPO
JOFL
GETT
FRST
EDAL
INDE
NERO
THKO
FRHI
FLNI
EISE
HAMP
FONE
FOMC
CEBE
THST
FRSP
GEWA
ASIS
GARI
SHEN
RICH
NERI
APCO
MAWA
COLO
BLUE
APPA
PETE
FOMR
BOWA
SAPA
PAGR
HAGR
GEGR
EDIS
STON
THRB
AFBG
CACL
FEHA
STLI
GOIS
GATE

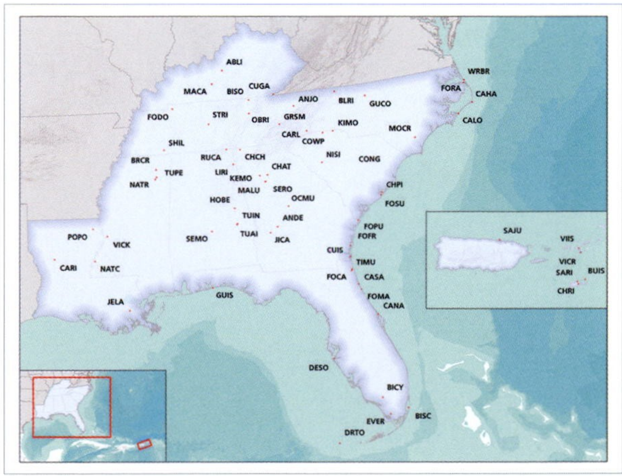
ABLI
WRBR
CUGA
FORA
MACA
BISO
CAHA
ANJO
BLRI
GUCO
FODO
STRI
OBRI
GRSM
KIMO
CALO
MOCR
CARL
COWP
SHIL
NISI
BRCR
RUCA
CHCH
CONG
CHAT
TUPE
LIRI
KEMO
NATR
MALU
SERO
CHPI
HOBE
OCMU
FOSU
TUIN
ANDE
FOPU
SEMO
TUAI
JICA
FOFR
POPO
VICK
SAJU
VIIS
CUIS
TIMU
CARI
NATC
VICR
SARI
BUIS
FOCA
CASA
CHRI
JELA
GUIS
FOMA
CANA
DESO
BICY
BISC
EVER
DRTO

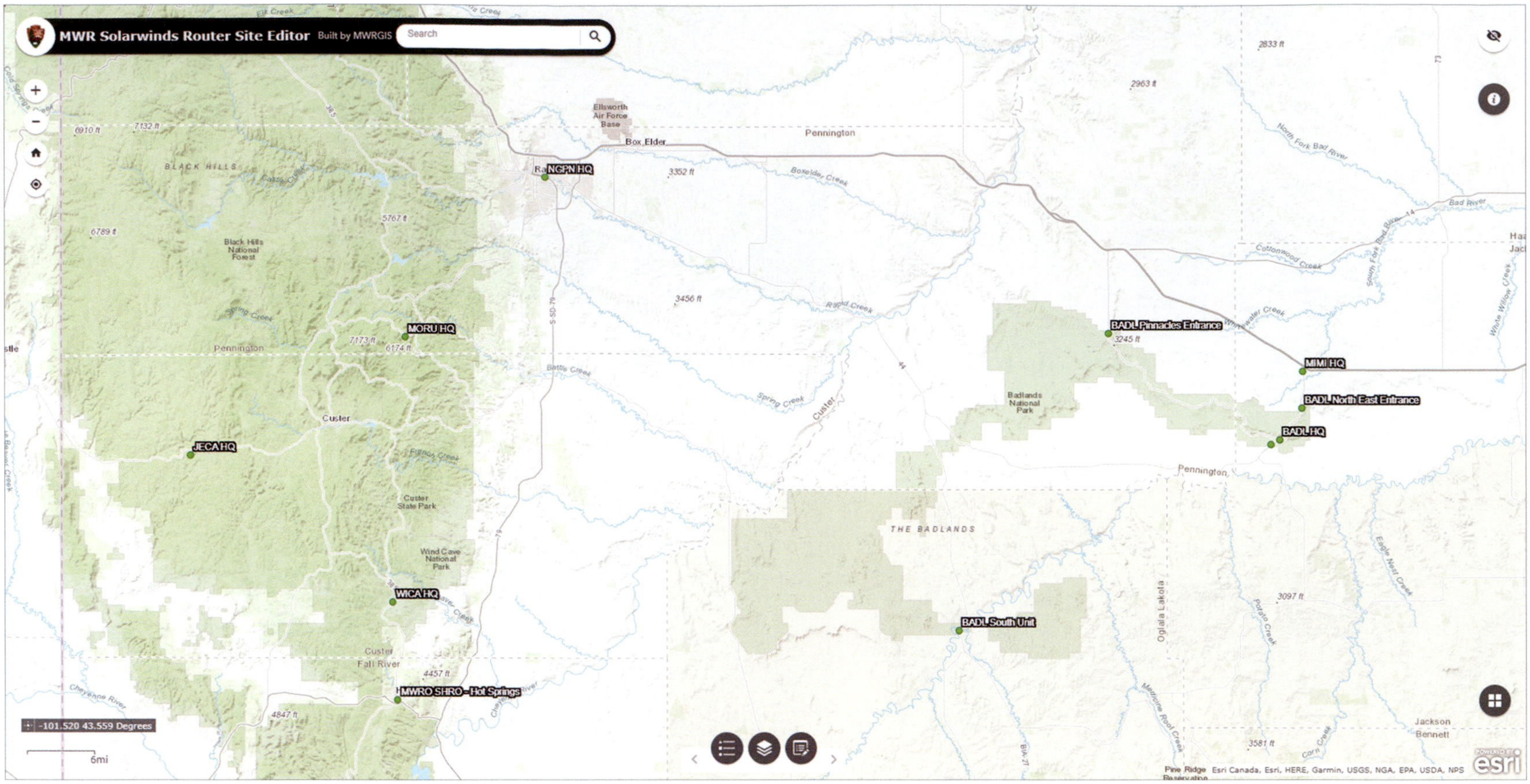

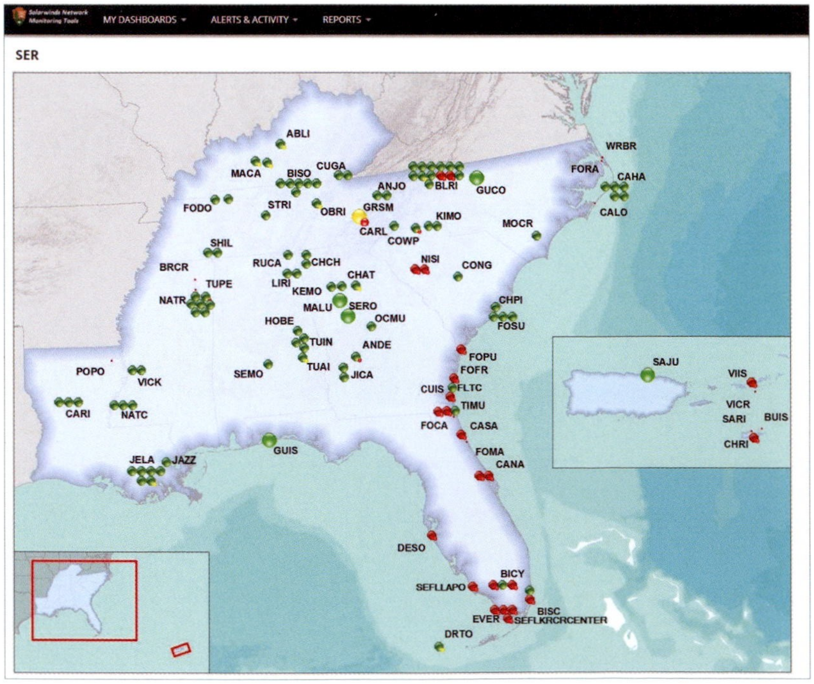

(*Above*) The NPS portal application is used by park and region staff to more accurately map network switch locations and update their names. These maps are then used within SolarWinds for detail beyond region level. (*Left*) This screenshot from SolarWinds was taken during Hurricane Maria. Data sources: NPS (Midwest Region Geospatial Resources and Land Resources Division), Esri.

MAPPING TRAVEL TIME IN SEQUOIA AND KINGS CANYON NATIONAL PARKS

ALEX EDDY, NPS SIERRA NEVADA INVENTORY AND MONITORING NETWORK
PAUL HARDWICK AND GREG FAUTH, SEQUOIA AND KINGS CANYON NATIONAL PARKS
JAMES TRICKER AND PETER LANDRES, ALDO LEOPOLD WILDERNESS RESEARCH INSTITUTE, US FOREST SERVICE

How long would it take a hiker to reach some of the most remote places in the Sierra Nevada? The NPS has partnered with the Aldo Leopold Wilderness Research Institute to develop a model that estimates travel time in Sequoia and Kings Canyon National Parks (SEKI). The model uses detailed geographic information to estimate how long it would take to hike from a road to any place in the parks.

Trail corridors provide hikers with access to the Sierra Nevada's spectacular mountain crests and valleys. NPS photo/Mandy Holmgren.

The parks' landscape is diverse and dramatic, with high mountain crests and deep river canyons. The travel time model captures the challenges of hiking in these national parks by including rivers, lakes, and steep slopes that delay or prevent travel. Trails enable faster travel, depending on the level of development or whether they include structures such as stairs or bridges. A vegetation map helps estimate how difficult or easy it would be to hike through cross-country areas, including woodlands, shrubland, and evergreen forests.

Based on the final map, a hiker could travel to many areas of the parks in less than an hour but might need up to 27 hours to reach the most remote areas. Park managers use this information to better understand visitor experience and make informed decisions on how to best manage wilderness character. They also use the map to help choose optimal locations for field project sites that contribute to protecting and preserving these remarkable places.

Some rivers and streams may significantly impede travel, while others are relatively easy to traverse. NPS photo/Wesley Meyers.

Bare ground and developed trails enable faster travel, even in remote high-elevation places. NPS photo/Linda Mutch.

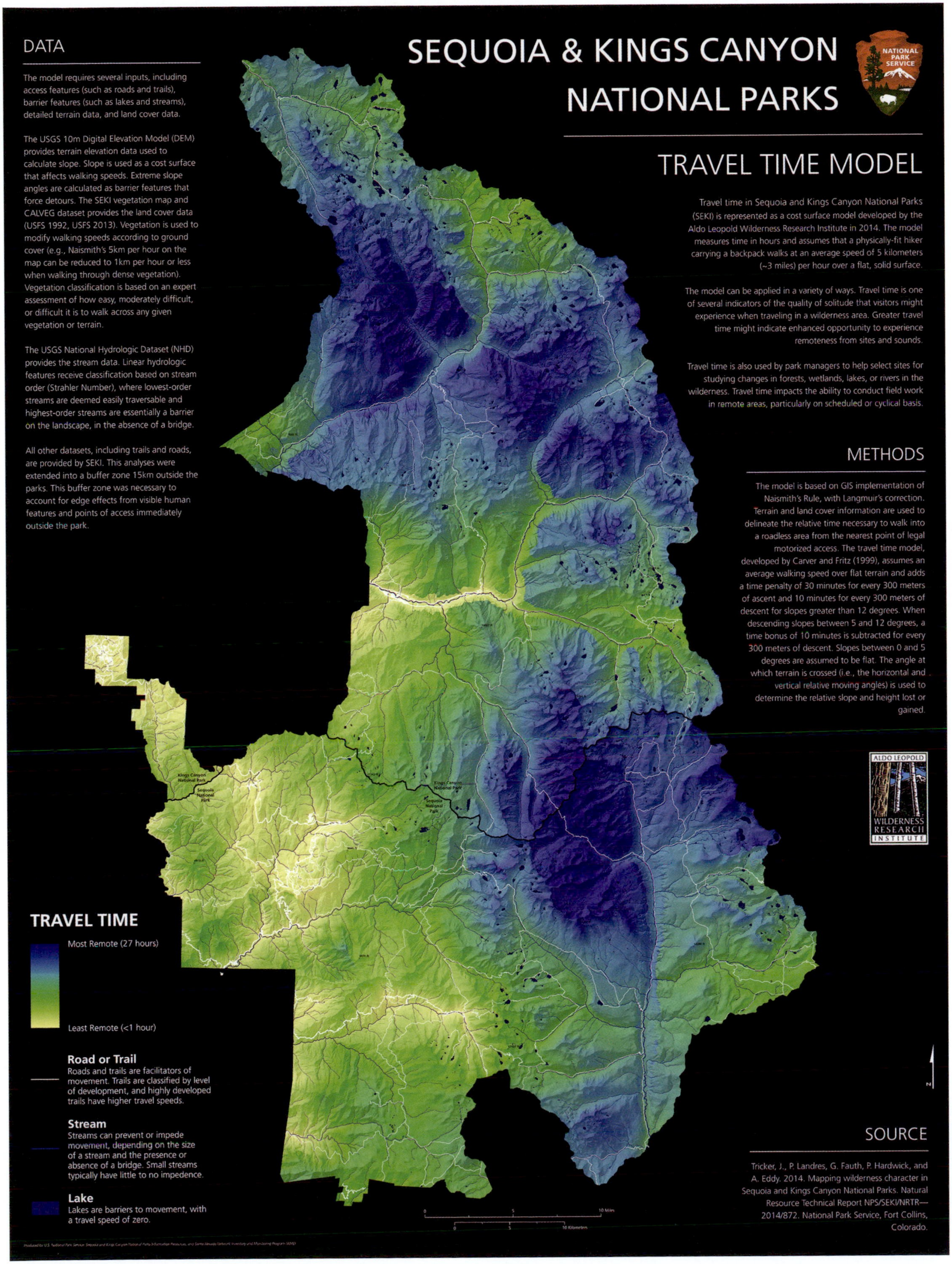
SEQUOIA & KINGS CANYON NATIONAL PARKS
NATIONAL PARK SERVICE
TRAVEL TIME MODEL
Travel time in Sequoia and Kings Canyon National Parks (SEKI) is represented as a cost surface model developed by the Aldo Leopold Wilderness Research Institute in 2014. The model measures time in hours and assumes that a physically-fit hiker carrying a backpack walks at an average speed of 5 kilometers (~3 miles) per hour over a flat, solid surface.
The model can be applied in a variety of ways. Travel time is one of several indicators of the quality of solitude that visitors might experience when traveling in a wilderness area. Greater travel time might indicate enhanced opportunity to experience remoteness from sites and sounds.
Travel time is also used by park managers to help select sites for studying changes in forests, wetlands, lakes, or rivers in the wilderness. Travel time impacts the ability to conduct field work in remote areas, particularly on scheduled or cyclical basis.
DATA
The model requires several inputs, including access features (such as roads and trails), barrier features (such as lakes and streams), detailed terrain data, and land cover data.
The USGS 10m Digital Elevation Model (DEM) provides terrain elevation data used to calculate slope. Slope is used as a cost surface that affects walking speeds. Extreme slope angles are calculated as barrier features that force detours. The SEKI vegetation map and CALVEG dataset provides the land cover data (USFS 1992, USFS 2013). Vegetation is used to modify walking speeds according to ground cover (e.g., Naismith's 5km per hour on the map can be reduced to 1km per hour or less when walking through dense vegetation). Vegetation classification is based on an expert assessment of how easy, moderately difficult, or difficult it is to walk across any given vegetation or terrain.
The USGS National Hydrologic Dataset (NHD) provides the stream data. Linear hydrologic features receive classification based on stream order (Strahler Number), where lowest-order streams are deemed easily traversable and highest-order streams are essentially a barrier on the landscape, in the absence of a bridge.
All other datasets, including trails and roads, are provided by SEKI. This analyses were extended into a buffer zone 15km outside the parks. This buffer zone was necessary to account for edge effects from visible human features and points of access immediately outside the park.
METHODS
The model is based on GIS implementation of Naismith's Rule, with Langmuir's correction. Terrain and land cover information are used to delineate the relative time necessary to walk into a roadless area from the nearest point of legal motorized access. The travel time model, developed by Carver and Fritz (1999), assumes an average walking speed over flat terrain and adds a time penalty of 30 minutes for every 300 meters of ascent and 10 minutes for every 300 meters of descent for slopes greater than 12 degrees. When descending slopes between 5 and 12 degrees, a time bonus of 10 minutes is subtracted for every 300 meters of descent. Slopes between 0 and 5 degrees are assumed to be flat. The angle at which terrain is crossed (i.e., the horizontal and vertical relative moving angles) is used to determine the relative slope and height lost or gained.
ALDO LEOPOLD WILDERNESS RESEARCH INSTITUTE
TRAVEL TIME
Most Remote (27 hours)
Least Remote (<1 hour)
Road or Trail
Roads and trails are facilitators of movement. Trails are classified by level of development, and highly developed trails have higher travel speeds.
Stream
Streams can prevent or impede movement, depending on the size of a stream and the presence or absence of a bridge. Small streams typically have little to no impedence.
Lake
Lakes are barriers to movement, with a travel speed of zero.
SOURCE
Tricker, J., P. Landres, G. Fauth, P. Hardwick, and A. Eddy. 2014. Mapping wilderness character in Sequoia and Kings Canyon National Parks. Natural Resource Technical Report NPS/SEKI/NRTR—2014/872. National Park Service, Fort Collins, Colorado.

The final poster shows that a hiker might need up to 27 hours to reach the most remote areas of the park. Data sources: NPS and Aldo Leopold Wilderness Research Institute.

ISLE ROYALE NATIONAL PARK CONCESSIONER-ASSIGNED FACILITIES

MATTHEW COLWIN, NPS MIDWEST REGION GEOSPATIAL RESOURCES

This series of maps depict facilities at Isle Royale National Park—a remote island cluster in Lake Superior near Michigan's border with Canada—and emphasize facilities that are concessioner-assigned, meaning that these facilities are operated by groups for commercial purposes. The maps were created as part of a project by the NPS Midwest Region Commercial Services team and provided to the project contractors for evaluating the concessioner-assigned facilities that would be included in a new concession contract. The largely grayscale format of the maps allowed them to be easily printed and reproduced, and the lack of non-facilities data or imagery basemaps helped keep the purpose of the maps clear. The maps use a consistent format that allows them to be easily altered for use in future commercial services projects.

Overview map shows Isle Royale National Park concessioner-assigned facilities. The map was created for NPS Midwest Regional Office (MWRO) Commercial Services team to evaluate the concessioner-assigned facilities that would be included in a new concession contract.

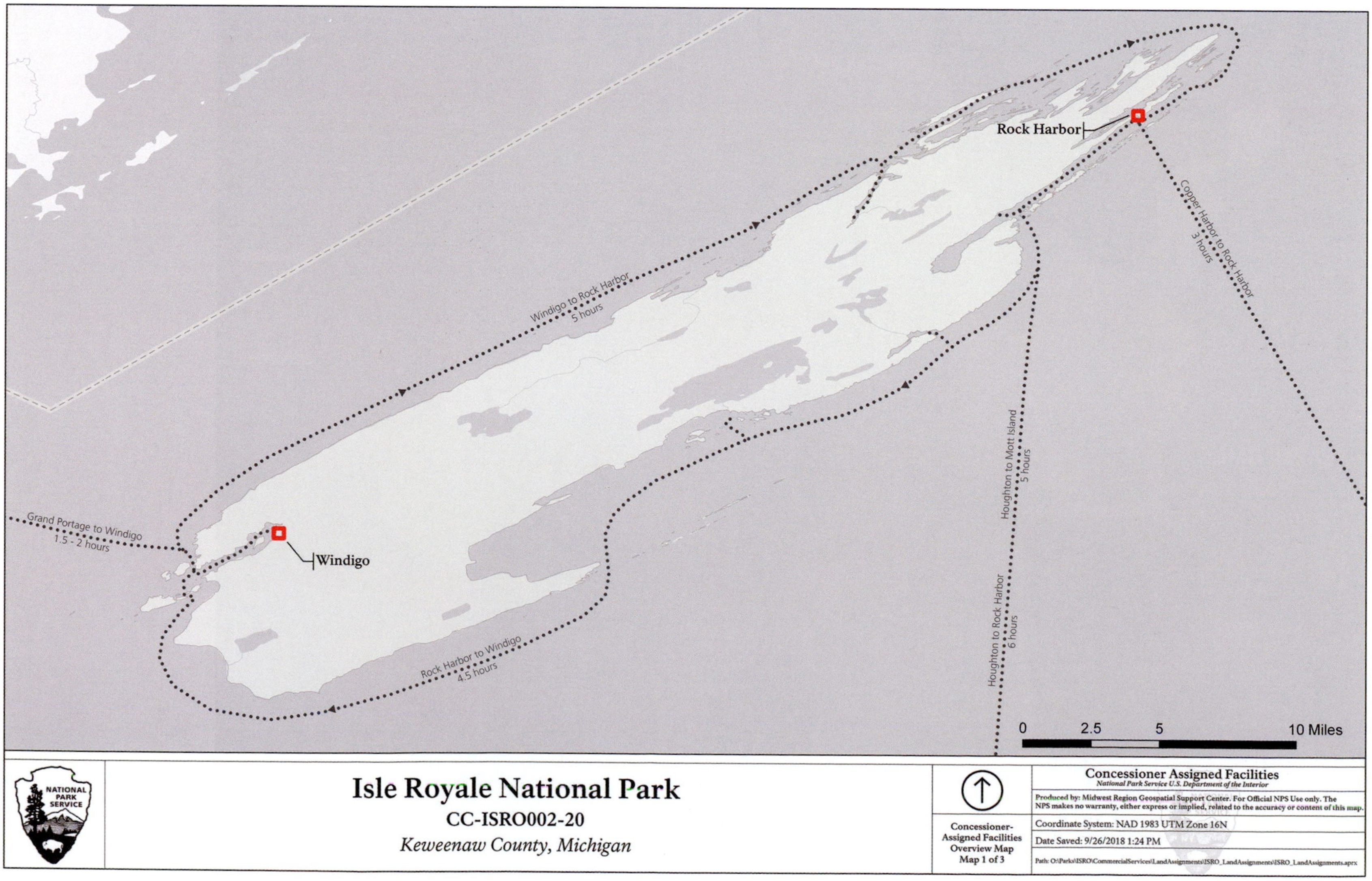

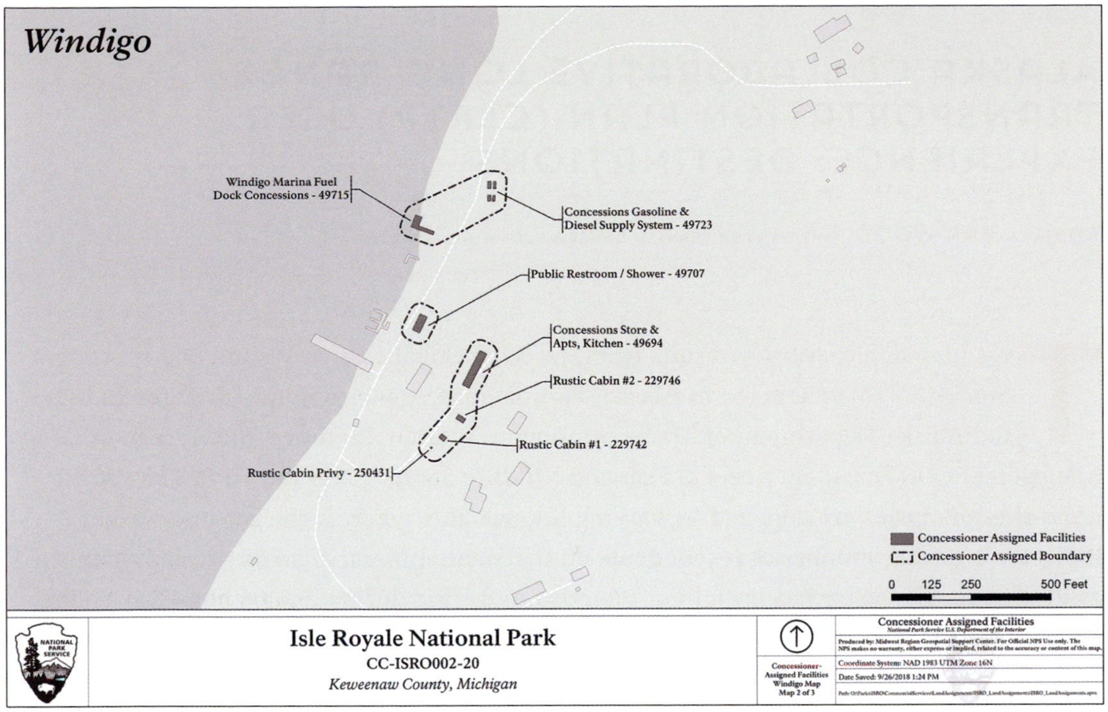

The Windigo map shows Isle Royale National Park concessioner-assigned facilities. The map was created for NPS MWRO Commercial Services team to evaluate the concessioner-assigned facilities that would be included in a new concession contract.

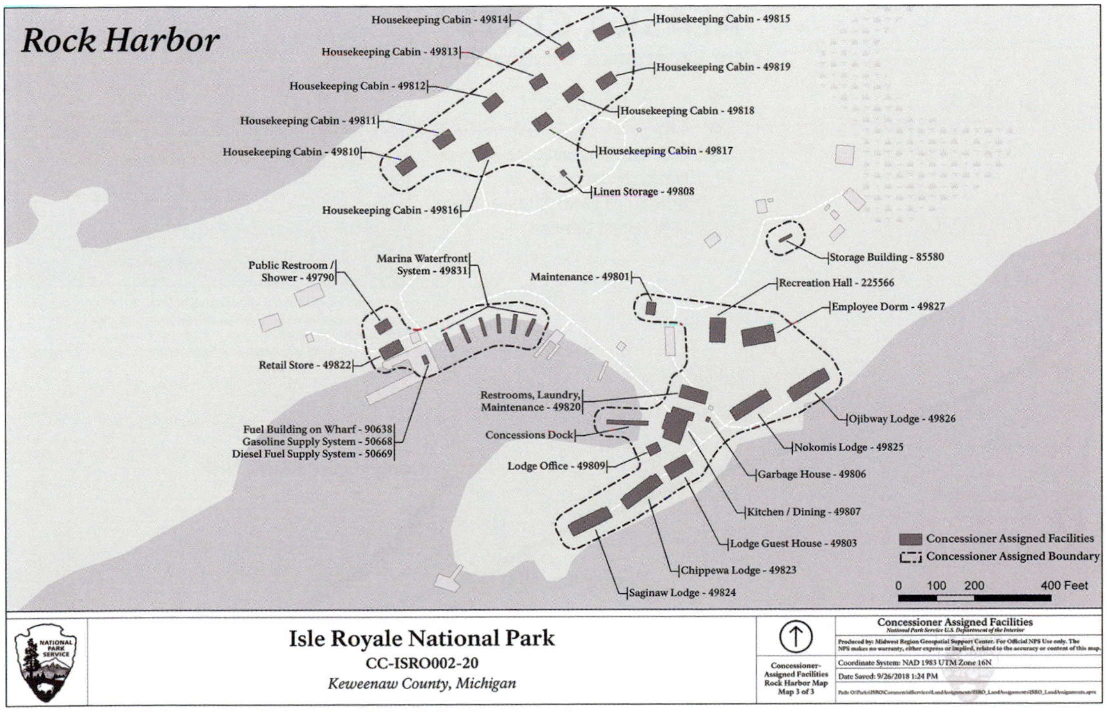

Isle Royale National Park concessioner-assigned facilities map shows Rock Harbor. The map was created for the NPS MWRO Commercial Services team to evaluate the concessioner-assigned facilities that would be included in a new concession contract. Data sources: NPS Midwest Region Geospatial Resources, NPS.

ALASKA COLLABORATIVE LONG-RANGE TRANSPORTATION PLAN (CLRTP) USER EXPERIENCE: DESTINATIONS

RAFAEL ZAK WOOD, NPS DENVER SERVICE CENTER PLANNING DIVISION

This map depicts survey results from the 2016 Collaborative Visitor Transportation Study (CVTS) conducted in Alaska. Participating agencies in the CVTS included the Alaska Department of Transportation and Public Facilities, the Bureau of Land Management US Forest Service, US Fish and Wildlife Service, and the NPS. The top 10 cross-site itineraries are depicted by varying line weights where a thicker line weight indicates a greater number of respondents on that same itinerary. Cross sites are identified as destinations a visitor has been to or intends on visiting during one trip to Alaska. The itineraries do not reflect the order of visitation. The map will be used in the Collaborative Long-Range Transportation Plan (CLRTP) analysis and document. The map contributes to better informed decision making on future transportation investments such as allocating funding based on visitation numbers at public lands and investing in transportation infrastructure used to connect common destinations.

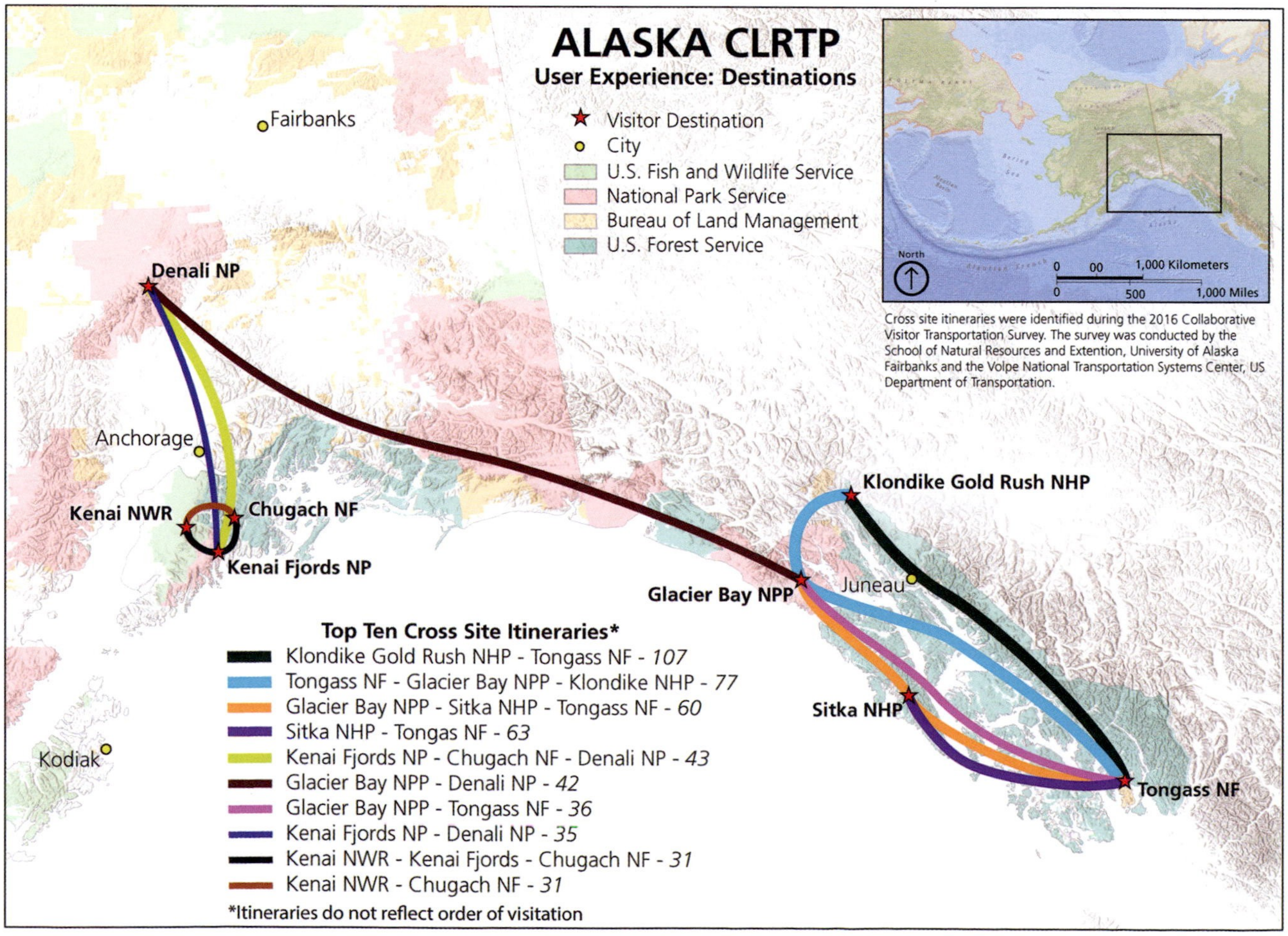

This map shows survey results of the 2016 Collaborative Visitor Transportation Study. Credit for visitor destinations layer and line segments depicting visitor flow across Alaska: Collaborative Visitor Transportation Survey: Results from summer 2016 Alaska Survey, by Peter Fix, Alisa Wedin, Jasmine Shaw, Karen Petersen, and Margaret Petrella. Data sources: Alaska Department of Transportation, Department of Community and Regional Affairs, US Department of the Interior, Bureau of Land Management, Alaska Department of Natural Resources, Peter Fix, Alisa Wedin, Jasmine Shaw, Karen Petersen, and Margaret Petrella; Esri Hillshade Basemap, National Geographic World Map Basemap.

8 WORKING WITH COMMUNITIES AND PARTNERS

ROBERT RATCLIFFE, NPS CONSERVATION AND OUTDOOR RECREATION DIVISION

CULTIVATING COMMUNITY AND PARTNER RELATIONSHIPS

The National Park Service (NPS) oversees much more than 88 million acres of its well-known system of more than 400 parks. The NPS is also responsible for a distinct and unique portfolio of over 54 congressionally authorized programs that assist communities across America with:

- historic preservation,
- redevelopment of historic sites,
- special area conservation and acquisition,
- recreation and improved access to public lands, and
- education and professional development in and outside national parks.

These programs include the popular Federal Historic Preservation Historic Tax Credit; Rivers, Trails, and Conservation Assistance; National Heritage Areas, National Trails, and Wild and Scenic Rivers Systems; and Land and Water Conservation Fund state assistance.

Through programs such as these, the NPS works directly with community and non-profit partners to leverage funding and support to protect and conserve resources as well as advance critical community and environmental needs. These collaborative relationships help to enhance and create valuable assets, such as trails and greenways, that boost local economies and community quality of life through environmental restoration, business development, tourism promotion, recreation access, and job creation.

Program partners include community organizations, local governments, nonprofits, and private entrepreneurs who align and share their knowledge and resources to collaborate on community and regional goals.

Geospatial mapping and analyses have become an important and essential tool in identifying, planning, monitoring, and managing these programs and resources. The analytical capabilities now available span a host of social, environmental, and economic considerations and resources. The ability to assess data across sectors enables programs to be implemented in a more efficient, effective, and accountable manner. The collective and aggregation assessment capabilities that geospatial mapping provides allow for an improved and integrative approach to understanding data and evaluating implications of planning and management actions. New mapping capabilities have allowed for various data analytics and considerations of community needs such as park deserts or recreation access limitations, viewsheds and soundscapes, and transportation in the implementation of NPS assistance programs.

Mapping capabilities and products have proven to be a vital component to working with our neighbors and affected communities. These capabilities and products are instrumental as public outreach and engagement tools. Maps and the data associated with them make for compelling public engagement, enhancing the public's understanding of the benefits and outcomes of these programs and projects. Maps and related data are essential in generating needed support of decision-makers as they prioritize investments in their communities and states. Maps present analytical information in a powerful way that allows stakeholders to visualize and understand challenges and interrelationships of resources as they seek solutions and identify opportunities. This chapter reviews a few of the uses and applications for the new generation of tools. The NPS has just begun to tap the full potential of mapping and data analytics for the collaborative management of its parks and programs.

INTERACTIVE WEB MAP FOR THE NATIONAL TRAILS SYSTEM

MATT ROBINSON, APPALACHIAN NATIONAL SCENIC TRAIL
PETER BONSALL, NPS CONSERVATION AND OUTDOOR RECREATION DIVISION

This web map shows the 11 National Scenic Trails and 19 National Historic Trails that make up the National Trails System in the United States. It is presented in an interactive format utilizing Esri's ArcGIS Online web application and allows users to explore individual routes and access additional information about each trail. It also includes information of the more than 1,200 National Recreation Trails and the National Trails 50th Anniversary events and incorporates data from the Protected Areas Database of the United States (PADUS). Visit www.pnts.org/new/map-interactive-gis-pnts.org to view the web map and other information about the National Trails System.

The National Trails System interactive web map includes all 30 National Scenic and National Historic Trails.

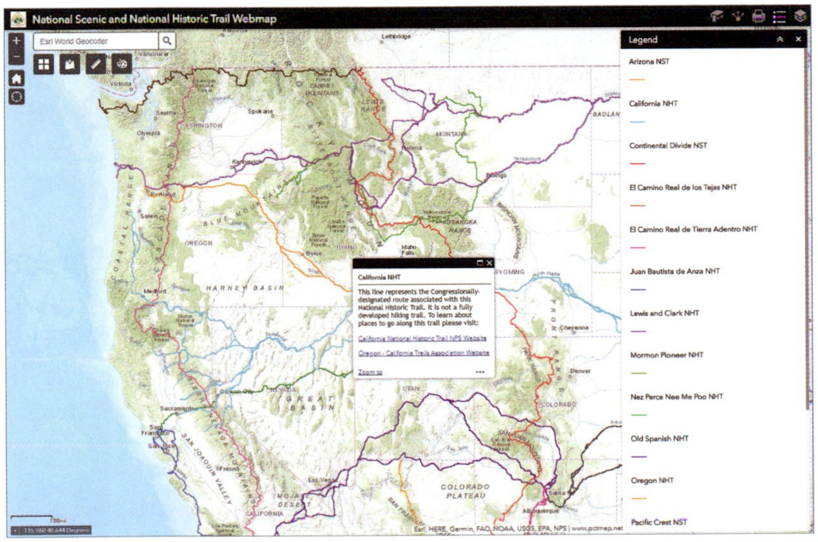

The National Trails System interactive web map includes a trail pop-up screen describing the California National Historic Trail.

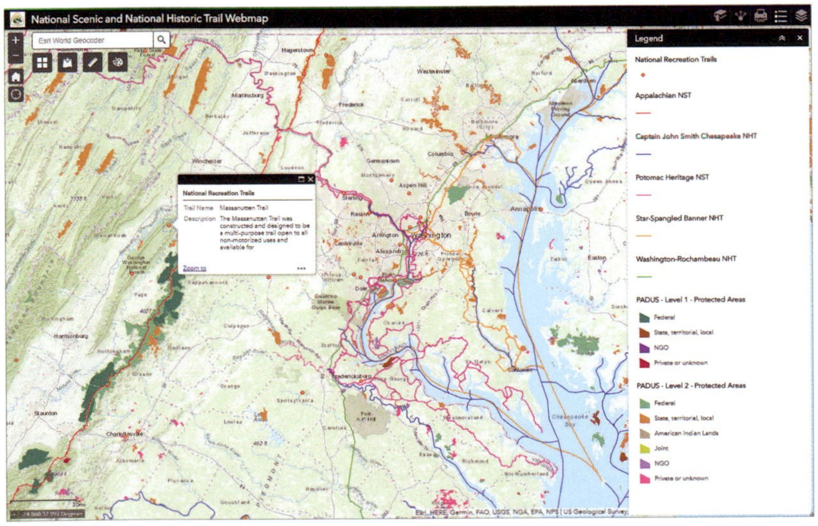

A larger scale view shows the web map with a trail pop-up and federal lands.

Credit American Trails, Mike Bullington: National Recreation Trails Point Feature Layer. Data sources: NPS, Bureau of Land Management, US Forest Service, Appalachian Trail Conservancy, Arizona Trail Association, Chesapeake Conservancy, US Forest Service Region 2, NPS Intermountain Region, Continental Divide Trail Coalition, Bear Creek Survey, Florida Trail Association, Florida Fish and Wildlife Conservation Commission, Trail Volunteers, Ice Age Trail Alliance, Connecticut Forest and Park Association, Appalachian Mountain Club, North Country Trail Association, Pacific Crest Trail Association, USGS, Gap Analysis Program (GAP), Mike Bullington—American Trails, University of Montana, and Esri Topographic Basemap.

MAP OF THE NATIONAL TRAILS SYSTEM'S 50TH ANNIVERSARY

MATT ROBINSON, APPALACHIAN NATIONAL SCENIC TRAIL

This map was produced for the 50th Anniversary of the National Trails System Act, enacted October 2, 1968. It shows the 11 National Scenic Trails and 19 National Historic Trails that make up the National Trails System as of October 2, 2018, and includes logos for each trail in the system.

This map shows the National Trails System. Data sources: NPS.

SIGN PLANNING FOR NATIONAL HISTORIC TRAILS

BRIAN DEATON AND SARAH RIVERA, NATIONAL TRAILS

Partners across the country collaborate with the National Trails office to plan signs for the nine National Historic Trails that the office administers. To initiate a plan, partners contact the office to discuss scope, location, sign types, and funding for the project. The partners or National Trails staff can then design a sign plan upon agreement. The sign plan is designed using ArcGIS Online group capability, where partners can join the sign-planning group and input signs into the map viewer. Together, partners and National Trails staff can collaborate to create a draft plan for the proposed area. Once sign locations for the plan are finalized, National Trails runs the sign plan tool in ArcGIS Pro from a Python script. This tool downloads the sign plan data from ArcGIS Online, queries the sign plan name, and utilizes the ArcGIS Pro reporting function and map series to generate a customized PDF sign plan report. This report contains maps, sign summaries, cost estimates, and sign specifications. Once National Trails staff, partners, and involved road jurisdictions approve the sign plan report, an approval form is signed by the road jurisdictions. The signs can then be ordered and installed based on the report. The sign-planning tool allows our office an opportunity to cooperate with partners across the country to protect, develop, and promote National Historic Trails for the enjoyment and recreational opportunities of the public.

Santa Fe Trail
Rock Creek Crossing
Santa Fe Trail
HISTORIC ROUTE
Santa Fe Trail
CROSSING

An example of installed signs near Council Grove, Kansas, and two examples of sign design graphics are shown.

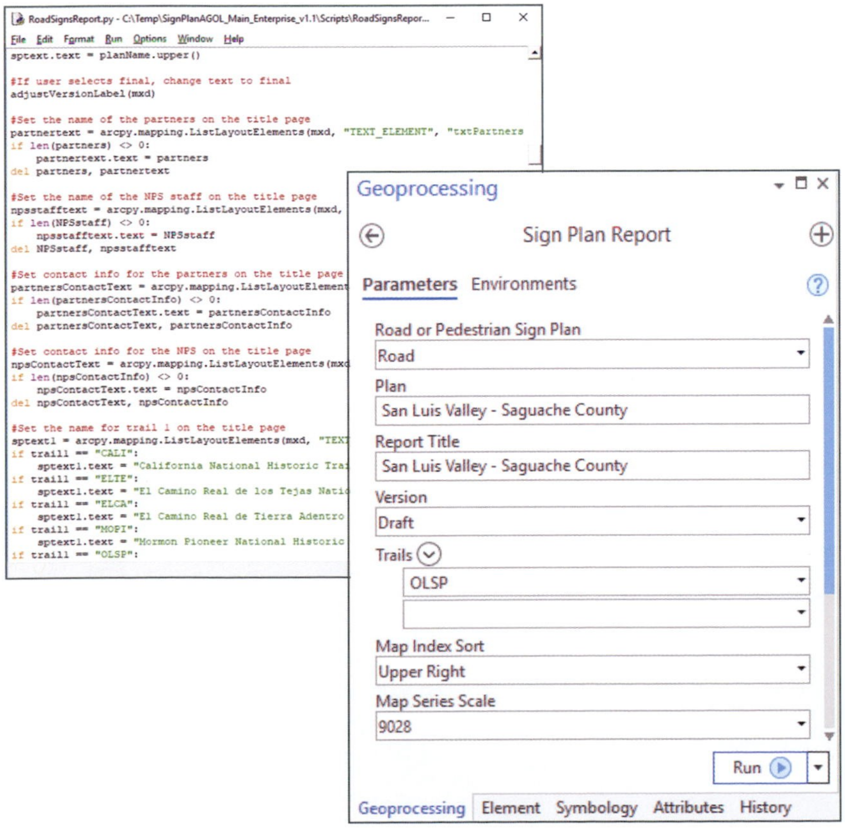
RoadSignsReport.py - C:\Temp\SignPlanAGOL_Main_Enterprise_v1.1\Scripts\RoadSignsRepor...
File Edit Format Run Options Window Help
sptext.text = planName.upper()
#If user selects final, change text to final
adjustVersionLabel(mxd)
#Set the name of the partners on the title page
partnertext = arcpy.mapping.ListLayoutElements(mxd, "TEXT_ELEMENT", "txtPartners
if len(partners) <> 0:
partnertext.text = partners
del partners, partnertext
#Set the name of the NPS staff on the title page
npsstafftext = arcpy.mapping.ListLayoutElements(mxd,
if len(NPSstaff) <> 0:
npsstafftext.text = NPSstaff
del NPSstaff, npsstafftext
#Set contact info for the partners on the title page
partnersContactText = arcpy.mapping.ListLayoutElement
if len(partnersContactInfo) <> 0:
partnersContactText.text = partnersContactInfo
del partnersContactText, partnersContactInfo
#Set contact info for the NPS on the title page
npsContactText = arcpy.mapping.ListLayoutElements(mx
if len(npsContactInfo) <> 0:
npsContactText.text = npsContactInfo
del npsContactText, npsContactInfo
#Set the name for trail 1 on the title page
sptext1 = arcpy.mapping.ListLayoutElements(mxd, "TEX
if trail1 == "CALI":
sptext1.text = "California National Historic Tra
if trail1 == "ELTE":
sptext1.text = "El Camino Real de los Tejas Nati
if trail1 == "ELCA":
sptext1.text = "El Camino Real de Tierra Adentro
if trail1 == "MOPI":
sptext1.text = "Mormon Pioneer National Historic
if trail1 == "OLSP":
Geoprocessing
Sign Plan Report
Parameters
Environments
Road or Pedestrian Sign Plan
Road
Plan
San Luis Valley - Saguache County
Report Title
San Luis Valley - Saguache County
Version
Draft
Trails
OLSP
Map Index Sort
Upper Right
Map Series Scale
9028
Run
Geoprocessing
Element
Symbology
Attributes
History

An example Python script code and ArcGIS Pro script tool interface are shown.

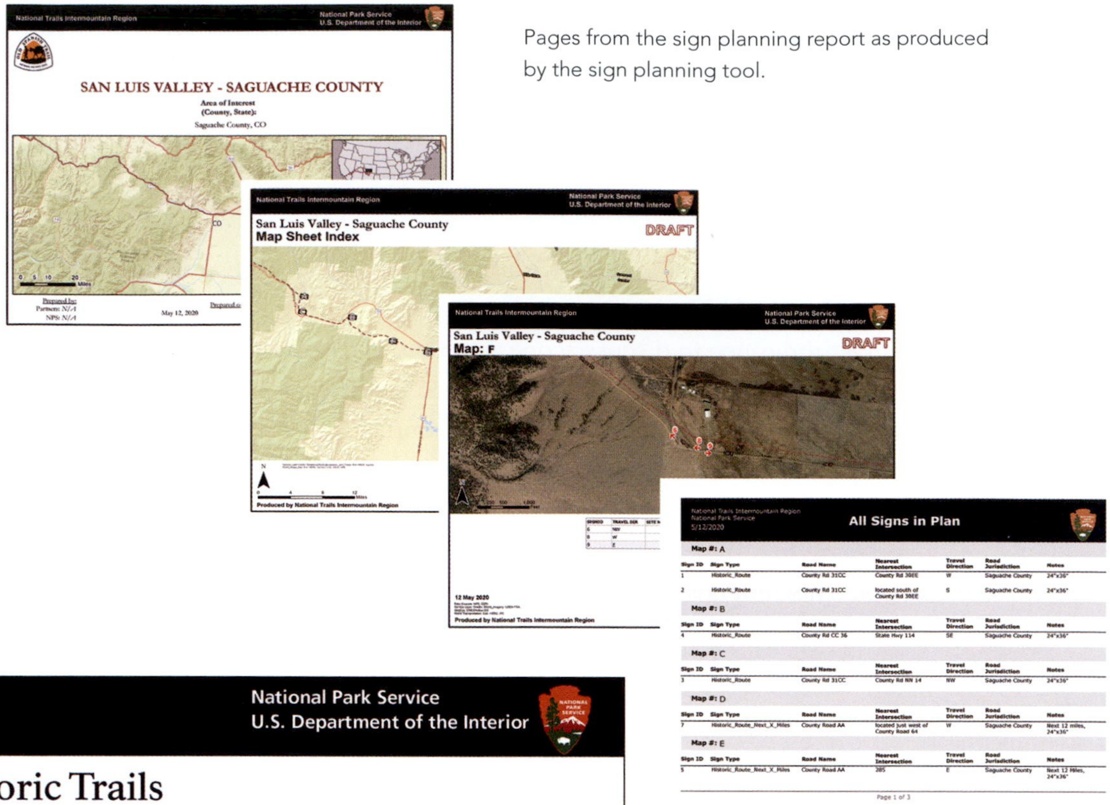

Pages from the sign planning report as produced by the sign planning tool.

This map displays several ongoing sign plans across the country.

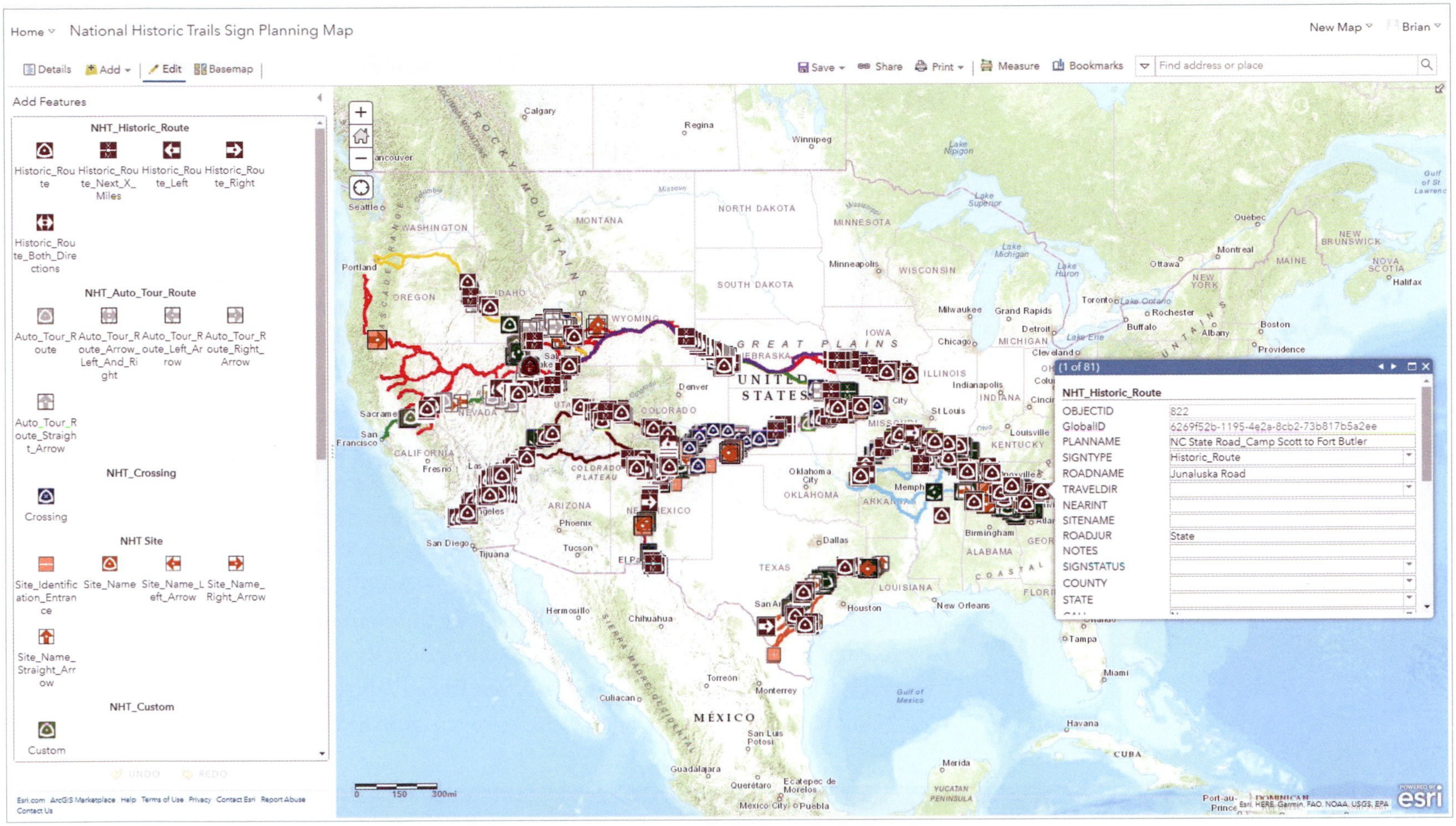

This National Historic Trails map is an ArcGIS Online web map used by partners and National Trails to develop sign plans.

National Historic Trails and Route 66 Corridor Preservation Program logos are shown. Data sources: National Trails-National Park Service, US Census Bureau, USGS National Atlas, World Ocean Basemap.

RIVERS, TRAILS, AND CONSERVATION ASSISTANCE PROGRAM STATE PAGES

PETER BONSALL, NPS CONSERVATION AND OUTDOOR RECREATION DIVISION

The NPS Rivers, Trails, and Conservation Assistance (RTCA) program supports community-led natural resource conservation and outdoor recreation projects across the nation. The RTCA State Pages use an ArcGIS application to create stories for highlighting annual projects occurring in each state across the country. Each project is given a point location to display the geographic spread of RTCA's engagement among communities in each state. Points are color-coded as success stories (red points) or current projects (green points). Success stories provide a brief narrative on an accomplishment associated with a previous year's project. Current projects provide basic project information such as project name, project goal, NPS role, project partner, NPS contact, and location. The images provided are sample RTCA State Pages for Hawai'i, Texas, and Florida. The RTCA State Pages are used as promotional tools to show the communities where RTCA is actively engaged and to attract potential new community partners to apply for future community assistance. The RTCA program helps carry out the NPS mission by supporting community-led outdoor recreation and conservation projects for the benefit of all Americans.

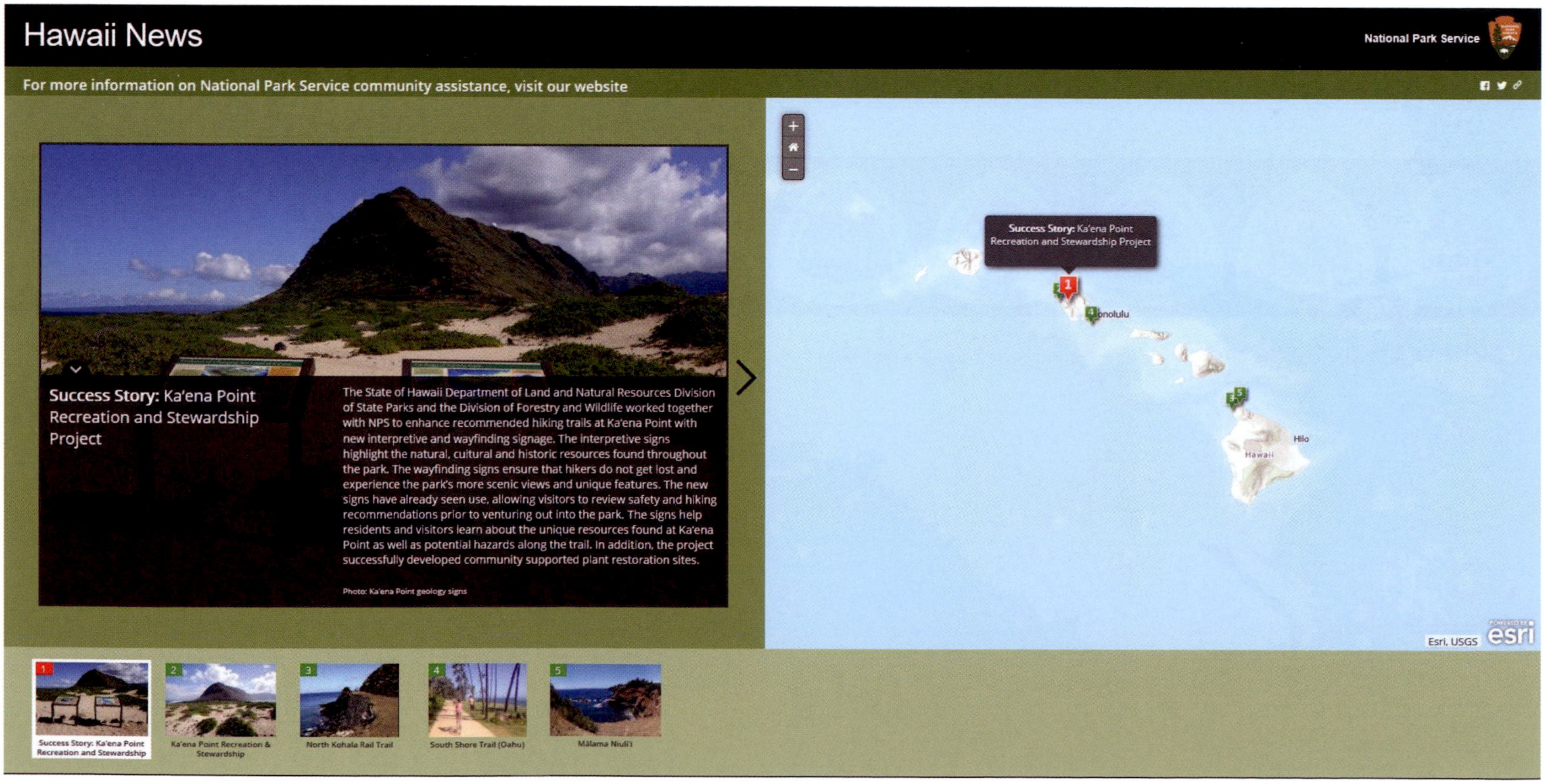
Hawaii News
National Park Service
For more information on National Park Service community assistance, visit our website
Success Story: Ka'ena Point Recreation and Stewardship Project
The State of Hawaii Department of Land and Natural Resources Division of State Parks and the Division of Forestry and Wildlife worked together with NPS to enhance recommended hiking trails at Ka'ena Point with new interpretive and wayfinding signage. The interpretive signs highlight the natural, cultural and historic resources found throughout the park. The wayfinding signs ensure that hikers do not get lost and experience the park's more scenic views and unique features. The new signs have already seen use, allowing visitors to review safety and hiking recommendations prior to venturing out into the park. The signs help residents and visitors learn about the unique resources found at Ka'ena Point as well as potential hazards along the trail. In addition, the project successfully developed community supported plant restoration sites.
Photo: Ka'ena Point geology signs
Success Story: Ka'ena Point Recreation and Stewardship Project
Honolulu
Hilo
Hawaii
Esri, USGS
esri
Success Story: Ka'ena Point Recreation and Stewardship
Ka'ena Point Recreation & Stewardship
North Kohala Rail Trail
South Shore Trail (Oahu)
Mālama Niuli'i

This RTCA state page uses an ArcGIS application to highlight annual projects in Hawai'i.

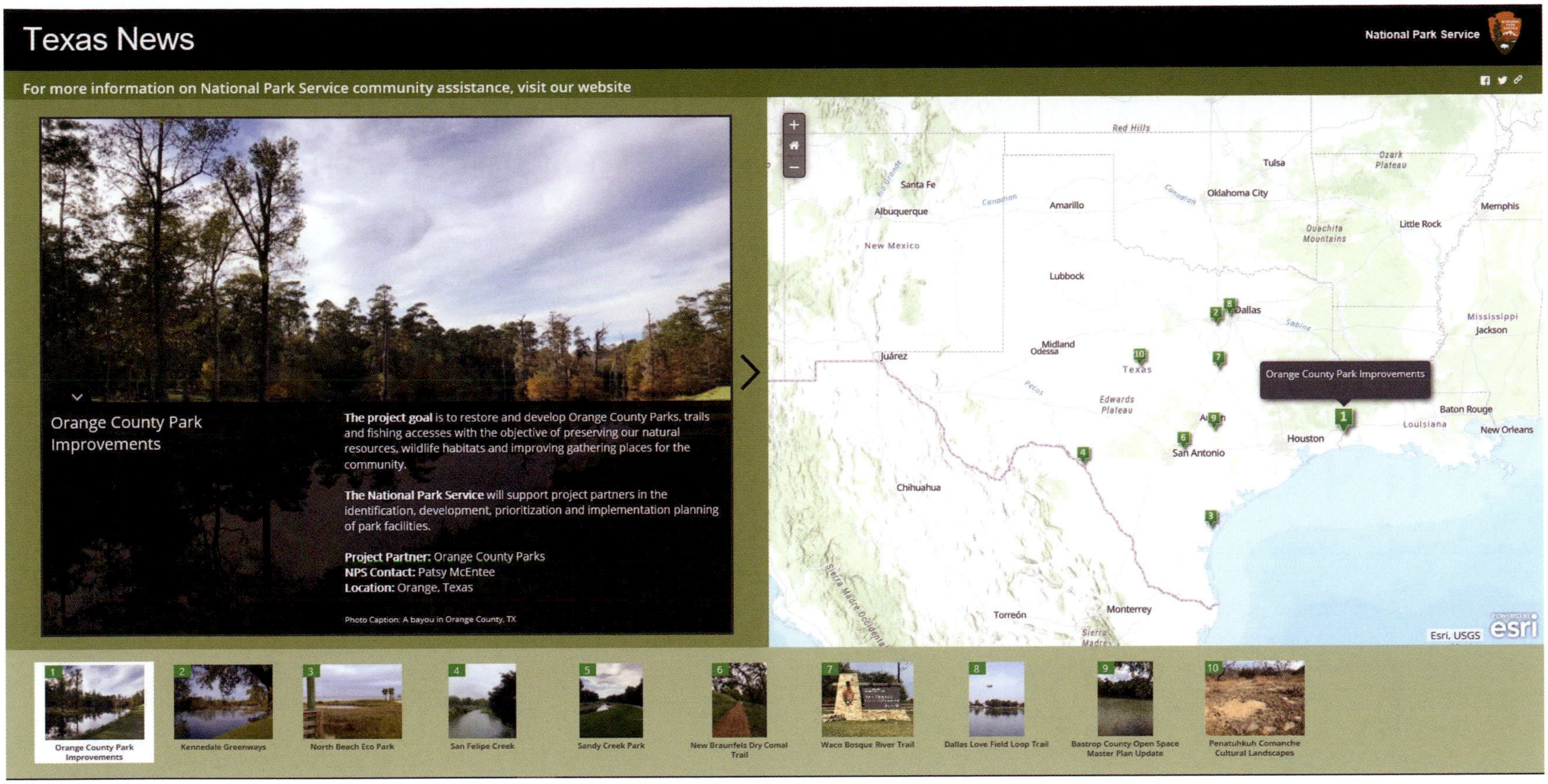
Texas News
National Park Service
For more information on National Park Service community assistance, visit our website
Orange County Park Improvements
The project goal is to restore and develop Orange County Parks, trails and fishing accesses with the objective of preserving our natural resources, wildlife habitats and improving gathering places for the community.
The National Park Service will support project partners in the identification, development, prioritization and implementation planning of park facilities.
Project Partner: Orange County Parks
NPS Contact: Patsy McEntee
Location: Orange, Texas
Photo Caption: A bayou in Orange County, TX
Orange County Park Improvements
Esri, USGS
Orange County Park Improvements
Kennedale Greenways
North Beach Eco Park
San Felipe Creek
Sandy Creek Park
New Braunfels Dry Comal Trail
Waco Bosque River Trail
Dallas Love Field Loop Trail
Bastrop County Open Space Master Plan Update
Penatuhkuh Comanche Cultural Landscapes

This RTCA state page uses an ArcGIS application to highlight annual projects in Texas.

Florida News
National Park Service
For more information on National Park Service community assistance, visit our website
Chattahoochee River Landing Park-Heritage Trail
The project goal is to review outdoor recreation amenities and recommend sustainable practices that protects the special archaeological history while providing camping, increased access, and greater outdoor recreation opportunity.
The National Park Service will assist the City of Chattahoochee in their recovery efforts by providing valuable public engagement, reviews, feedback, and recommendations on how to renovate and upgrade the park infrastructure while protecting their historical archaeological assets.
Project Partner: Chattahoochee Mainstreet
Project Location: Chattahoochee, Florida
NPS Contact: Deidre Hewitt & Russell Clark
Photo: Docking and kayak launch site along the Chattahoochee River in Chattahoochee, Florida.
Chattahoochee River Landing Park-Heritage Trail
Esri, USGS
Chattahoochee River Landing Park-Heritage Trail
Salinas Park - Gulfside
Mexico Beach Restoration Basins
Parker Parks Redevelopment
Springfield Parks Redevelopment
Bay County Parks
Marianna Redevelopment

This RTCA state page uses an ArcGIS application to highlight annual projects in Florida. Data sources: NPS Rivers, Trails, and Conservation Assistance Program.

JAPANESE AMERICAN REMEMBRANCE TRAIL

ARISA NAKAMURA, JAPANESE CULTURAL AND COMMUNITY CENTER OF WASHINGTON
STEPHANIE STROUD, RIVERS, TRAILS, AND CONSERVATION ASSISTANCE PROGRAM

The Japanese American Remembrance Trail is a walking route in Seattle, Washington, connecting over 40 historic sites in the present-day Chinatown-International District. It was created in 2017, the year that marked the 75th Anniversary of the Day of Remembrance, when President Franklin D. Roosevelt signed Executive Order 9066. The signing of EO 9066 resulted in the forced removal and incarceration of 120,000 Japanese and Japanese Americans (Nikkei) from the West Coast into US concentration camps. The Remembrance Trail serves to celebrate the rich culture of Japanese Americans during this time and throughout Seattle's history, promote businesses and organizations that work in the community today, and share the important message, "Never again."

The trail benefits Seattle residents and visitors alike and strives to be accessible for all ages and abilities, connecting generations. Along the trail is the Klondike Gold Rush National Historical Park, an NPS unit that was once owned by a Japanese American family, and The Wing, an Affiliated Area of the NPS. The trail offers a unique opportunity to leave the buildings of these two sites and explore the surrounding neighborhoods, offering a deeper connection to the stories of the community and creating a sense of place through healthy outdoor recreation. The trail was established as part of a grassroots community effort led by the voices of residents with planning assistance from the NPS Rivers, Trails, and Conservation Assistance Program (RTCA). Community members selected all the sites and RTCA provided GIS maps as tools to explore different safe and accessible walking routes for the trail over a series of iterative exercises. Local artist Arisa Nakamura was able to take the results of the mapping exercises and create her own unique map with added illustrations that matched the character of the historic sites along the trail.

Using the printed map, visitors can take a self-guided tour or can take a tour with a guide through The Wing. Along the trail, visitors can explore historic hotels, restaurants, businesses, community organizations, sculpture, greenspace, dojos, and much more.

The WING
Wing Luke Museum of the Asian Pacific American Experience

The logo for the Wing Luke Museum of the Asian Pacific American Experience.
Wing Luke Museum of the Asian Pacific American Experience.

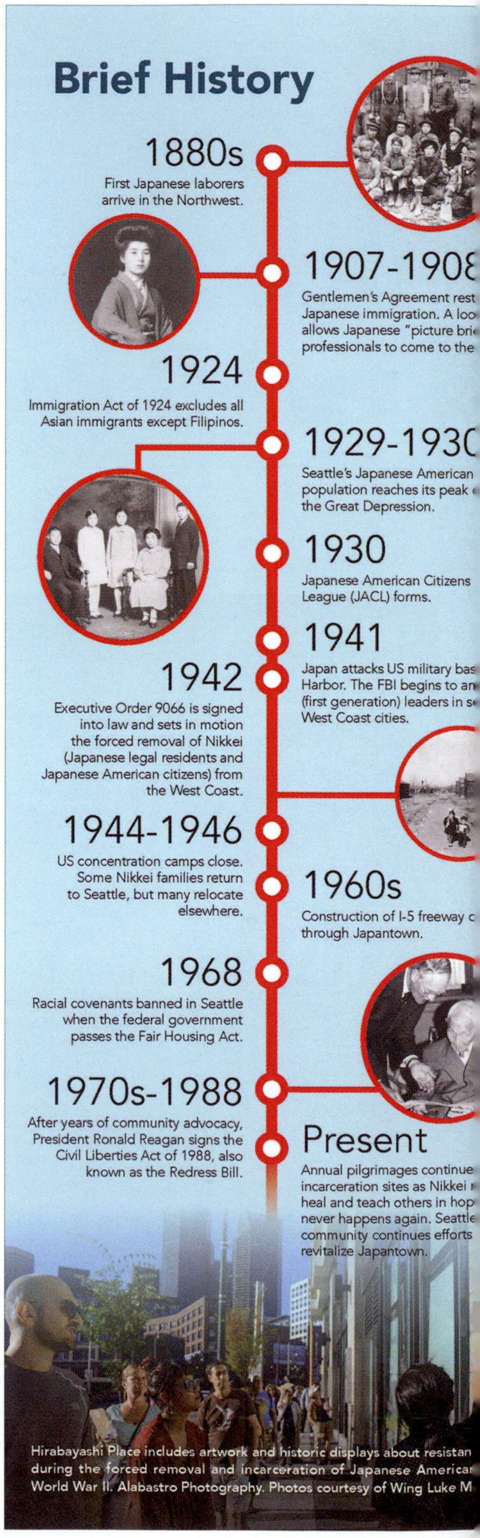
Brief History
1880s
First Japanese laborers arrive in the Northwest.
1924
Immigration Act of 1924 excludes all Asian immigrants except Filipinos.
1930
Japanese American Citizens League (JACL) forms.
1941
1942
Executive Order 9066 is signed into law and sets in motion the forced removal of Nikkei (Japanese legal residents and Japanese American citizens) from the West Coast.
1944-1946
US concentration camps close. Some Nikkei families return to Seattle, but many relocate elsewhere.
1960s
through Japantown.
1968
Racial covenants banned in Seattle when the federal government passes the Fair Housing Act.
1970s-1988
After years of community advocacy, President Ronald Reagan signs the Civil Liberties Act of 1988, also known as the Redress Bill.
Present

Hirabayashi Place includes artwork and historic displays about resistan during the forced removal and incarceration of Japanese Americar World War II. Alabastro Photography. Photos courtesy of Wing Luke M

The Japanese American Remembrance Trail Map (page 1) shows a timeline. Illustrations by Arisa Nakamura.

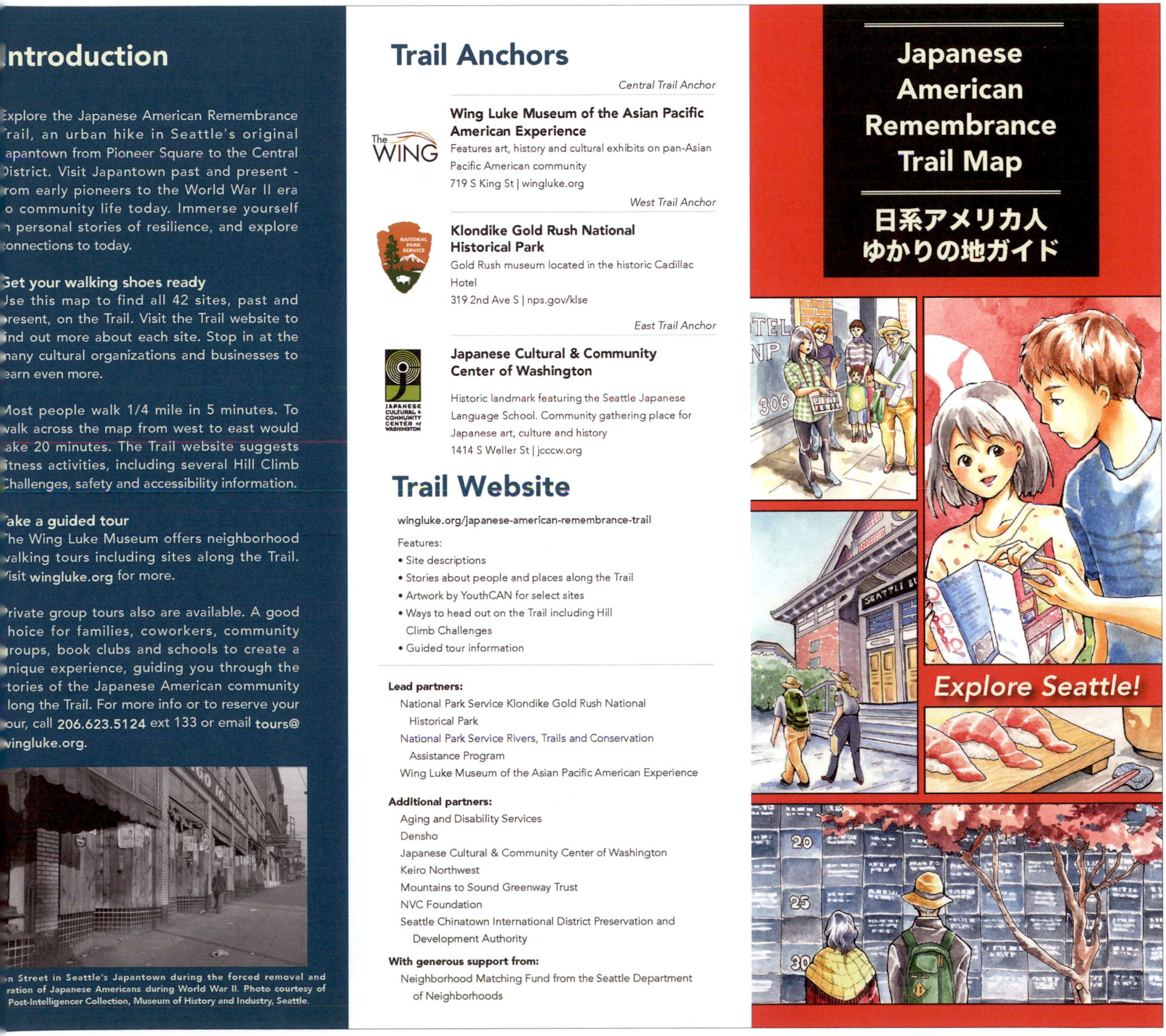

ntroduction

xplore the Japanese American Remembrance rail, an urban hike in Seattle's original apantown from Pioneer Square to the Central istrict. Visit Japantown past and present - rom early pioneers to the World War II era o community life today. Immerse yourself n personal stories of resilience, and explore onnections to today.

et your walking shoes ready
Jse this map to find all 42 sites, past and resent, on the Trail. Visit the Trail website to ind out more about each site. Stop in at the nany cultural organizations and businesses to earn even more.

Most people walk 1/4 mile in 5 minutes. To valk across the map from west to east would ake 20 minutes. The Trail website suggests itness activities, including several Hill Climb hallenges, safety and accessibility information.

ake a guided tour
he Wing Luke Museum offers neighborhood valking tours including sites along the Trail. isit **wingluke.org** for more.

rivate group tours also are available. A good hoice for families, coworkers, community roups, book clubs and schools to create a nique experience, guiding you through the tories of the Japanese American community long the Trail. For more info or to reserve your our, call **206.623.5124** ext 133 or email **tours@ vingluke.org**.

n Street in Seattle's Japantown during the forced removal and ration of Japanese Americans during World War II. Photo courtesy of Post-Intelligencer Collection, Museum of History and Industry, Seattle.

Trail Anchors

Central Trail Anchor

Wing Luke Museum of the Asian Pacific American Experience
Features art, history and cultural exhibits on pan-Asian Pacific American community
719 S King St | wingluke.org

West Trail Anchor

Klondike Gold Rush National Historical Park
Gold Rush museum located in the historic Cadillac Hotel
319 2nd Ave S | nps.gov/klse

East Trail Anchor

Japanese Cultural & Community Center of Washington
Historic landmark featuring the Seattle Japanese Language School. Community gathering place for Japanese art, culture and history
1414 S Weller St | jcccw.org

Trail Website

wingluke.org/japanese-american-remembrance-trail

Features:
- Site descriptions
- Stories about people and places along the Trail
- Artwork by YouthCAN for select sites
- Ways to head out on the Trail including Hill Climb Challenges
- Guided tour information

Lead partners:
National Park Service Klondike Gold Rush National Historical Park
National Park Service Rivers, Trails and Conservation Assistance Program
Wing Luke Museum of the Asian Pacific American Experience

Additional partners:
Aging and Disability Services
Densho
Japanese Cultural & Community Center of Washington
Keiro Northwest
Mountains to Sound Greenway Trust
NVC Foundation
Seattle Chinatown International District Preservation and Development Authority

With generous support from:
Neighborhood Matching Fund from the Seattle Department of Neighborhoods

Japanese American Remembrance Trail Map

日系アメリカ人ゆかりの地ガイド

Explore Seattle!

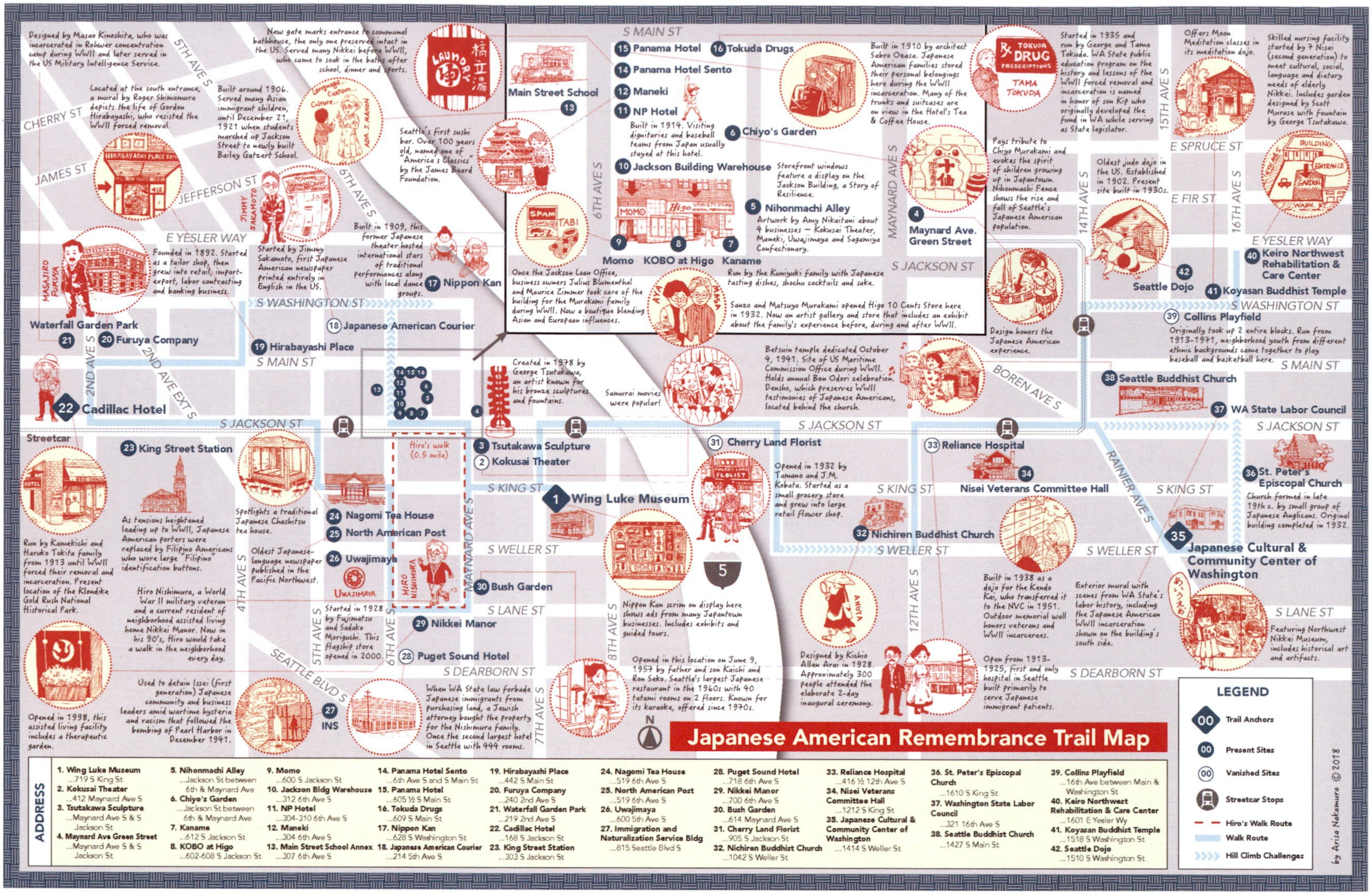
Japanese American Remembrance Trail Map
ADDRESS
1. Wing Luke Museum ...719 S King St
2. Kokusai Theater ...412 Maynard Ave S
3. Tsutakawa Sculpture ...Maynard Ave S & S Jackson St
4. Maynard Ave Green Street ...Maynard Ave S & S Jackson St
5. Nihonmachi Alley ...Jackson St between 6th & Maynard Ave
6. Chiyo's Garden ...Jackson St between 6th & Maynard Ave
7. Kaname ...612 S Jackson St
8. KOBO at Higo ...602-608 S Jackson St
9. Momo ...600 S Jackson St
10. Jackson Bldg Warehouse ...312 6th Ave S
11. NP Hotel ...304-310 6th Ave S
12. Maneki ...304 6th Ave S
13. Main Street School Annex ...307 6th Ave S
14. Panama Hotel Sento ...6th Ave S and S Main St
15. Panama Hotel ...605 ½ S Main St
16. Tokuda Drugs ...609 S Main St
17. Nippon Kan ...628 S Washington St
18. Japanese American Courier ...214 5th Ave S
19. Hirabayashi Place ...442 S Main St
20. Furuya Company ...240 2nd Ave S
21. Waterfall Garden Park ...219 2nd Ave S
22. Cadillac Hotel ...168 S Jackson St
23. King Street Station ...303 S Jackson St
24. Nagomi Tea House ...519 6th Ave S
25. North American Post ...519 6th Ave S
26. Uwajimaya ...600 5th Ave S
27. Immigration and Naturalization Service Bldg ...815 Seattle Blvd S
28. Puget Sound Hotel ...718 6th Ave S
29. Nikkei Manor ...700 6th Ave S
30. Bush Garden ...614 Maynard Ave S
31. Cherry Land Florist ...905 S Jackson St
32. Nichiren Buddhist Church ...1042 S Weller St
33. Reliance Hospital ...416 ½ 12th Ave S
34. Nisei Veterans Committee Hall ...1212 S King St
35. Japanese Cultural & Community Center of Washington ...1414 S Weller St
36. St. Peter's Episcopal Church ...1610 S King St
37. Washington State Labor Council ...321 16th Ave S
38. Seattle Buddhist Church ...1427 S Main St
39. Collins Playfield ...16th Ave between Main & Washington St
40. Keiro Northwest Rehabilitation & Care Center ...1601 E Yesler Wy
41. Koyasan Buddhist Temple ...1518 S Washington St
42. Seattle Dojo ...1510 S Washington St
LEGEND
Trail Anchors
Present Sites
Vanished Sites
Streetcar Stops
Hiro's Walk Route
Walk Route
Hill Climb Challenges
by Aria Nakamura © 2018

Facing page, clockwise from top:
The Japanese American Remembrance Trail Map (page 2) shows addresses of locations. Illustrations by Arisa Nakamura.

The wall of Chiyo's Garden, a site along the walking trail, which commemorates the Japanese American population in Seattle over time. Alabastro Photography/Alan Alabastro.

Visitors on a pilot walk for the Japanese American Remembrance Trail view a handheld historic image. Alabastro Photography/Alan Alabastro.

This page, clockwise from top:
A tour guide shares stories along the Japanese American Remembrance Trail. Alabastro Photography/Alan Alabastro.

Modern-day Japantown in the Chinatown-International District in Seattle. Alabastro Photography/Alan Alabastro.

A modern-day retail shop is paired with a historic sign from a Japanese American–owned variety store along the trail. Alabastro Photography/Alan Alabastro.

(*Above*) The planning group used a series of maps in their process to select sites, make connections, and create a safe walking route for the trail. Alabastro Photography/Alan Alabastro.

(*Below and right*) Visitors along the Japanese American Remembrance Trail participate in a pilot walk to test and provide feedback for the trail. Alabastro Photography/Alan Alabastro.

9 PARK MANAGEMENT

PATRICK GREGERSON, NPS PARK PLANNING AND SPECIAL STUDIES

SUPPORTING PARK MANAGEMENT

The National Park Service (NPS) has been going though growing pains in various ways. The NPS recently celebrated its 100th anniversary, bringing with it unprecedented attention and visitation to most of its parks. This attention presents many challenges that parks are unprepared to handle without careful planning and management support to provide for a quality visitor experience while protecting valuable resources.

Planning is key to addressing the needs of the park, and it guides informed and insightful decisions that provide relevant and timely direction to park management. An NPS foundation document serves as the underlying guidance for all management and planning decisions for a national park unit by describing the core mission of the park unit and identifying the purpose, significance, fundamental and important resources and values, interpretive themes, assessment of planning and data needs, special mandates and administrative commitments, and the unit's setting in the regional context.

The park atlas was developed as a fundamental component of every foundation document. Each park atlas is a compilation of that park unit's unique baseline GIS data presented on an interactive web mapping site.

For over 30 years, GIS has been used as an integral part of the planning process. Today, GIS is incorporated into every management aspect in the NPS. Over 280 parks have identified the need for GIS data in the park's foundation document through the assessment of planning and data needs. These include, but are not limited to, natural and cultural resources, scenic resources, visitor opportunities, and regional land use. GIS specialists now join park planners to develop alternatives that protect resources while allowing visitors to experience their parks and learn about nature. The evolution of GIS technology offers exciting possibilities as NPS planners strive to understand our nation's dynamic ecosystems and protect them for generations to come.

IMPROVING COMMUNICATION WITH WEB MAPS IN THE MID-ATLANTIC

JUSTIN SHEDD, MAKIKO SHUKUNOBE, NICOLE INGLIS, DR. MEGAN SKRIP, AND DR. JELENA VUKOMANOVIC, CENTER FOR GEOSPATIAL ANALYTICS, NORTH CAROLINA STATE UNIVERSITY

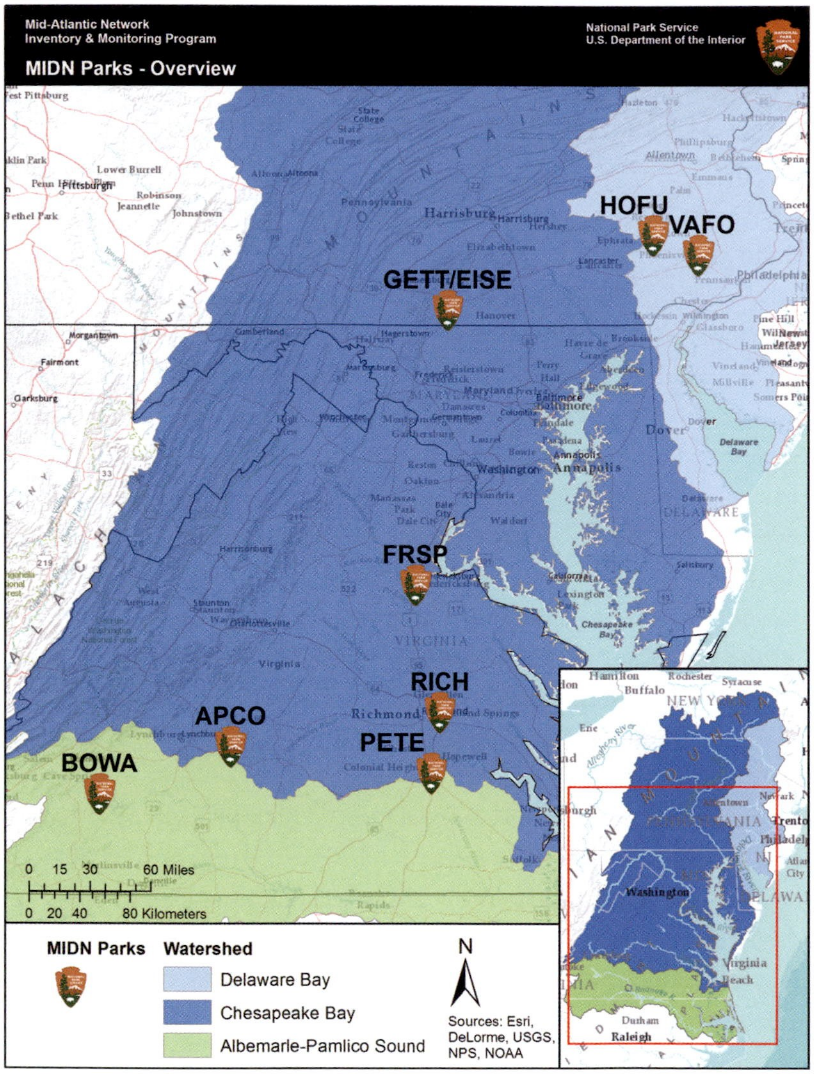

This map shows the NPS Mid-Atlantic Inventory and Monitoring Network. Map by Nathan Dammeyer, NPS.

Sprawling invasive vines, flammable weeds that overgrow delicate marshland communities, and sapling trees that encroach on bird-filled meadows are all challenges that parks face in the NPS Mid-Atlantic Inventory and Monitoring Network. To overcome these challenges, teams within the network use prescribed fire, herbicides, and mowers to remove problem plants that threaten native ecosystems. Maintaining the health and safety of workers and optimizing time and resources are top priorities for these teams, so seamless communication is essential. With so many people working on specialized projects across the parks, a widely accessible data-sharing system is particularly important. What tools can best meet this need? Enter customized ArcGIS Online web maps, which now allow park service staff and collaborators to view and upload a wide variety of data, even while out in the field. Created by researchers from the Center for Geospatial Analytics at North Carolina State University, these web maps display the where, when, and what of management treatments across nine national parks from central Virginia to southeast Pennsylvania, providing a complete picture of current and past management and allowing new information to be continually added. In recognition for its sweeping impact on improving communication, this mapping project received the 2017 Inventory and Monitoring Division Science Management Partner Award, honoring team members from the Mid-Atlantic Inventory and Monitoring Network, network parks, NPS Wildland Fire, NPS Exotic Plant Management Teams, NPS Natural Resource Stewardship and Science Northeast Regional Office, and the Center for Geospatial Analytics.

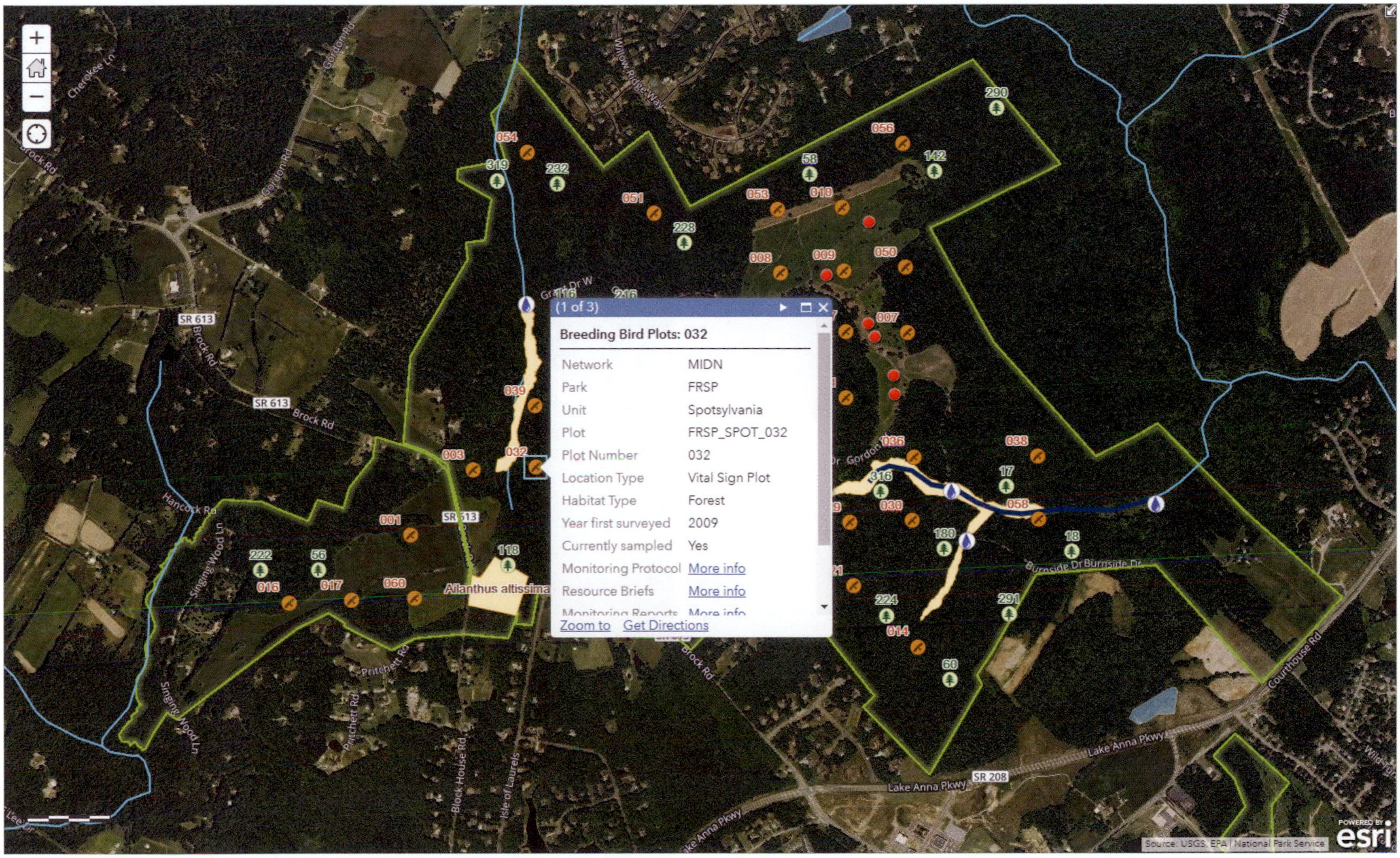

An example of one web map developed for Fredericksburg & Spotsylvania National Military Park in Virginia shows inventory and monitoring plot locations (breeding bird, forest vegetation, and aquatic monitoring), fire effects plots (red circles), and Exotic Plant Management Team treatments. Image provided by NC State University. Data sources: NPS, USGS, Esri.

Clockwise from top left:

An NPS wildland firefighter helps implement a prescribed fire. Photo credit: Grace Crawford, NPS.

An NPS Exotic Plant Management Team member conducts chemical treatments to remove exotic species. Photo credit: Nathan Wender, NPS.

Forest vegetation monitoring in Valley Forge National Historic Park. Photo credit: Mark Johnson, NPS.

The blacknose dace (*Rhinichthys atratulus*) is one of the fish species inventoried during water quality monitoring in Valley Forge National Historical Park. Photo credit: Mark Johnson, NPS.

INTEGRATED PARK IMPROVEMENTS AT CHACO CULTURE NATIONAL HISTORICAL PARK

LAURA BABCOCK, NPS DENVER SERVICE CENTER PLANNING DIVISION

Integrated park improvements (IPI) is a strategic process to foster an integrated and comprehensive approach to facility planning at a park or district level. Through the IPI process, a strategy is developed to guide improvement of a park unit or district with identified resource, visitor experience, and facility improvement needs. The process is comprehensive in that, where possible, projects at the park or district level are combined to guide the rehabilitation, maintenance, and funding of priority assets and resources. Priority assets are identified by mapping optimizer bands, in which individual assets are given a band value ranging from one to five that represents the level of maintenance it should receive. Optimizer bands are one of many attributes listed in the facility management software system (FMSS) that contains a wealth of asset information to help parks plan for future work. The IPI process also identifies assets that have project management information system (PMIS) projects by mapping PMIS statement numbers to each individual asset. Assets with several PMIS projects are identified as high priority for future improvement. Combining optimizer band information with PMIS project locations helps inform the strategic and coordinated IPI process.

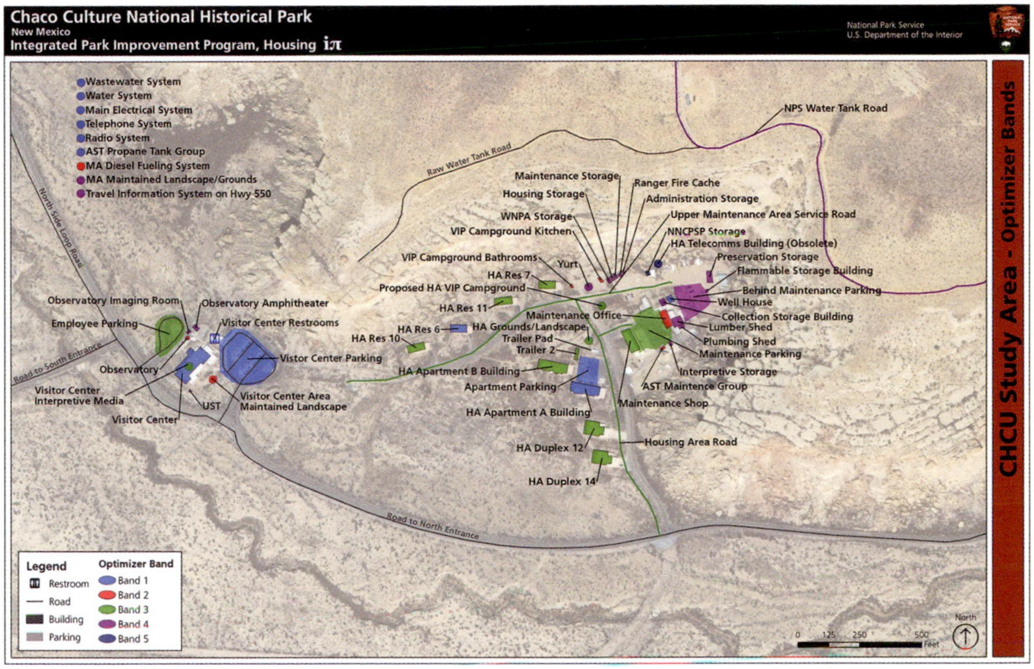

Park assets are represented by associated optimizer bands from the facility management software system (FMSS).

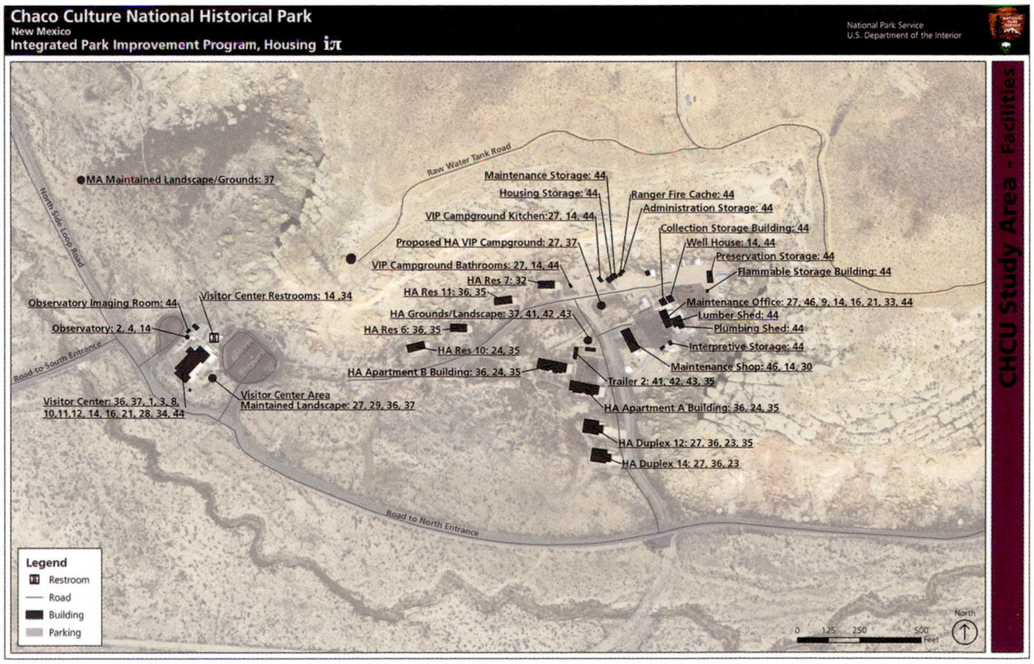

Park facility assets are identified at Chaco Culture National Historical Park in New Mexico in the project management information system (PMIS). Data sources: NPS, Esri.

WIND TURBINE VISIBILITY AT HOMESTEAD NATIONAL MONUMENT OF AMERICA

MATTHEW COLWIN, NPS MIDWEST REGION GEOSPATIAL SERVICES

The map in this section depicts how visible proposed wind turbines would be from locations in the area surrounding Homestead National Monument of America (HOME) in Nebraska. The analysis was performed at multiple important viewpoints around the park, and this map shows percent visibility from the park's native plant display outside the Education Center. The analysis was specifically designed to test how visible a specific proposed wind turbine would be, but the analysis can be generalized to show the range of the potential visual impacts from any location on the landscape. This map is one example of a growing trend of projects in the NPS Midwest Region related to wind power. As the number of wind farm projects increases, many new vertical elements are being introduced into a largely horizontal landscape, so parks have an increased need to know how these turbines will affect their viewsheds. These maps have been used by HOME to help enact changes to the placement of wind turbines to minimize visual intrusions.

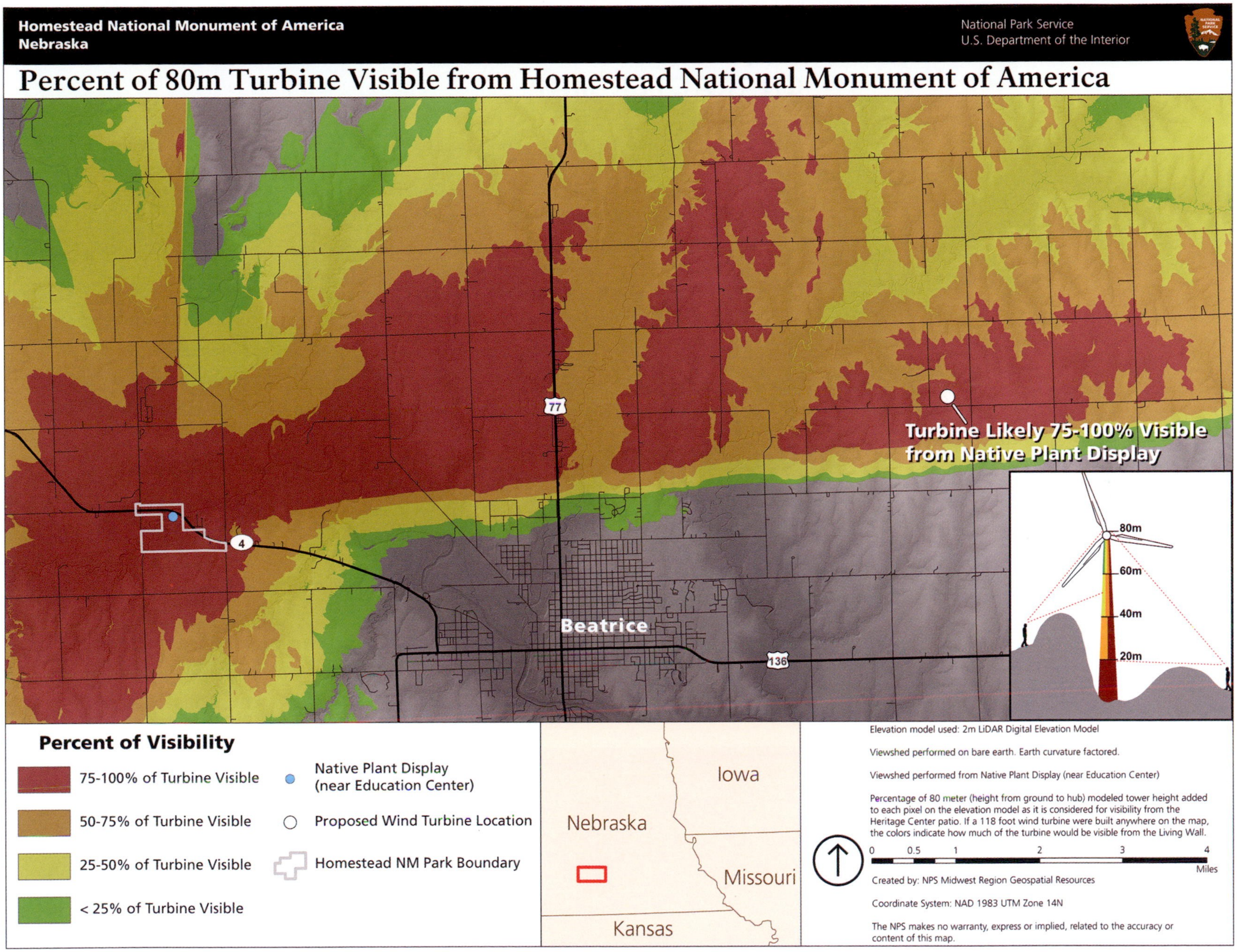

This map shows proposed turbine location and visibility viewshed analysis from the Homestead National Monument of America in Nebraska. Data sources: NPS (Land Resources Division and Midwest Region Geospatial Resources), USGS, Esri.

SEQUOIA AND KINGS CANYON WILDERNESS CHARACTER MAP

JAMES TRICKER AND PETER LANDRES, ALDO LEOPOLD WILDERNESS RESEARCH INSTITUTE, US FOREST SERVICE
GREGG FAUTH AND PAUL HARDWICK, SEQUOIA AND KINGS CANYON NATIONAL PARKS
ALEX EDDY, NPS INVENTORY AND MONITORING DIVISION, SIERRA NEVADA NETWORK

Sequoia and Kings Canyon National Parks (SEKI) and the US Forest Service's Aldo Leopold Wilderness Research Institute worked together to map the current condition of wilderness across the parks based on the wilderness characteristics of natural, untrammeled, undeveloped, and on opportunities for solitude described in the interagency strategy, Keeping It Wild (Landres et al. 2008). The project included these objectives:

- Show the current condition of wilderness character and how it varies across SEKI's 837,806 acres of wilderness.
- Provide a baseline from which future monitoring can show changes in wilderness character.
- Allow the parks to analyze the potential impacts of management actions on wilderness character.
- Identify areas within the wilderness where resource managers should make an effort to control or mitigate impacts.
- Improve internal and external communication about wilderness-related issues.

A backpacker hikes through foxtail pine at Little Claire Lake, Sequoia National Park. Photo credit: Linda Mutch, NPS.

SEKI staff identified a set of measures for each indicator to capture negative impacts on wilderness. Project participants used 79 spatial datasets to measure and delineate wilderness character. Variation in the scale, accuracy, and completeness in the data placed limitations on how the map products were developed. Measures were assigned under an appropriate indicator and weighted. The weighted measures for each indicator were added together, and maps were developed for each quality. The four different maps for each quality were added together to produce a composite wilderness character map.

Interpreting this map requires a clear understanding of the methods and data used. To understand why these areas are degraded, one must drill down into the individual qualities, indicators, and measures.

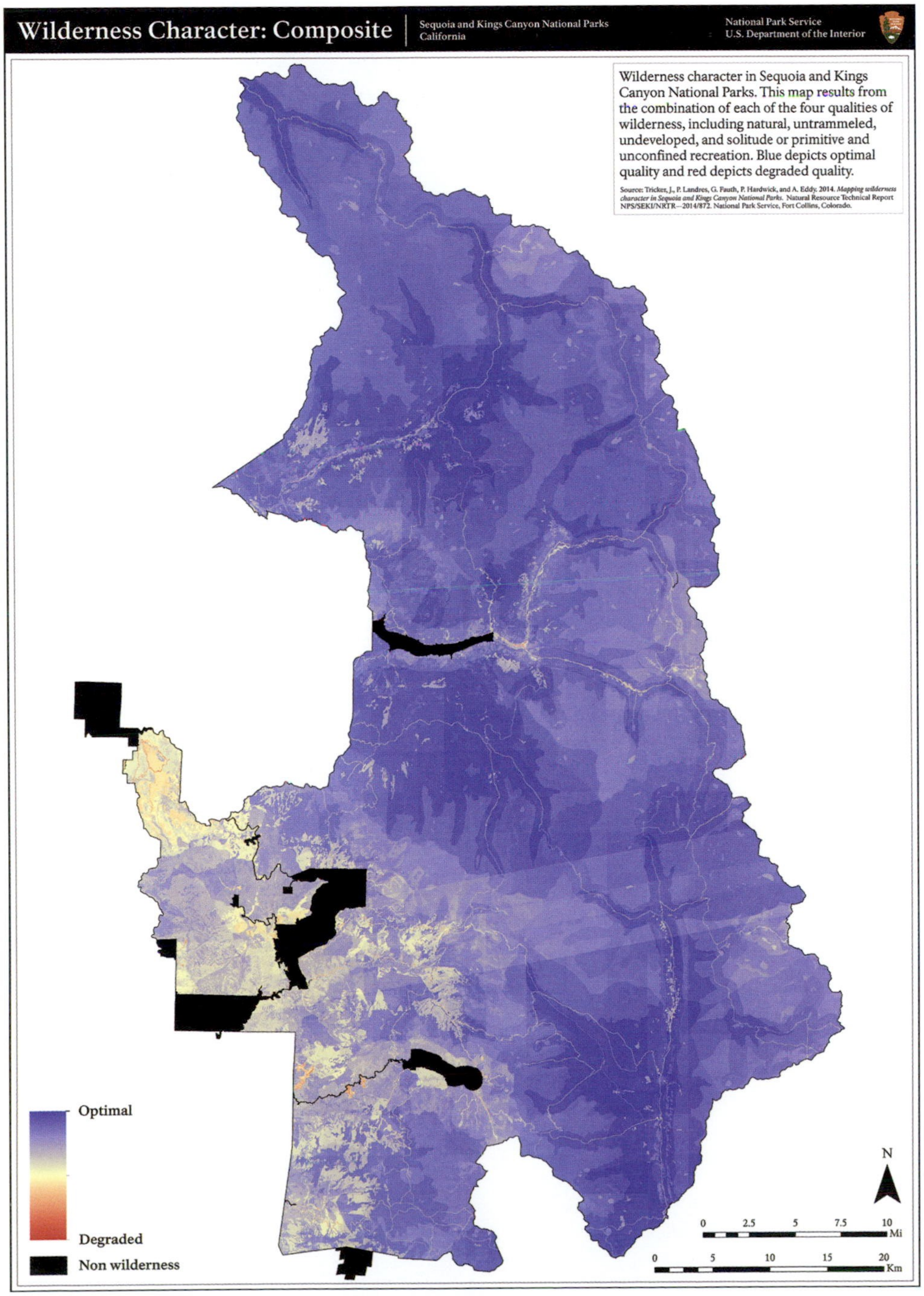
Wilderness Character: Composite
Sequoia and Kings Canyon National Parks
California
National Park Service
U.S. Department of the Interior
Wilderness character in Sequoia and Kings Canyon National Parks. This map results from the combination of each of the four qualities of wilderness, including natural, untrammeled, undeveloped, and solitude or primitive and unconfined recreation. Blue depicts optimal quality and red depicts degraded quality.
Source: Tricker, J., P. Landres, G. Fauth, P. Hardwick, and A. Eddy. 2014. Mapping wilderness character in Sequoia and Kings Canyon National Parks. Natural Resource Technical Report NPS/SEKI/NRTR—2014/872. National Park Service, Fort Collins, Colorado.
Optimal
Degraded
Non wilderness
N
0 2.5 5 7.5 10 Mi
0 5 10 15 20 Km

A derived Sequoia and Kings Canyon National Parks composite wilderness character basemap.

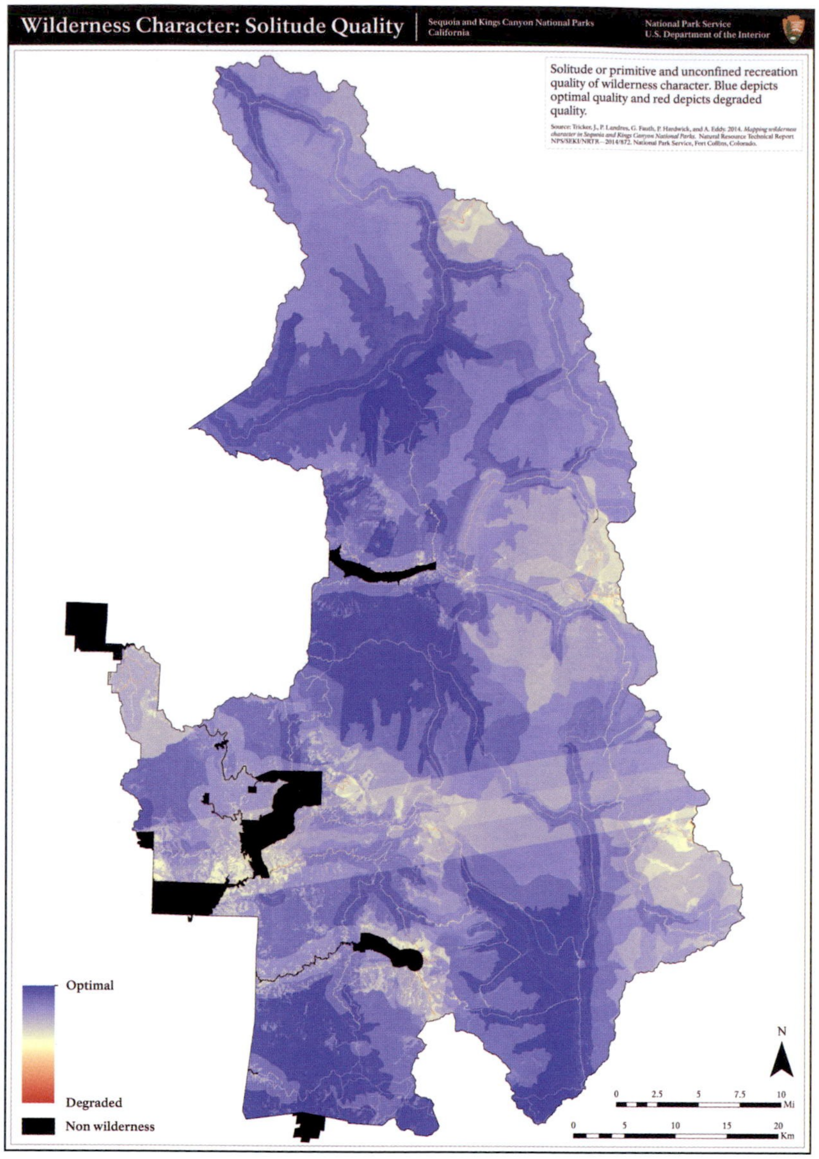
Wilderness Character: Solitude Quality
Sequoia and Kings Canyon National Parks
California
National Park Service
U.S. Department of the Interior
Solitude or primitive and unconfined recreation quality of wilderness character. Blue depicts optimal quality and red depicts degraded quality.
Source: Tricker, J., P. Landres, G. Fauth, P. Hardwick, and A. Eddy. 2014. Mapping wilderness character in Sequoia and Kings Canyon National Parks. Natural Resource Technical Report NPS/SEKI/NRTR—2014/872. National Park Service, Fort Collins, Colorado.
Optimal
Degraded
Non wilderness
N
0 2.5 5 7.5 10 Mi
0 5 10 15 20 Km

A derived Sequoia and Kings Canyon National Parks wilderness solitude wilderness character basemap.

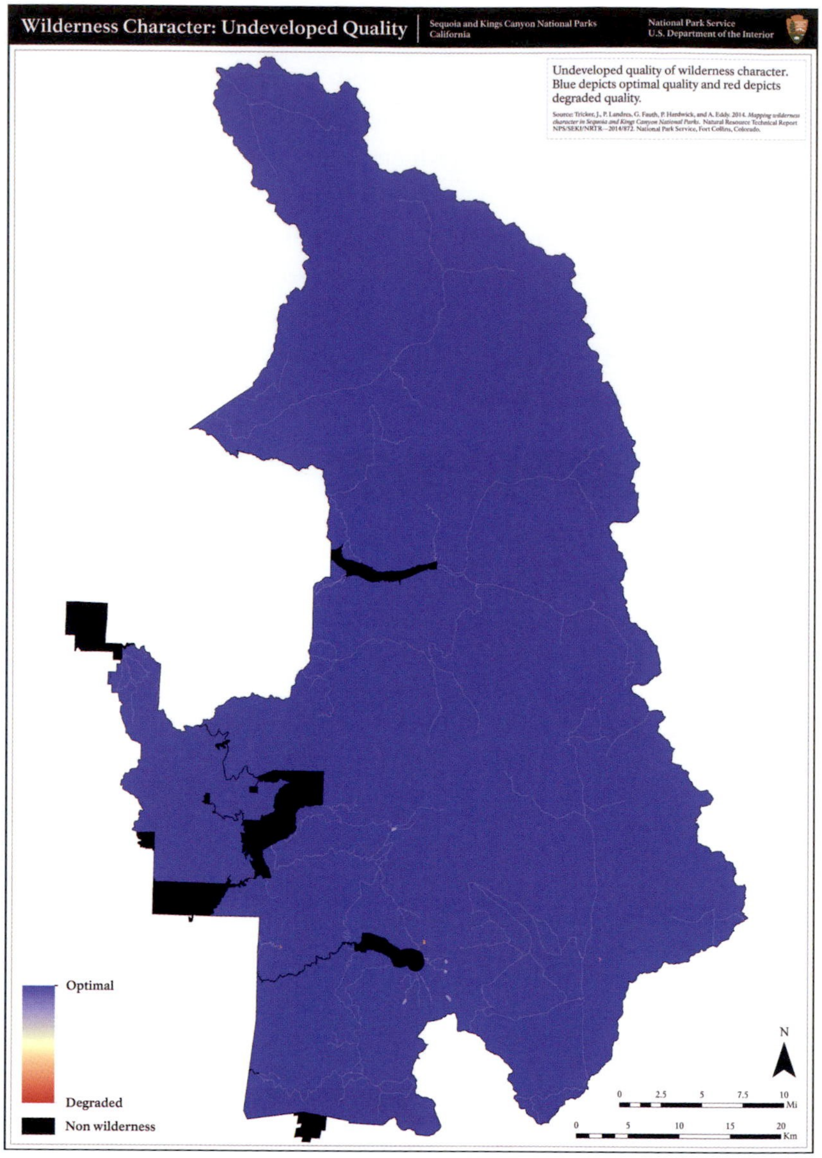
Wilderness Character: Undeveloped Quality
Sequoia and Kings Canyon National Parks
California
National Park Service
U.S. Department of the Interior
Undeveloped quality of wilderness character. Blue depicts optimal quality and red depicts degraded quality.
Source: Tricker, J., P. Landres, G. Fauth, P. Hardwick, and A. Eddy. 2014. Mapping wilderness character in Sequoia and Kings Canyon National Parks. Natural Resource Technical Report NPS/SEKI/NRTR—2014/872. National Park Service, Fort Collins, Colorado.
Optimal
Degraded
Non wilderness
N
0 2.5 5 7.5 10 Mi
0 5 10 15 20 Km

A derived Sequoia and Kings Canyon National Parks wilderness undeveloped wilderness character basemap.

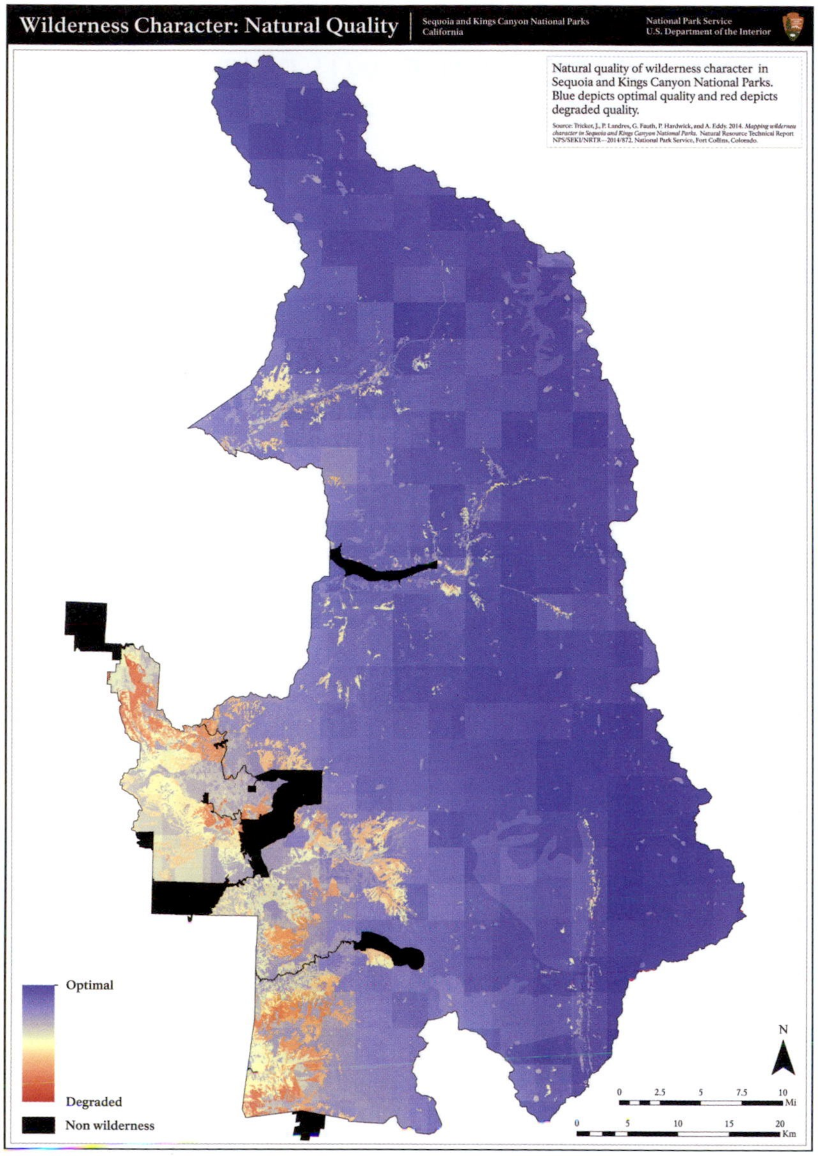
Wilderness Character: Natural Quality
Sequoia and Kings Canyon National Parks
California
National Park Service
U.S. Department of the Interior
Natural quality of wilderness character in Sequoia and Kings Canyon National Parks. Blue depicts optimal quality and red depicts degraded quality.
Source: Tricker, J., P. Landres, G. Fauth, P. Hardwick, and A. Eddy. 2014. Mapping wilderness character in Sequoia and Kings Canyon National Parks. Natural Resource Technical Report NPS/SEKI/NRTR—2014/872. National Park Service, Fort Collins, Colorado.
Optimal
Degraded
Non wilderness
N
0 2.5 5 7.5 10 Mi
0 5 10 15 20 Km

A derived Sequoia and Kings Canyon National Parks wilderness natural wilderness character basemap.

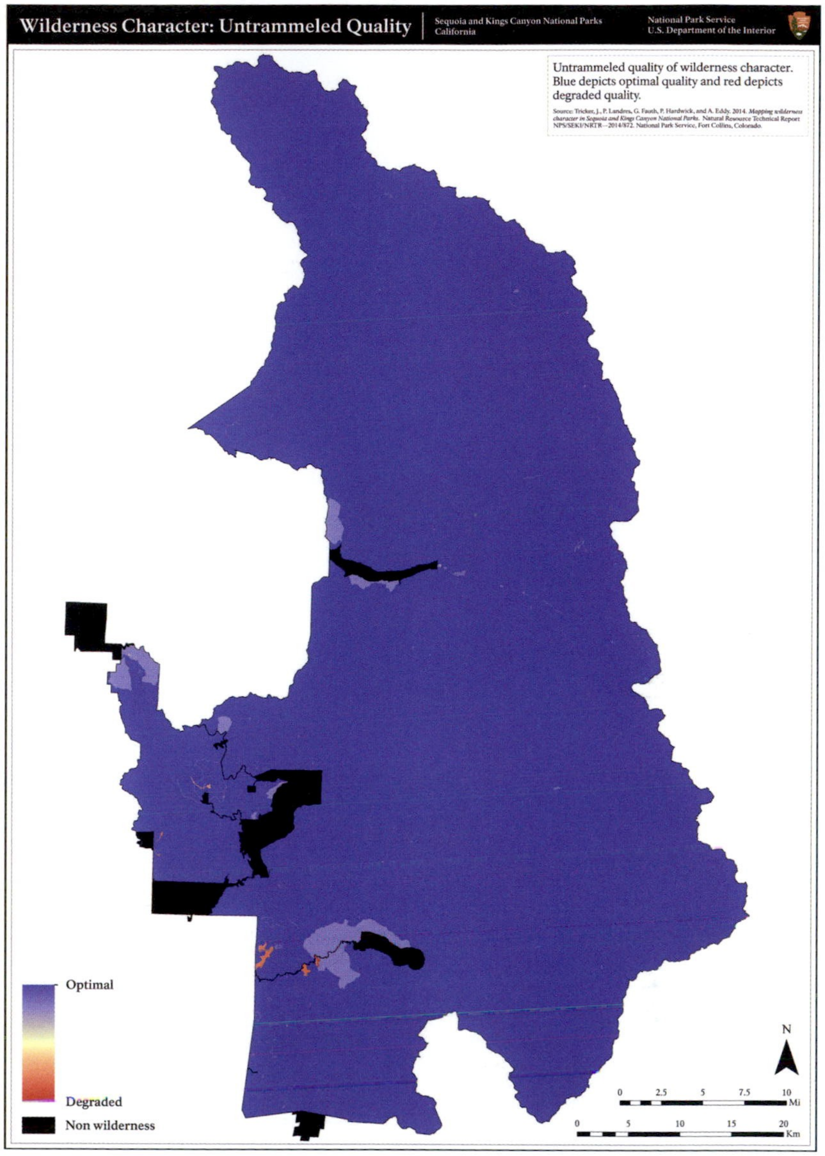
Wilderness Character: Untrammeled Quality
Sequoia and Kings Canyon National Parks
California
National Park Service
U.S. Department of the Interior
Untrammeled quality of wilderness character. Blue depicts optimal quality and red depicts degraded quality.
Source: Tricker, J., P. Landres, G. Fauth, P. Hardwick, and A. Eddy. 2014. Mapping wilderness character in Sequoia and Kings Canyon National Parks. Natural Resource Technical Report NPS/SEKI/NRTR—2014/872. National Park Service, Fort Collins, Colorado.
Optimal
Degraded
Non wilderness
N
0 2.5 5 7.5 10 Mi
0 5 10 15 20 Km

A derived Sequoia and Kings Canyon National Parks wilderness untrammeled wilderness character basemap. Data sources: Sequoia and Kings Canyon National Parks, US Forest Service Aldo Leopold Wilderness Research Institute.

MAPPING ELIGIBLE WILDERNESS IN ALASKA NATIONAL PARK UNITS

NPS ALASKA REGION GIS TEAM

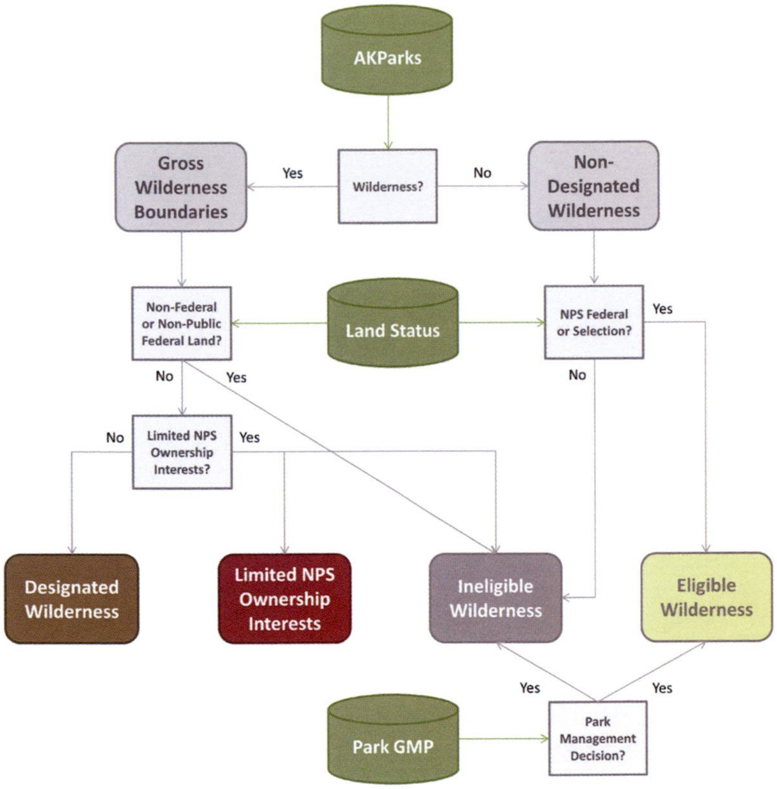

(*Above*) The overall workflow of the Alaska Wilderness Model is shown. (*Below*) The Alaska Wilderness Model is a programmatic sequence of geospatial-related commands that automates the otherwise onerous manual process.

The basis for much of Alaska's wilderness eligibility comes from regulations and definitions from the 1980s. The eligibility determinations made in Alaska at that time create challenges for managers today. Wilderness eligibility review content in general management plans (GMPs) from the 1980s was non-digital. These maps provide insufficient detail due to both scale and technical accuracy. Improvements in geospatial software for mapping and spatial analysis have significantly improved the accuracy for locating features and extent on the ground. In some locations, the wilderness eligibility determinations are no longer representative of the current boundary or land conditions.

Since the 1980s, park and wilderness boundaries have changed due to changes in land ownership or use. In response, the Wilderness Eligibility Mapping project was initiated to increase the accuracy of wilderness information by developing digital representations of NPS Alaska wilderness areas. Through this effort, the spatial extent of congressionally designated wilderness and eligible wilderness areas are accurately mapped in a manner that reflects the original intent of the GMP wilderness reviews, but within current land status and management conditions. The primary tool developed and used to update wilderness area boundaries and acreages is the Alaska Wilderness Model, an automated ModelBuilder™ routine that uses a controlled set of source inputs, structured data queries, and geoprocessing tools executed in a set order to accomplish specific tasks for NPS Alaska wilderness mapping. The updated wilderness data and maps facilitate wilderness preservation and management and create a more accurate accounting of Alaska's wilderness areas.

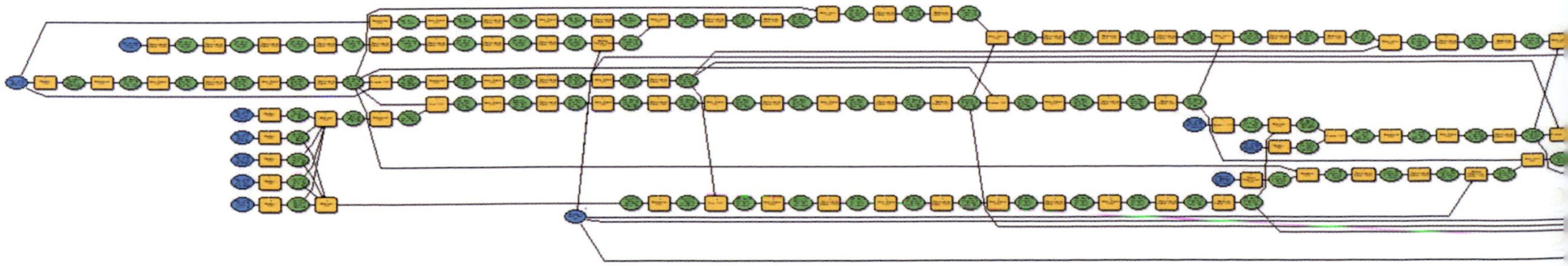

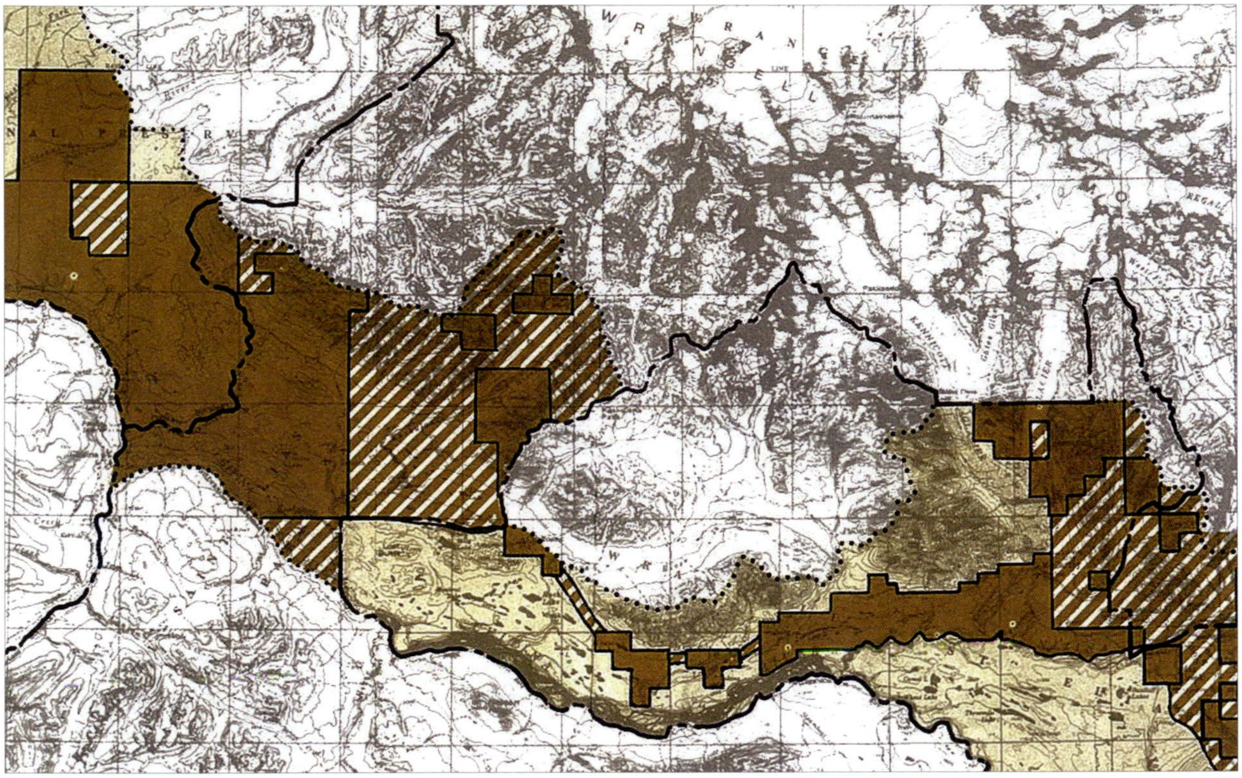

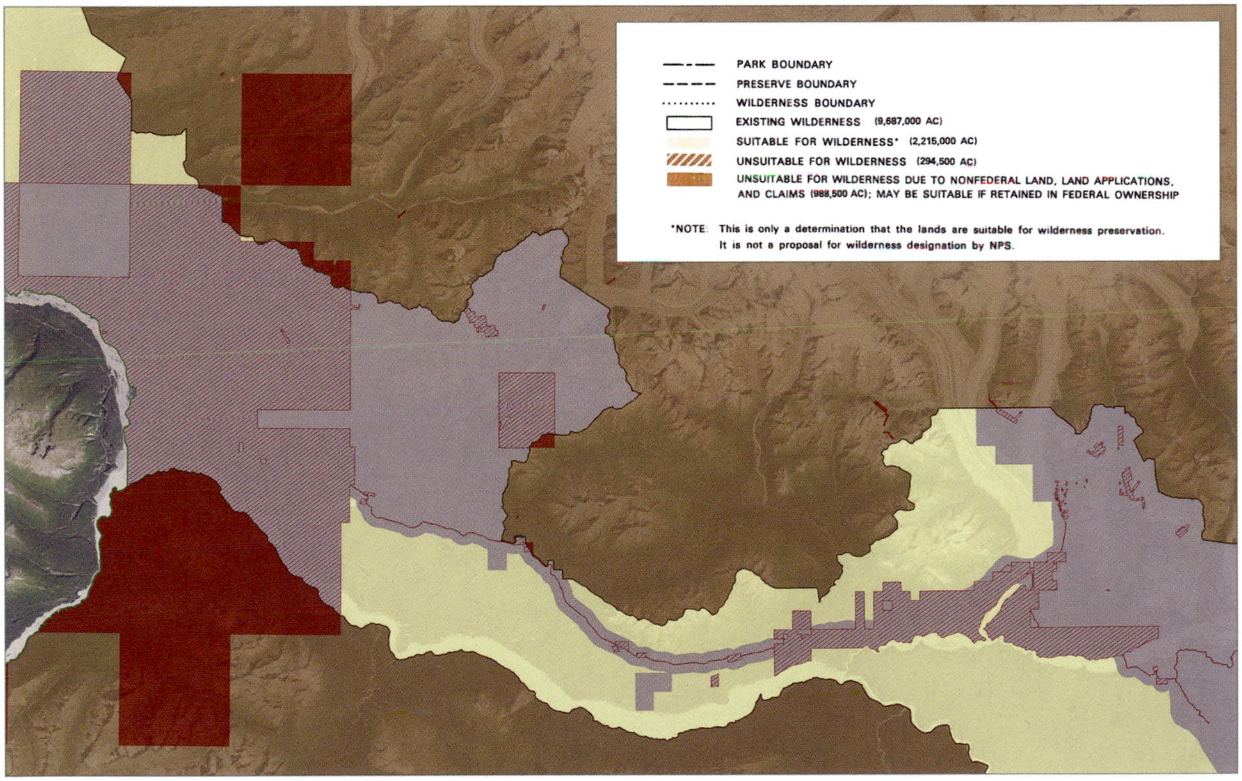

A comparison of wilderness eligibility in Wrangell-St. Elias National Park and Preserve. The top image depicts wilderness eligibility per the 1986 Wrangell-St. Elias National Park and Preserve General Management Plan, while the bottom image depicts current wilderness eligibility mapping in the same area. Data sources for the current map: NPS (NPS) Alaska Region GIS Team; Alaska Department of Natural Resources, Information Resource Management; and USGS (USGS) (locator map only).

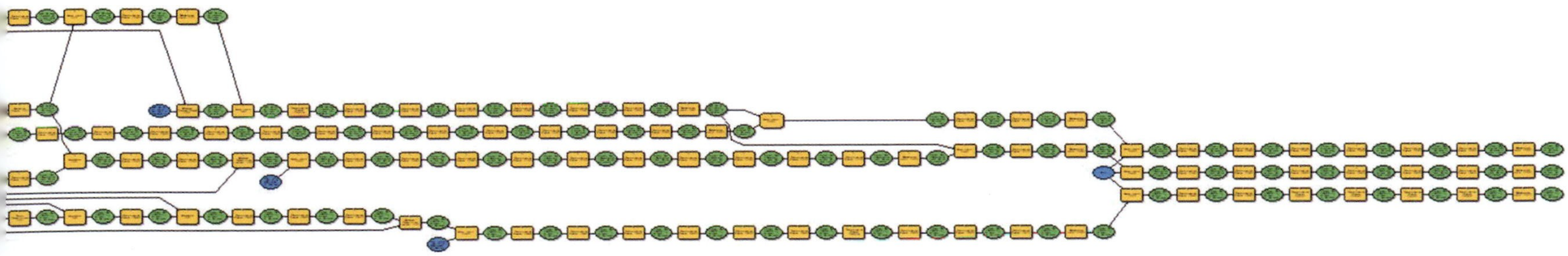

ICE AGE COMPLEX GLACIATED AREA

DOUGLAS WILDER, NPS DENVER SERVICE CENTER PLANNING DIVISION
MATTHEW COLWIN, NPS MIDWEST REGION GEOSPATIAL RESOURCES

This map was created as part of the planning process for the Ice Age Complex at Cross Plains, Wisconsin, a piece of land acquired by the Ice Age National Scenic Trail. This map depicts what the Ice Age Complex would have looked like during the maximum extent of glaciation. The stark contrast in topography between the glaciated and driftless areas can be easily seen on the map, which was beneficial to trail staff as they planned various alternative management strategies for the Ice Age Complex. Though it was created during the planning process, a version of the map is still used by trail staff to help interpret the Ice Age Complex and the larger story of the Ice Age National Scenic Trail. It is used as a way to help visitors understand how glaciation affected Wisconsin and to demonstrate how its effects can still be seen on the landscape today.

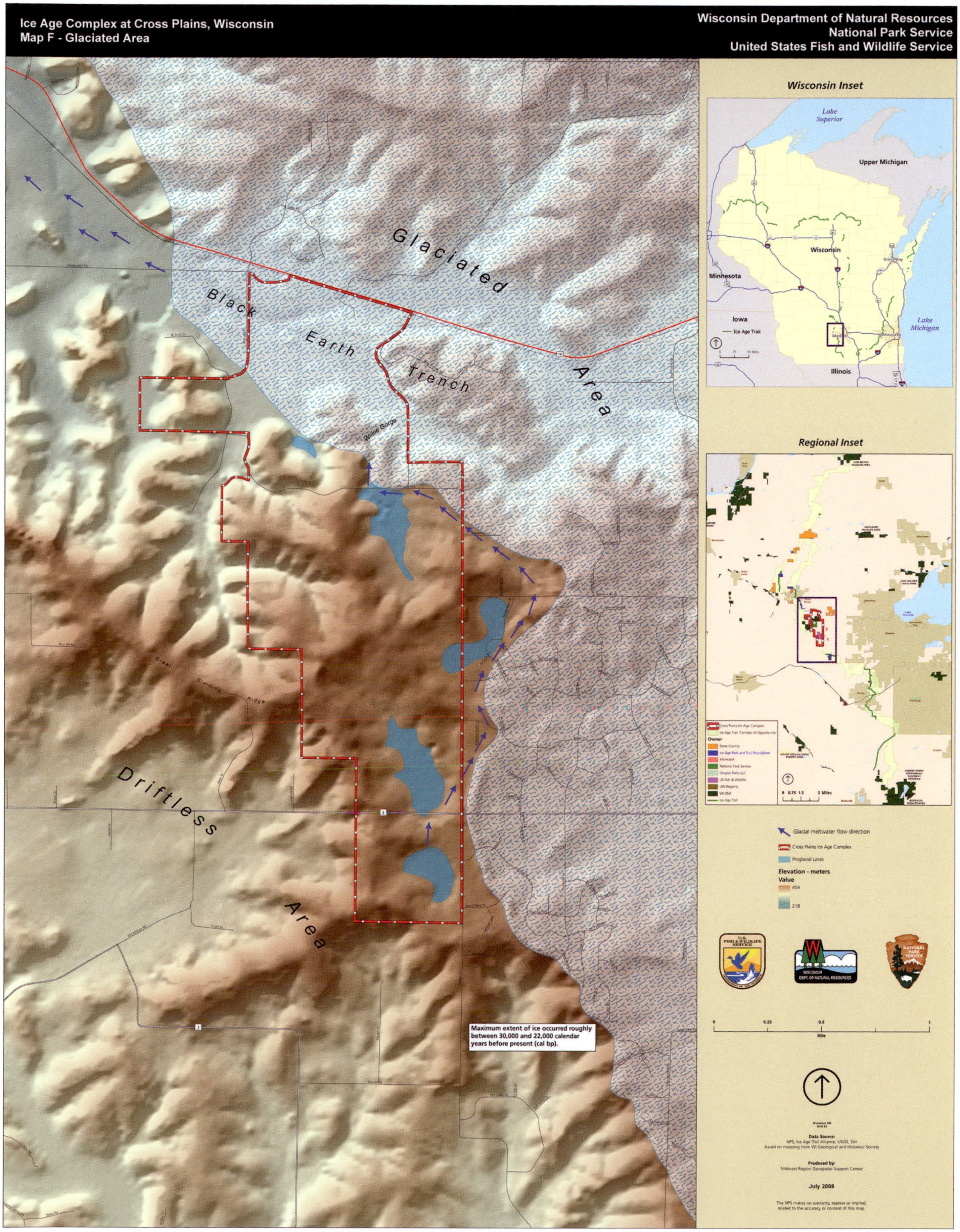

The Ice Age Complex is shown at Cross Plains, Wisconsin, from July 2008.

Data sources: NPS , USGS, Ice Age Trail Alliance, Esri, based on mapping from the Wisconsin Geological and Historical Society.

INDEX

V

W

Y